Proceedings of the 2007 National Conference on Environmental Science and Technology

Godfrey A. Uzochukwu

Proceedings of the 2007 National Conference on Environmental Science and Technology

Springer

Editor
Godfrey A. Uzochukwu
Waste Management Institute
North Carolina A&T State University
261 Carver Hall Annex
Greensboro NC 27411
USA
uzo@ncat.edu

ISBN 978-0-387-88482-0 e-ISBN 978-0-387-88483-7
DOI 10.1007/978-0-387-88483-7

Library of Congress Control Number: 2008937810

Printed on acid-free paper

springer.com

Preface

The purpose of the Third National Conference on Environmental Science and Technology, which was held in Greensboro, North Carolina on September 12–14, 2007 was to address pollution prevention, solutions, and research needs and foster relationships that could result in partnerships needed to protect and sustain the environment and improve the quality of life. The following topics are included in this book: Pollution Prevention, Fate and Transport of Contaminants, Bioremediation, Bio-processing, Innovative Environmental Technologies, Global Climate Change, and Environmental Justice and Ethics.

Several discussions about Global Climate Change, Pollution Prevention, Environmental Justice and Ethics among Godfrey A. Uzochukwu (Waste Management Institute, North Carolina A & T State University), Sherry Southern and Jeffrey Allison (DOE—Savannah River Site), Thomas Parker (CDM), Glennis Nelson (CDM), Jason Callaway (Allied Waste), Steve Roland (O'Brien & Gere), Marv Richardson (O'Brien & Gere) and Rick Crume (US Environmental Protection Agency) set the stage for the Third National Conference on Environmental Science and Technology. The following persons served on the Executive Conference Committee: G.B. Reddy (Professor of Environmental Microbiology), Shoou-Yuh Chang (Professor of Environmental Engineering), Vinayak Kabadi (Professor of Chemical Engineering), Keith Schimmel (Associate Professor of Chemical Engineering), Emmanuel Nzewi (Professor and Director of Civil and Environmental Engineering), Stephanie Luster-Teasley (Assistant Professor of Environmental Engineering) and Godfrey A. Uzochukwu (Professor and Director, Waste Management Institute). These individuals approved the conference theme – Environmental Science and Technology. The authors of presentations accepted for the Conference were asked to prepare a six-page (maximum) paper, following the guidelines developed by Springer. Full papers were peer reviewed and recommended for publication in the proceedings.

The following agencies are thanked for their financial support of the conference; NOAA Interdisciplinary Environmental Technology Cooperative Science Center); NSF Science & Technology Center; Allied Waste Services, O'Brien & Gere; US Department of Energy (Savannah River Site); US Environmental Protection Agency; North Carolina A&T State University (Waste Management Institute, DOE Chair of Excellence in Environmental Disciplines, School of Agriculture and

Environmental Sciences and College of Engineering. In addition, the hard work of Cheryl Bowers, Allyson Alston, Sheena Shearin, Cecilia Worthington, Patricia White, Patricia O'Connor, Gloria Hughes, Furmose Gomez and Jesse Davis is gratefully acknowledged. We would like to express our thanks to Springer staff for assembling this book.

Special thanks go to the following conference keynote speakers: Stanley F. Battle, Chancellor, North Carolina A&T State University; Gregory Green, US Environmental Protection Agency, Research Triangle Park; Sherry Southern, United States Department of Energy and Thomas Parker, CDM.

The conference co-sponsors, Springer and other supporting organizations did not review the published materials in this book. Their support for the conference is not an endorsement of the content.

Greensboro, NC, USA — Godfrey A. Uzochukwu
Greensboro, NC, USA — Keith Schimmel
Greensboro, NC, USA — Shoou-Yuh Chang
Greensboro, NC, USA — Vinayak Kabadi
Greensboro, NC, USA — Stephanie Luster-Teasley
Greensboro, NC, USA — Gudigopuram B. Reddy
Greensboro, NC, USA — Emmanuel Nzewi

Contents

Part III Environmental Justice and Ethics

Contributors

Amer AbuGhazaleh Southern Illinois University, Carbondale, IL, USA

Blakely Adair North Carolina Central University, Durham, NC, USA

Siham A. Ahmed Department of Family and Consumer Sciences, North Carolina A&T State University, Greensboro, NC 27411-1064, USA

Souod R. Alani Department of 4H Youth Development and Family & Consumer Sciences, North Carolina State University, Raleigh, NC 27695, USA

Farouk F. Al-Nouri Food Sciences & Biotechnology Department, College of Agriculture, Baghdad University, Iraq

Mustafa A. Alsharafani Ministry of Science & Technology, P.O. Box 765, Baghdad, Iraq

Yolanda B. Anderson North Carolina Central University, Durham, NC, USA

Olatunde Animashaun Department of Medical Technology, Morgan State University, Baltimore, Maryland, 21251, USA

Aaron Ball Western Carolina University, Cullowhee, NC, USA

John J. Bang North Carolina Central University, Durham, NC, USA

Daniel Barnhurst Washington Savannah River Corporation, Aiken, SC, USA

Marcia Bradshaw North Carolina Central University, Durham, NC, USA

Zarhelia Carlo-Rojas Environmental Science and Engineering PhD (ESE) Program UTEP, El Paso, TX, USA

David M. Cass O'Brien & Gere, Blue Bell, Pennyslvania, USA

Shoou-Yuh Chang North Carolina A&T State University, Greensboro, NC, 27411, USA

S. Chen Department of Civil and Environmental Engineering, MRC 319, Rensselaer Polytechnic Institute, 110 8th Street, Troy, NY 12180, USA

Roger Clapp WATR, Bryson City, NC, USA

Cathy Connor University of Alaska Southeast, Juneau, Alaska, USA

Richard V. Crume U.S. EPA, Research Triangle Park, NC, USA

Yoko S. Crume NC A&T State University, Greensboro, NC, USA

Johnsely S. Cyrus Department of Natural Resources and Environmental Design, North Carolina A&T State University, Greensboro, NC 27411, USA

Devra Davis University of Pittsburgh Cancer Institute, Center for Environmental Oncology, Pittsburgh PA, USA; University of Pittsburgh, Graduate School of Public Health, Pittsburgh PA, USA

Saundra F. DeLauder North Carolina Central University, Durham, NC, USA

Chris Donaldson WCU, Cullowhee, NC, USA

Maryann Donovan University of Pittsburgh Cancer Institute, Center for Environmental Oncology, Pittsburgh PA, USA; School of Medicine, University of Pittsburgh, Pittsburgh PA, USA

Patricia Eagon University of Pittsburgh Cancer Institute, Center for Environmental Oncology, Pittsburgh PA, USA; Veterans Research Foundation, Veterans Affairs Medical Center, Pittsburgh PA USA; School of Medicine, University of Pittsburgh, Pittsburgh PA, USA

Peter Egeghy U.S. EPA National Exposure Research Laboratory, RTP, NC, USA

Albert Esterline NCA&T, Greensboro, NC, USA

Rob Fatland Vexcel/Microsoft Geospatial Solutions, Boulder, CO, USA

George Ford Western Carolina University, Cullowhee, NC, USA

Roy Fortmann U.S. EPA National Exposure Research Laboratory, RTP, NC, USA

Timothy Foster North Carolina A&T State University, Greensboro, NC, USA

Edward Fosu North Carolina A&T State University, Greensboro, NC, USA

Angela M. Fraser Department of 4H Youth Development and Family & Consumer Sciences, North Carolina State University, Raleigh, NC 27695, USA

Austin Frazier North Carolina A&T State University, Greensboro, NC, USA

Kawana O. Fuller Department of Computer Science, North Carolina A&T State University, Greensboro, NC 27411, USA

Noelle Garvin University of Georgia SREL, Aiken, SC, USA

Phillip W. Gassman Iowa State University, Ames, IA, USA

Carresse Gerald North Carolina A&T State University, Greensboro, NC, USA

Curry I. Guinn University of North Carolina Wilmington, USA

Gennadyi Gurman Queens Botanical Garden, Flushing, NewYork, USA

Tracy Lee Hansen NOAA Earth System Research Lab, Boulder, CO, USA

Matt Heavner University of Alaska Southeast, Juneau, Alaska, USA

Talal El Hefnawy University of Pittsburgh Cancer Institute, Center for Environmental Oncology, Pittsburgh PA, USA

Vincent C. Henrich The University of North Carolina at Greensboro, Greensboro, NC, USA

Eran Hood University of Alaska Southeast, Juneau, Alaska, USA

Frank Houghton Veterans Research Foundation, Veterans Affairs Medical Center, Pittsburgh PA, USA

Brian P. Howell Western Carolina University, Cullowhee, NC, USA

Salam A. Ibrahim Department of Family and Consumer Sciences, Food and Nutritional Sciences, North Carolina A&T State University, Greensboro, NC 27411-1064, USA

Wahid Islam Department of Chemical Engineering, North Carolina A&T State University, Greensboro, NC, USA

A. Jin North Carolina A&T State University, Greensboro, NC 27411, USA

Ibipo Johnston-Anumonwo State University of New York, Cortland, New York, USA

Vinayak Kabadi Department of Chemical Engineering, North Carolina A&T State University, Greensboro, NC, USA

Chyer Kim Department of Family and Consumer Sciences, North Carolina A&T State University, Greensboro, NC 27411-1064, USA

Parakalan Krishnamachari North Carolina A&T State University, Greensboro, NC, USA

Lyubov A. Kurkalova North Carolina A&T State University, Greensboro, NC, USA

Marion Lamberth North Carolina Central University, Durham, NC, USA

Sikdar M.I. Latif North Carolina A&T State University, Greensboro, North Carolina 27411, USA

Wen-Yee Lee ESE Program; Chemistry Department, UTEP, El Paso, TX, USA

Tom LeFebvre NOAA Earth System Research Lab, Boulder, CO, USA

Diana Lenzner University of Pittsburgh, Graduate School of Public Health, Pittsburgh PA, USA

Xiaopeng Li North Carolina A&T State University, Greensboro, North Carolina 27411, USA

Yebo Li The Ohio State University/Ohio Agricultural Development Center, Wooster, OH, USA

Jianzhong Lou Department of Chemical Engineering, North Carolina A&T State University, Greensboro, North Carolina 27411, USA

Stephanie Luster-Teasley North Carolina A&T State University, Greensboro, NC, USA

Rakesh Malhotra North Carolina Central University, Durham, NC, USA

Michael M. Marshall The University of North Carolina at Greensboro, Greensboro, NC, USA

Jeff Marston WCU, Cullowhee, NC, USA

Brent McCoy WCU, Cullowhee, NC, USA

William McDaniel Western Carolina University, Cullowhee, NC, USA

Gwendolyn McFadden-Wade North Carolina A&T State University, Greensboro, NC, USA

Faustina Meheux North Carolina Central University, Durham, NC, USA

Gary Mills University of Georgia SREL, Aiken, SC, USA

M.C. Nyman Department of Civil and Environmental Engineering, MRC 319, Rensselaer Polytechnic Institute, 110 8th Street, Troy, NY 12180, USA

Ethelbert Odo Orient Home Care Services Inc., Grand Prairie TX, USA

Gbekeloluwa B. Oguntimein Department of Civil Engineering, Morgan State University, Baltimore, Maryland, 21251, USA

Ivie Okpere Department of Civil Engineering, Morgan State University, Baltimore, Maryland, 21251, USA

Somphavanh Phetsomphou North Carolina A&T State University, Greensboro, NC, USA

David Price North Carolina A&T State University, Greensboro, NC, USA

G.B. Reddy Department of Natural Resources and Environmental Design, North Carolina A&T State University, Greensboro, NC 27411, USA

D.J.R. Reeves University of North Carolina Wilmington, USA

Diana R. Robinson North Carolina A&T State University, Greensboro, NC, USA

Parke A. Rublee The University of North Carolina at Greensboro, Greensboro, NC, USA

Moussa M.E. Salem National Research Center, Dokki, Cairo, Egypt

Mary Sue Schultz NOAA Earth System Research Lab, Boulder, CO, USA

Silvia Secchi Iowa State University, Ames, IA, USA

Chung W. Seo North Carolina A&T State University, Greensboro, NC, USA

Abolghasem Shahbazi Department of Natural Resources and Environmental Design, North Carolina A&T State University, Greensboro, NC 27411-1064, USA

Myra M. Shird North Carolina Agricultural and Technical State University, Greensboro, NC, USA

Parikhit Sinha O'Brien & Gere, Blue Bell, Pennyslvania, USA

Garriet W. Smith University of SC, Aiken, SC, USA

Marie-Ange Smith University of SC, Aiken, SC, USA

Xianzhi Song Western Carolina University, Cullowhee, NC, USA

Sherry Southern U.S. DOE, Aiken, SC, USA

Sara E. Steudler Elon University, Elon, NC, USA

Katherine Suda North Carolina A&T State University, Greensboro, NC, USA

Nancy Sussman University of Pittsburgh, Graduate School of Public Health, Pittsburgh PA, USA

Kathleen E. Sykes U.S. EPA, Washington, DC, USA

Terrence Thomas North Carolina A&T State University, Greensboro, NC, USA

Brant W. Touchette Elon University, Elon, NC, USA

Leonard Uitenham North Carolina A&T ST.U., Greensboro, NC, USA

Leonard Uitenham North Carolina A&T State University, Greensboro, NC, USA

Godfrey A. Uzochukwu North Carolina A&T State University, Greensboro, NC, USA

Mary E. Uzochukwu North Carolina A&T State University, Greensboro, North Carolina

Renay Van Wyk Masters Environmental Health, University of Johannesburg, South Africa

Conrad Daniel Volz Department of Environmental and Occupational Health, Graduate School of Public Health, University of Pittsburgh, Pittsburgh, PA, USA; Center for Environmental Oncology, Co-Director Exposure Assessment, University of Pittsburgh Cancer Institute, Pittsburgh, PA, USA; Scientific Director, Center for Healthy Environments and Communities, Pittsburgh, USA

Rashawn K. Washignton Chemical Engineering, North Carolina A&T State University, Greensboro, NC 27411, USA

Jean T. Wells Howard University, Washington, DC, USA

Donald Whitaker U.S. EPA National Exposure Research Laboratory, Research Triangle Park, NC, USA

John B. Williams Department of Biological & Physical Sciences, South Carolina State University, Orangeburg, SC, USA

Ronald Williams U.S. EPA National Exposure Research Laboratory, RTP, NC, USA

Regina M. Williams North Carolina Agricultural and Technical State University, Greensboro, NC, USA

Vereda Johnson Williams Economics, Transportation and Logistics Department, North Carolina A&T State University, Greensboro, NC 27411, USA

Robert C. Wingfield Department of Chemistry, Fisk University, Nashville, TN 37208, USA

Dereje Worku North Carolina A&T State University, Greensboro, NC, USA

Mulumebet Worku North Carolina A&T State University, Greensboro, NC, USA

Maxine Wright-Walters Department of Environmental and Occupational Health, Graduate School of Public Health, University of Pittsburgh, Pittsburgh, PA, USA

Jizhong Yan North Carolina A&T State University, Greensboro, NC, USA

Thomas M. Yanoschak HDR, Raleigh, North Carolina, USA

Osei Yeboah North Carolina A&T State University, Greensboro, NC, USA

Jian Zhang University of Kansas, Greensboro, NC, USA

Part I
Bioprocessing

Synthesis of Poly(L(+) Lactic Acid) by Polycondensation Method in Solution

Jian Zhang, Parakalan Krishnamachari, Jianzhong Lou, and Abolghasem Shahbazi

Abstract The disposal problem due to non-degradable petroleum-based plastics has raised the demand for biodegradable polymers. Poly(lactide) (PLA) is a biodegradable aliphatic polyester derived from 100% renewable resources, such as corn and sugar beets. Moreover, it has unique physical properties that make it useful in diverse applications including paper coating, fibers, films, and packaging. In this work, the results of the synthesis of high molar weight poly(L(+) lactic acid) (PLA) which is able to thermally crystallize have been described. The synthesis of poly(L(+) lactic acid) was carried out in a solution of p-xylene. Tin(II) chloride was used as catalysts in a quantity of 1.0 wt% calculated on the monomer. Polycondensation was carried out over a period from ten to over 20 h. The resulting poly(L(+) lactic acids) were characterized by FT-IR, DSC and the GPC. The molecular weight of PLLA reached as high as 26000. FTIR spectra of PLLA were analyzed to determine the configuration of the PLLA. The main byproduct of condensation polymerization is cyclo-polyester. The glass transition temperature of PLA after is 46.25°C. PLA after the recrystallization does not show glass transmission temperature because the crystallinity is more than 40%.

Introduction

Poly(L-lactic acid) (PLLA) is synthesized from renewable resources and it has received much interest in recent years. It is degraded by hydrolytic cleavage of the ester bonds to produce lactic acid and its oligomers, which can be metabolized by many microorganisms. Application of PLLA includes biomedical purposes such as surgical sutures, fractured bone fixation, tissue engineering, and drug delivery systems. Cargill Dow Co. began to produce PLLA on a commercial scale under the trade name of Nature Works^{TM}. The price of PLLA has been reduced greatly

J. Lou (✉)
Department of Chemical Engineering, North Carolina A&T State University, Greensboro, North Carolina 27411, USA

G.A. Uzochukwu et al. (eds), *Proceedings of the 2007 National Conference on Environmental Science and Technology*, DOI 10.1007/978-0-387-88483-7_1,

owing to the technology developed by Cargill Dow Co. by combining agricultural processes with biological and chemical technologies. (Richard and Bhanu 2002; Ray et al., 2000)

As far as we know, Cargill Dow Co. produces PLLA through ring opening polymerization of L-lactide, which is formed by catalytic depolymerization of low molecular weight PLLA prepolymer synthesized by direct polycondensation polymerization of l-lactic acid (LA). Cargill Dow Co. employs this route mainly because direct condensation polymerization of LA in bulk state gives only low- to intermediate-molecular weight polymers due to the low equilibrium constant of the condensation polymerization reaction. A solvent with high boiling point, such as xlyene, which is compatible with PLLA, is used for the removal of dissociated water by means of the so-called azeotropic distillation technique to shift the equilibrium state to the polymer side. PLLA with high molecular weight can be produced through this solution polymerization route. This route requires large reactor volume and facilities for evaporation and recovery of the solvent, which increases the production cost of PLLA. Recently, Mitsui Chemicals opened a new process based on direct polycondensation of low-molecular weight PLLA in solid state without the use of an organic solvent11 to produce high-molecular-weight PLLA.

Experimental Section

Materials. L-Lactic acid (LA) from Aldrich was a 85% aqueous solution of the monomer. The excess water was removed before use by distillation under reduced pressure at 100°C. The following products were used without further treatment: Tin(II) chloride, Methanol, Chloroform and p-xylene all (HPLC grade) from Aldrich.

Polymerization. 80 g of the product of distillation L(+) lactic acid, 0.8 g Tin(II) chloride and 200 mL of xylene were added into a 600 mL Parr 4590 series stirred reactor and the reaction mixture was heated. The water was azeotropically distilled off, and the reaction was continued at 160°C for 24 h under reduced pressure (~400 mm Hg). Then the Tin(II) chloride was filtered off, and the resulting polymer was dissolved in chloroform, and then was precipitated twice in excess methanol. The product was dried in a vacuum oven at 60°C until constant weight was attained. (Hiltunen et al., 1997)

Characterizations. The DSC measurements were carried out under a nitrogen atmosphere with samples sealed into the standard DSC aluminum sample kits using DSC Q100 (TA instruments). The following thermal cycle was employed: First, the sample was heated up to a temperature 110°C for 30 min; next, cooled down to 20°C; then, sample was heated up to 200°C (1st scan) and then cooled down to 10°C; finally, heated to 200°C again (2nd scan). The samples investigated had weights within the range of 5–10 mgs. Both heating and cooling rates were 20°C/min. By integrating the normalized area of the melting exotherm, determining the heat involved, and rating it to the reference 100% crystalline polymer (93.6 J/g), the relative crystallinity of PLA was assessed. Infrared spectra were recorded with a Pike Miracle ATR, Ge crystal, single bounce on a Bruker Tensor 37 FTIR and 4 wave number resolution scans for sample and background. GPC measurements were performed on a Waters model 410 differential refractometer equipped with Waters Styragel HMW 6E THF column (7.8 × 300 mm) with effective molecular weight range 5000 to 1 × 107. Operating conditions were solvent THF; solute PLA and polystyrenes (standards of Shodex company); injected volume 20 μL; sample concentration of the order of 1 mg/ml. Flow rate of 1 mL/min. The sensitivity of the refractometer was set to 128 and a scale factor of 100 was adopted. (Aou et al., 2005)

Results and Discussion

Figure 1 presents the FTIR spectra of synthesized PLA. The 1758 cm-1 of FTIR spectra presents the C=O stretching region. And C-O-C stretching mode was observed at 1190 cm-1 as a IR band. 1459, 1392 and 2997 cm-1 regions are CH3 and CH bending region and are further confirmed the existence of CH and CH3. 1264 and 1099 cm-1 present the C-O stretching of cyclo-polyester. And this is the main byproduct of condensation polymerization. (Kister et al., 1998)

Figure 2 shows the DSC thermograms of the PLA. The results of thermal characterization are shown in Table 1. The first scan values for Tg, Tm and crystallinity denote the properties of PLA just after the recrystallization at 110°C for 30 min. The second scan values are results after heated to 200°C cycle. The glass transition

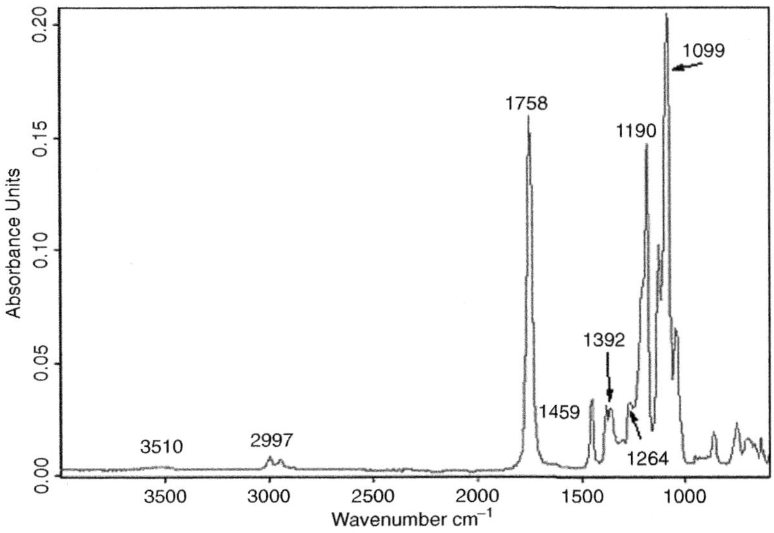

Fig. 1 FTIR-spectrum of PLA

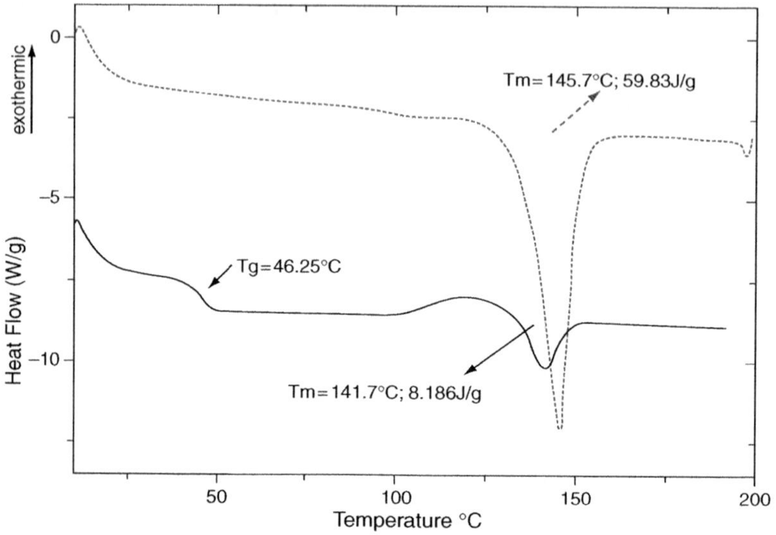

Fig. 2 DSC thermograms of PLA: 1st scan (after recrystallization at 110°C, *top dash line*); 2nd scan (after first cycle, *bottom solid line*)

Table 1 Thermal characterization results

PLA	T_g, °C	T_m, °C	ΔH, J/g	Crystallinity, %
1st Scan	–	145.7	59.83	63.9
2nd Scan	46.25	141.7	8.168	8.7

temperature of PLA after is 46.25°C. PLA after the recrystallization does not show glass transmission temperature. because the crystallinity is more than 40%.

Polydisperse polymers in solution are fractionated according to size or hydrodynamic Volume during GPC, which is also known as size exclusion chromatography. Molecular weight is related to the hydrodynamic volume. In GPC a dilute polymer solution is injected into a solvent stream which then flows through a series of columns packed with porous gel beads. Smaller molecules pass through and around the beads while larger molecules are excluded from all but the largest pores. Thus fractionation occurs with the largest molecules eluting first. The molecular weight of an eluting polymer molecule varies exponentially with eluting volume, the latter of which is proportional to time under constant flow rate conditions. To obtain molecular weight data and convert the GPC chromatogram into a molecular weight distribution, the relation between molecular weight and elution time is obtained from a series of polymer standards of known molecular weight. In these study Polystyrene standards made by the Shodex company was used. Three different molecular weight standards was used, of the order of 104,105 and 106 respectively. Dependence between molecular size and molecular weight can be described by Mark Houwink equation. The parameters K and Alpha as found in literature for the both polymers in THF solvent is 0.011 ml/g, 0.725 and 0.00174 ml/g, 0.736 for Polystyrene and PLA respectively. This is used in the universal calibration method. We can calculate molecular mass of unknown polymer by using a calibration for another (more common) polymer.

$$[\eta] = K\ Ma \tag{1}$$

where η is the intrinsic viscosity and M is the molecular weight of the polymer. By linearizing the plot of Log (K*Ma+1) versus the retention volume for different standards of polystyrene we can a get a slope and intercept of the fit. Which can be used back in fitting a straight line to already known values of K and a of PLA in THF solvent. This helps us in finding the unknown molecular weight of PLA from the equation (Fig. 3) and GPC Chromatogram data of PLA (Fig. 4).

$$\text{Log}(K^*\ Ma + 1) = (\text{Retention time of PLA})^*\text{Slope} + \text{Intercept} \tag{2}$$

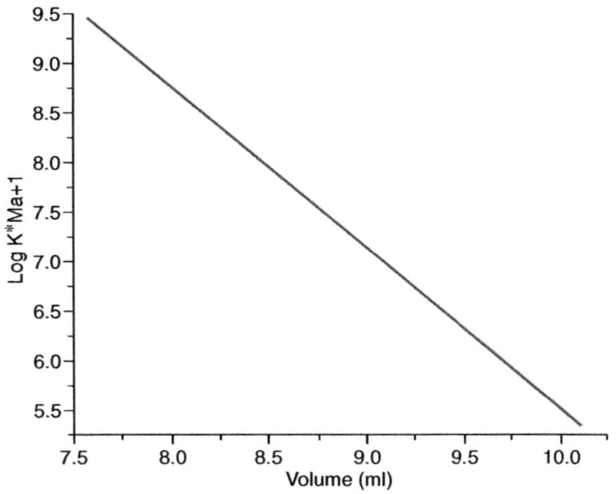

Fig. 3 Polystyrene calibration

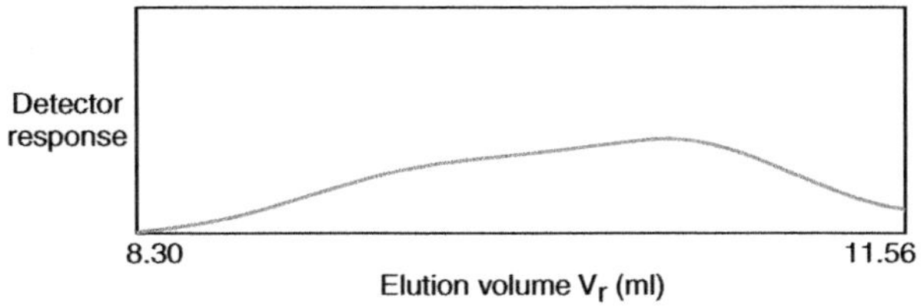

Fig. 4 Chromatogram showing the broad peak of PLA

After substituting the appropriate values in equation (2) i.e. K = 0.00174 ml/g, a = 0.736 and the retention volume of 10.38, Molecular weight of PLA can be calculated as: Log (0.00174 × M1.736) = –1.61302(10.38) + 21.652, M = 26.11 × 10^3. Therefore the Molecular weight of PLA is 26000.

Conclusion

PLLA was synthesized by the direct bulk condensation polymerization using Tin(II) chloride as a catalyst under reduced pressure (~400 mm Hg). The molecular weight of PLLA reached as high as 26000. FTIR spectra of PLLA were analyzed to determine the configuration of the PLLA. The main byproduct of condensation polymerization is cyclo-polyester. The glass transition temperature of PLA after is 46.25°C. PLA after the recrystallization does not show glass transmission temperature because the crystallinity is more than 40%.

Acknowledgments Funding for this project was provided by the USDA; Mrs. Nicole Whiteman of NatureWorks provided technical advice.

References

Aou, K., Kang, S., Hsu, S.L. 2005. Morphological study on thermal shrinkage and dimensional stability associated with oriented poly(lactic acid). *Macromolecues* 38:7730–7735.

Hiltunen, K., Seppala, J.V., Harkonen, M. 1997. Effect of catalyst and polymerization conditions on the preparation of low molecular weight lactic acid polymers. *Macromolecules* 30:373–379.

Kister, G., Cassanas, G., Vert, M. 1998. Effects of morphology, conformation and configuration on the IR and Raman spectra of various poly(lactic acid)s. *Polymer* 39:267–273

Ray, E.D., Patrick, R.G., David, E.H. 2000. Polylactic acid technology. *Advanced Material* 12:1841.

Richard, A.G., Bhanu, K. 2002. Biodegradable polymers for the environment. *Science* 297:803.

The Feasibility of Using Cattails from Constructed Wetlands to Produce Bioethanol

Katherine Suda, Abolghasem Shahbazi, and Yebo Li

Abstract This project investigates the feasibility of harvesting the cattails in the constructed wetlands of the North Carolina A&T Farm to be converted into ethanol. Using the cattails to produce renewable energy will add value to the land as well as reduce emissions of greenhouse gases by replacing petroleum products. Pretreatment of the dried cattails with dilute NaOH was followed by solid-liquid separation and enzymatic hydrolysis and fermentation of the solids. Two trials gave an average conversion efficiency of 43.4% for the pretreated solids alone which, in conjunction with the crop yield for the cattails, would give up to 4,012 L ethanol/ha, a favorable comparison with corn stover's 1,665 L/ha at a 60% conversion rate. Given the high potential – 9,680 L/ha at 60% conversion efficiency for solid and liquid streams – and the social and environmental benefits gained by adding value to the waste management system and reducing carbon emissions otherwise made by gasoline, it is recommended that further studies be made using cattails as a feedstock for bioethanol.

Introduction

Renewable transportation fuels are being developed to lower emissions of greenhouse gases and to enhance energy security. Biodiesel and starch-based ethanol predominate, but the technology required for conversion of cellulose to ethanol is improving; some pilot plants are currently in operation, and the first full-scale plants are due to come on line in 2007 and 2008. Cellulosic ethanol has a better environmental profile than starch-based ethanol, with a 90% reduction in carbon emissions over gasoline as compared to a 29% reduction using starch-based ethanol (Wang, 2005), and it has the advantages of not requiring fertile agricultural land and of adding value to marginal farmland. Another benefit of cellulosic ethanol is the

K. Suda (✉)
North Carolina A & T State University, Greensboro, NC, USA

G.A. Uzochukwu et al. (eds), *Proceedings of the 2007 National Conference on Environmental Science and Technology*, DOI 10.1007/978-0-387-88483-7_2,

supply distribution afforded since it can be converted from a variety of cellulosic materials according to growing conditions and availability around the country.

One possible source of cellulosic feedstock is the cattails used for phytoremediation in constructed wetlands. Whereas harvesting plants from natural wetlands is a sensitive issue because of possible environmental harm, harvesting the above-water portion of cattails once a year from a constructed wetland is a sustainable practice which will not interfere with their role in cleaning the wastewater. There is no energy input required to grow them, as they receive their water and nutrients from the wastewater. Using the cattails as a feedstock for bioethanol is one way to add value to the land with little environmental impact.

This project examines the feasibility of using the cattails from constructed wetlands to produce bioethanol. The focus is on pretreatment, followed by enzymatic hydrolysis and fermentation of the solid fraction. Environmental impact, in terms of emissions reductions, is calculated, and economic factors are considered. The preliminary results are presented.

Methods

The basic steps to be followed in producing ethanol from cellulosic plant material include harvesting, cleaning and shredding the feedstock, pretreatment, enzymatic hydrolysis, fermentation, distillation and drying. This paper focuses on pretreatment of the feedstock with sodium hydroxide followed by enzymatic hydrolysis and yeast fermentation.

Feedstock Characteristics. Cattails were cut and chopped with pruning shears, dried at 70°C for 5 days, and ground in a Wiley mill to 2 mm mesh size. The energy content for cattails was determined using a Parr bomb calorimeter. Dried samples were sent to Dairy One Forage Lab (Ithaca, NY) to be analyzed for cellulose, hemicellulose, and lignin content by ANKOM 200 Fiber Analyzer.

Pretreatment of the feedstock. Alkaline pretreatment was carried out in order to render the cellulose more accessible to the cellulase for enzymatic hydrolysis. About 500 g of the dried, ground cattails were stirred into 5 L of a 1.0 N NaOH solution and left at room temperature overnight. The mixture was then centrifuged for 20 min, after which, the liquid was poured off, the solids rinsed with water and re-suspended in the water two times with centrifuging and decanting of the supernatant after each rinse. Moisture content of the pretreated mix was determined by drying a sample at 60°C until weight was constant. Another sample was sent to Dairy One Labs for cellulose, hemicellulose and lignin analysis.

Culturing the yeast for fermentation. *Saccharomyces cerevisiae* (ATCC 24858) and *Yamadazyma stipitis* (ATCC 58784) were used to ferment the sugars. *Y. stipitis* breaks xylose down producing ethanol, and *S. cerevisiae* produces ethanol from glucose. Aseptic technique was used to rehydrate the freeze-dried yeasts. They were incubated at 30°C (*S. cerevisiae)* and 25°C (*Y.* stipitis) and stepped up until 1.0 L was obtained.

Enzymatic hydrolysis and fermentation. Enzymatic hydrolysis was carried out in two batches in a BioFLo 110 Fermentor/BioReactor (New Brunswick Scientific, Edison, NJ) at 50°C, 4.8 pH and 200 rpm agitation. Pretreated cattails were mixed with deionized water to a concentration of 50 g/L. Cellulase enzymes (Spezyme CP, Genencor International, Rochester, NY) were added at 82 GCU/g dry matter. The pH was monitored and adjusted with buffer solutions of 5% NaOH and 5% HCl. Samples were taken throughout the process for later analysis by high pressure liquid chromatography (HPLC) using a 717 Plus Autosampler (Waters, Milford, MA). Hydrolysis proceeded for 24 h before yeasts were added. At 24 h, the temperature was reduced to 37°C and the yeasts, *S. cerevisiae and Y. stipitis,* were added. A condenser was attached to the fermentation vessels to allow cooling water to condense the evaporating ethanol back into the vessel. Fermentation proceeded at 37°C, 250 rpm agitation, and pH of 4.8, and the process was allowed to run for five days.

Results and Discussion

Feedstock Characteristics. Two years of data from the A&T Farm indicate that the average yield of cattails from the constructed wetlands is 16.1 mton/ha with a maximum of 42.7 mton/ha (Dr. G.B. Reddy, personal communication, October 2006). The harvested plants had average moisture content (w.b.) of 78.6%. The energy content at 5.4% moisture content (w.b.) was found to be 17.4 MJ/kg. Cellulose, hemicellulose and lignin content for the dry material were determined as 28.7%, 23.4%, and 10.1%, respectively. Because a large part of the energy in the process is used to dry the feedstock, further studies should look at the possibility of using it without drying.

Pretreatment. Supernatant from the pretreated material was analyzed by HPLC to determine the concentration of xylose, glucose and arabinose in the liquid portion of the hydrolyzate. The only sugar detected was xylose at 0.156% which was decided to be too little to pursue further. However, later it was questioned exactly how much 0.156% might have been as the supernatant had been diluted with the rinse water. It may be possible to use suction filtration or a reverse osmosis membrane to separate the liquids and solids and get higher yields from the solid stream. Future plans are to track and ferment the liquid fraction separately.

After pretreatment, cellulose content had increased from 28.7% to 46.2% total dry solids. Calculating the cellulose content based on the mass of pretreated dry solids shows that 28.7% of the cellulose was actually lost during pretreatment. Likewise, hemicellulose and lignin were reduced in the solids, with lignin diminishing by 51.4%, and hemicellulose by 67.5%. Evaporation of the supernatant showed that there were suspended solids, presumably the lost cellulose. Hemicellulose is alkaline soluble, and most was expected to be dissolved by the NaOH, while lignin has alkaline-soluble fractions (Reith et al., 2002; Walthers et al., 2001).

Enzymatic Hydrolysis and Fermentation. Enzymes began to convert the biomass to sugars within 2 min of being added (Fig. 1). The initial glucose

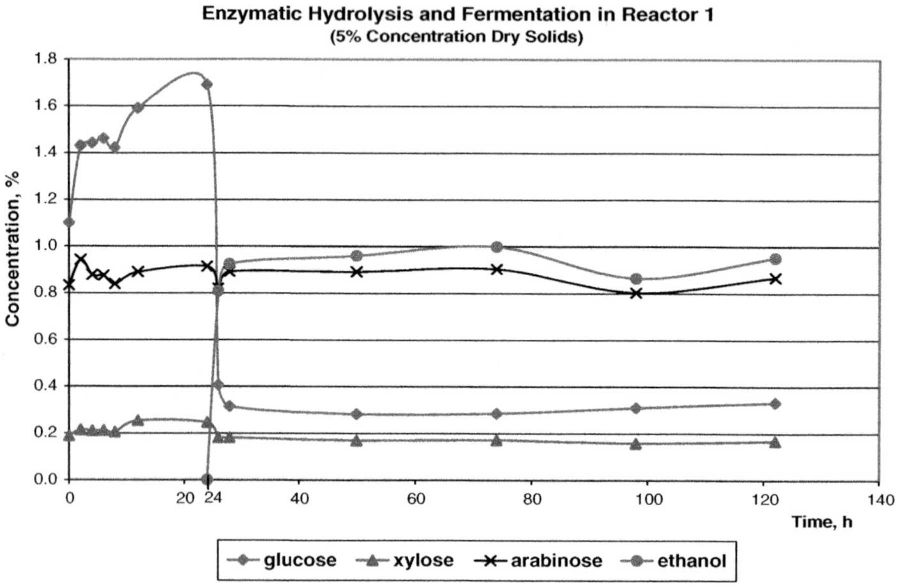

Fig. 1 Enzymatic hydrolysis and fermentation in reactor 1

concentration measured 1.100% and 1.041% respectively, in reactor 1 and reactor 2. Just before addition of the yeast, at 24 h, the glucose concentration had reached 1.691% and 1.615%. Xylose showed an increase in concentration from 0.187% and 0.158% to 0.245% and 0.229% in reactors 1 and 2 respectively. The xylose in two reactors and the glucose in one had actually started to decline in concentration before the yeast was added. Possibly they were further degrading into compounds that were not being measured.

Arabinose concentration during enzymatic hydrolysis and fermentation was steady at about 0.8%. The enzymes and yeast had no effect on it. The arabinose must have been released during pretreatment. With addition of arabinose-fermenting yeasts, the ethanol yield could be increased.

Results of the HPLC showed ethanol yield in this study was 0.997% and 0.951% in the two reactors, which yields an average efficiency of 43.4%. This is equal to 94 L ethanol/mton cattail. Badger (2002) reports a reasonable efficiency of 57% using acid hydrolysis to give 227 L ethanol/mton dry corn stover. Using enzymatic hydrolysis was expected to give a higher yield. One reason for the low yield could be that the enzymes used in this study, Spezyme CP, are tailor-made for corn stover. Supplemental enzymes might improve yields. Better fermenting organisms have been developed and a way of obtaining them for future studies is being studied. The two yeasts used have different optimum temperatures. It is supposed that the ethanol produced came from the glucose because of the higher than optimum temperature for *Y. stipitis*.

Tying the alcohol yield in with the yield of cattails from the wetlands gives an average of 1,521 L ethanol/ha cattails, and a maximum of 4,018 L ethanol/ha cattails. Based on current DOE estimates of 60% conversion efficiency (334 L/mton), and using the sustainable harvest amounts of corn stover for current tillage practices, we have an ethanol yield from corn stover of 804 L/ha, and for no-till farming, 1,665 L/ha (Perlack et al., 2005). Assuming that further trials with cattails will approach 60% efficiency, the yield of ethanol per acre of cattails could reach as much as 3,917 L/ha.

Environmental Analysis

The preliminary study did not address the environmental impact of harvesting cattails. There will be no need for machinery, except for the once yearly sustainable harvest in late summer, so the detrimental environmental impact should be less than for feedstocks requiring fertilization and irrigation. In addition, the cattails are being used for phytoremediation of wastewater, a benefit to the environment in itself.

The major benefit to the environment, if cattails were to be used as a feedstock for bioethanol, is the reduction in greenhouse gas emissions. Using a life cycle analysis, pure bioethanol emits 90% less carbon dioxide than an equivalent energy amount of gasoline (Wang, 2005). Pure ethanol has two thirds the energy of gasoline liter per liter, and one liter of gasoline, over its life, emits 2.32 kg carbon dioxide. To determine the reductions in emissions achieved by using bioethanol, the following formula can be used,

$$0.67 \times 2.32 \times 0.90 \times \text{EtOH} = \text{E}_\text{S} \qquad (1)$$

where EtOH is the amount of bioethanol in L/ha, and E_S is the emissions (kg/ha) saved by using the bioethanol. By using this formula, it can be seen that using the bioethanol from corn stover with current tillage practices will save 1,115 kg CO_2/ha. Cattails at average yield will save 2,128 kg CO_2/ha and cattails at maximum production will save 5,621 kg CO_2/ha. Once 60% conversion efficiency for cattails to ethanol is achieved, the savings will rise to 5.2–13.4 mtons of carbon dioxide per hectare.

Economic Analysis

A complete economic analysis needs to be carried out including the cost of harvesting. Because of the wet nature of the cattail environment, special machinery must be designed to cut the plants without sinking into the soft soil. In the case where the wetlands are constructed with ethanol production in mind, the dimensions can be designed to accommodate existing long-armed machinery such as those used to cut grass on highway banks. Sizing, storing, and processing costs are assumed to be approximately equal to those of corn stover.

Based on corn stover costs, a plant must have about 453–907 mtons/day to stay in operation (Fraser, 2006). Converting this weight of corn stover to a volume of ethanol gives a required 151,000–302,000 L/day. Assuming the plant operates 330 days a year, the amount of land required per crop per year can be determined from the following formula,

$$(302,800 \times 330)/(\text{EtOH}) = \text{A} \tag{2}$$

where EtOH is the amount of bioethanol in L/ha, and A is the area of land required in hectares. Using this formula gives a land requirement of about 60,000–124,000 ha/yr for corn stover and 25,000–66,000 ha/yr of cattails at the preliminary 43% conversion rate.

Although cattails are abundant in North Carolina, particularly in the eastern part of the state, there are not very many constructed wetlands. However, some mid-size municipalities across the country have found them to be cost effective as part of their wastewater treatment (Gelt, 1997). In addition, cattails have been used to clean some industrial wastewater streams. An integrated system which uses mixed feedstock including the cattails from constructed wetlands in the community should be investigated as a promising scenario.

Conclusion

Although the initial trials using cattails to make bioethanol achieved only 43.4% conversion efficiency, it is believed that further trials will obtain better results by using more advanced conversion organisms, and by combining the liquid and solid fractions. The low conversion rate is more than made up for by the high productivity of the wetlands, and by the reduction in emissions in the life cycle analysis of the fuel produced. With advances in technology, it is not unlikely that ethanol plants will soon be able to use a variety of feedstocks. Cattails from constructed wetlands could be a valuable part of an integrated system with their role as cleaning agents and clean fuel feedstock.

Acknowledgments Many thanks are extended to Kevin Jenkins and Crystal Biddle for their research assistance.

References

Badger, P. 2002. "Ethanol from Cellulose: a General Review." In: J. Janick and A. Whipkey (Eds.), *Trends in New Crops and New Uses*. ASHS Press, Alexandria, VA.

Fraser, J. 2006. Best Kept Secret in the Ethanol Industry: BlueFire Ethanol (BFRE). Seeking Alpha: Stock Market Opinions and Analysis. Online at http://seekingalpha.com.

Gelt, J. 1997. "Constructed Wetlands: Using Human Ingenuity, Natural Processes to Treat Water, Build Habitat." *Arroyo* , March 1997, 9(4).

Perlack, R., L. Wright, A. Turhollow, R. Graham, B. Stokes, and D. Erbach. 2005. "Biomass as Feedstock for a Bioenergy and Bioproducts Industry: the Technical Feasibility of a Billion-ton

Annual Supply." A joint study sponsored by US DOE and USDA. Oak Ridge, TN: Oak Ridge National Laboratory, 15 December 2005.

Reith, J., H. den Uil, H. van Veen, W. de Laat, J. Niessen, E. de Jong, H. Elbersen, R. Weusthuis, J. van Dijken, L. Raamsdonk. 2002. "Co-production of Bioethanol, Electricity and Heat from Biomass Residues." In: Twelfth European Conference and Technology Exhibition on Biomass for Energy, Industry and Climate Protection, Amsterdam, The Netherlands.

Walthers, T., P. Hensirisak, and F. Agblevor. 2001. "Model Compound Studies: Influence of Aeration and Hemicellulosic Sugars on Xylitol Production by Candida Tropicalis." *Applied Biochemistry and Biotechnology* 91–93:423–435.

Wang, M. 2005. "The Debate on Energy and Greenhouse Gas Emissions Impacts of Fuel Ethanol." Argonne National Laboratory. Energy Systems Division Seminar. Chicago, IL, 3 August 2005.

Significance of Bile Salt Tolerant *Lactobacillus reuteri*

Siham A. Ahmed, Salam A. Ibrahim, Chyer Kim, and Abolghasem Shahbazi

Abstract Probiotic supplements have received worldwide attention due to their health benefits. One of the most commonly used probiotics is *Lactobacillus reuteri*, which has known to help prevent and treat both viral and bacterial diarrhea enhancing the body's resistance to gastrointestinal disease. In order to survive and colonize in the gastrointestinal tract, *L. reuteri* should express high tolerance to bile salt. This study was therefore undertaken to determine the effect of bile salt on the growth and ß-galactosidase activities of *L. reuteri*. The effect of Tween 80 on the promotion of *L. reuteri* tolerance against bile salt was also determined. Five strains (CF2F, DSM20016, MM2-3, and MM7) of *L. reuteri* used for this study were subjected to MRS broth supplemented with various concentrations (0.0, 0.1, 0.2, 0.3, 0.4, 0.5, 1.0, 2.0, and 4.0%) of bile salt. Bacterial growth was then monitored using a spectrophotometer (610 nm) for 10 h at 37°C. Results showed that 0.3% bile salt significantly ($P<0.05$) reduced the growth of *L. reuteri* strains. Strain MM2-3 showed higher survival and ß-galactosidase activity (800 and 1500 Miller units) than other strains tested. The addition of Tween 80 enhanced the bile salt tolerance of *L. reuteri*. The results of this study demonstrate that bile salt and Tween 80 has an influence on the biochemical properties of *L. reuteri* and should be considered for useful applications in enhancing the survival of Lactobacillus.

Introduction

The use of probiotic supplements is becoming increasingly popular in the world due to their health benefits such as adherence to epithelial cells and exclusion or reduction of pathogenic adherence. They are able to persist and multiply in intestinal or

S.A. Ahmed (✉)
Department of Family and Consumer Sciences, North Carolina A&T State University, Greensboro, NC 27411-1064, USA

G.A. Uzochukwu et al. (eds), *Proceedings of the 2007 National Conference on Environmental Science and Technology*, DOI 10.1007/978-0-387-88483-7_3,

urogenital tracts and produce acids, hydrogen peroxide and bacteriocins antagonistic to pathogens (Goldin and Gorbach et al., 1984). The most common probiotics include *Lactobacillus bulgaricus*, *Lactobacillus casei*, *Lactobacillus acidophilus*, and bifidobacteria. Probiotic organisms have great benefit because, like antibiotics, they are alternative methods for the treatment of disease. Moreover, consuming dairy products containing probiotic organisms can lead to improvement of lactose intolerance symptoms, improvement of gastrointestinal motility, control of intestinal disorders attributed to infection, anti-carcinogenic condition, and lowering of serum cholesterol (Havenaar and Huis, 1992).

The symptom lactose malabsorption (lactose intolerance), which generally occurs by individuals with reduced ability in digesting lactose, leads to cramps, flatulence and diarrhea, and is induced by the enzyme lactase (β-galactosidase). Since β-galactosidase hydrolyzes disaccharide (lactose) to glucose and galactose, its production by probiotic bacteria has been gaining attention in improving their potential to treat lactose intolerance. Gilliland and Kim (1984) indicated that microbial β-galactosidase survived gastric digestion and improved lactose digestion.

An important criterion when selecting a probiotic strain is its ability to survive the acidic conditions of the gastric tract. Kaila (1992) observed that pH of stomach fluid could fall to as low as 1.5 and a salt content not less than 0.5% (w/v). The intrinsic tolerance of *Lactobacilli* to pH 1 and 4 has been examined in vitro using solutions such as acidified broths, buffers and complex media consisting of pancreatic enzymes, or fresh human gastric fluids (Conway et al., 1987). Although *Lactobacillus* species are considered intrinsically resistant to acid, they generally demonstrate increased sensitivity at pH values below 3 (Kaila, 1992). The acid tolerance of *Lactobacilli* is attributed to their ability to maintain a constant pH gradient between the pH of the medium and their cytoplasmic pH. When the internal pH reaches a threshold value, cellular function is inhibited and the cell dies (Kaila, 1992).

Lactobacillus reuteri is one of the few indigenous *Lactobacillus* species in the gastrointestinal (GI) tract of man and animals (Mitsuoka, 1992). Dobrogosz and Lindgren (1988) found that under anaerobic conditions and in the presence of glycerol, *L. reuteri* produces and secretes a substance that inhibits the growth of harmful bacteria in the gastrointestinal tract. This could explain the protective role that *L. reuteri* provides against a number of illnesses, by producing a substance called Reuterin. *L. reuteri* has been used as a probiotic dietary supplement for many years. It provides numerous advantages such as lowering pH value in the gastro-intestinal tract, creating an inhospitable local environment that inhibits the proliferation of pathogenic micro-organisms. In addition, *L. reuteri* has been shown to modulate immune responses in a series of animal models (Casas et al., 1998).

Objective. The objectives of this research were (1) to study the effect of bile salt on the survival and growth of *L. reuteri*, (2) to study the influence of bile salt on ß-galactosidase activities of *L. reuteri*, and (3) to examine the effect of Tween 20 80 on the bile salt tolerance of *L. reuteri*.

Materials and Methods

Bacterial Strains

Four strains of *Lactobacillus reuteri*-CF2F (child fecal isolate), DSM20016 (mother's milk isolate), MM2-3 (mother's milk isolate), and MM7 (mother's milk isolate)-obtained from Biogaia Biologics, Inc. (Raleigh, NC) were used in this study. The strains were propagated three times at 37°C for 16 h before they were used for the experiment. A 24-h culture of each strain with cell density (A_{610}) of 1.2 containing 1.2×10^6 CFU/ml was centrifuged (Sorvall RC5B *plus*, Beckman Instruments, Fulerton, CA) for 15 min at 4°C (7,500 × g). The pellets were washed with sterile 0.1% peptone water. These suspended cells were used for the study.

Growth in Bile Salt

Ten ml batches of MRS broth (55 g/L, DifcoTM Lactobacilli MRS broth, Becton Dickson and Co., Sparks, MD) containing bile salt (Sigma Chemicals, St. Louis, MO) at concentrations of 0.0, 0.1, 0.2, 0.3, 0.4, 0.5, 1.0, 2.0, and 4.0% (w/v) were sterilized by autoclaving at 121°C for 15 min. One hundred micro liters of each suspended culture were then transferred to fresh MRS broth containing different concentrations of bile salts. The samples were then mixed by vortex and then placed in anaerobic jar. The jar was then placed in an incubator at 37°C and the extend of growth was measured using spectrophotometer using a Spectronic 21 Milton Roy Spectrophotometer (Thermo Electron Scientific Co., Madison, WI) at 610 nm at 2 h intervals during 10 h incubation period.

ß-galactosidase Study

A 24-h culture was grown in modified M17 (Becton Dickson and Co., Sparks, MD) with glucose or with lactose as a sole source of carbohydrate to achieve a mild log phase (Ibrahim and O'sullivan, 2000). The growth was monitored by measuring the turbidity at 610 nm using the Spectronic 21. Aliquot of culture was added to a Z buffer to make a final volume of 10 ml. To each tube, 10-μl chloroform was added, mixed immediately and then subjected to reaction at 37°C for 30 min. At the end of the reaction time, 0.2 ml of O-Nitrophenyl-ß-galactopyranoside (ONPG, 4 mg/ml) was added and the time needed to develop a strong yellow color was recorded. The chemical reaction was stopped by adding 0.5 ml of Na_2Co_3 (10.6 g/100 ml, Sigma Sigma-Aldrich Co., St. Louis, MO) and absorbance (OD) was recorded at both 420 and 550 nm by using the Spectronic 21. The ß-galactosidase activity (Miller unit) was calculated as follows:

$$100\,[\text{OD at } 420\,\text{nm} - 175 \times \text{OD at } 550\,\text{nm}/(\text{T} \times \text{V} \times \text{OD at } 600\,\text{nm})]$$

where T is the time and V is the volume.

Enhancement of Bile Tolerance in L. reuteri *by Tween 80*

In this study modified trypticase peptone yeast extract (TPY) medium (Scardovi, 1986) was used. One hundred ml batches of TPY broth supplemented with combinations of bile salt (0.03 g) and Tween 80 (either 1 or 10 ml) were aseptically prepared and defined as treatment media. One hundred microliters of each suspended culture were then subjected to the treatment media. The treatment media were placed in anaerobic jar and incubated at 37°C. The extend of growth was measured as described above.

Statistical Analysis

All data were analyzed using statistical analysis system (SAS) with general linear model procedure to determine significant differences among treatments ($P < 0.05$).

Results and Discussion

The survival and growth of *L. reuteri* in various concentrations of bile salt with incubation time is shown as in the turbidity (OD reading) of MRS broth (Fig. 1 a–d). During the initial incubation period, little growth of *L. reuteri* was observed with all the treated samples. All *L. reuteri* strains subjected to control samples reached to maximum growth within 10 h of incubation. Figure 1a–d indicated that bacterial strains tested in this study were somewhat tolerant to 0.1% bile salt. When bile salt concentrations were increased to 0.2% or higher, a significant growth inhibition was observed in all treated strains. OD readings of control samples reached to approximately 1.0 after 8 h incubation at 37°C while broth containing 0.3% bile salt had 0.6. In other words, high populations (as represented in OD readings) of microorganism were observed in the broth without bile salt whereas broth with bile salt inhibited the growth of all tested strains.

Table 1 show the effect of bile salt on the β-galactosidase activities of *L. reuteri* strains grown in modified M17 broth. Addition of bile salt significantly ($P<0.05$) enhanced the production of β-galactosidase 20% and 88% for strains DSM20016 and MM2-3, respectively. However, there was no significant difference of β-galactosidase production for strains CF2F and MM7 due to the addition of bile salt. Several scientists (Ibrahim and O'Sullivan, 2000) indicate that β-galactosidase production by probiotic bacteria is beneficial in improving their potential to treat symptoms of lactose malabsorption.

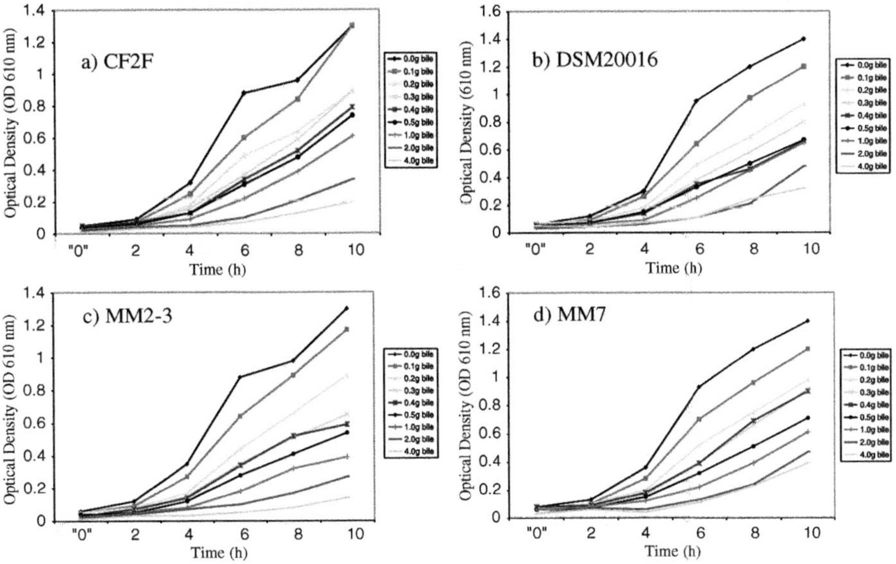

Fig. 1 **(a–d)** Survival and growth of *L. reuteri* strains in MRS broth due to the concentrations of bile salt

Table 1 Effect of bile salts on the β-galactosidase activities of *L. reuteri* strains in modified M17 broth

	β-galactosidase activity (Miller unit)	
Strain	Without 0.3% bile salt	With 0.3% bile salt
CF2F	700	800
DSM20016	1000	1200
MM2-3	800	1500
MM7	1100	1000

Effect of Tween 80 on the bile tolerance of microorganisms in TPY broth is shown in Table 2. OD reading of strain DSM20016 subjected to TPY broth was the same (0.94) whether bile salt was added or not indicating that this strain was bile resistant. However, growth rate of strains CF2F, MM2-3 and MM7 were inhibited by the presence of bile salt. Moreover, OD readings of broth were increased from 0.94 to 1.10 and 1.30 with the addition of 1 and 10 ml Tween 80, respectively, enhancing the growth of strain DSM20016. These results suggest that more addition of Tween 80 would allow for higher growth of DSM20016. These results clearly indicate that the addition of Tween 80 helps bacterial growth and increasing the amount of Tween 80 from 1% to 10% enhanced bile salt tolerance of the microorganisms.

Table 2 Effect of supplements on the growth of *L. reuteri* strains in TPY broth

Strain	TPY medium		OD reading
	0.3% Bile salt%	Volume of Tween 80 (ml)	
CF2F	0.0	0	1.00
	0.3	0	0.86
	0.3	1	0.94
	0.3	10	1.20
	0.0	1	1.20
	0.3	10	1.20
DSM20016	0.0	0	0.94
	0.3	0	0.94
	0.3	1	1.10
	0.3	10	1.30
	0.0	1	1.10
	0.0	10	1.30
MM2-3	0.0	0	1.10
	0.3	0	0.73
	0.3	1	1.00
	0.3	10	1.30
	0.0	1	1.20
	0.0	10	1.30
MM7	0.0	0	1.10
	0.3	0	0.60
	0.3	1	0.82
	0.3	10	1.20
	0.0	1	1.20
	0.0	10	1.20

Conclusions

Growth rate of all *Lactobacillus* strains tested was affected by the presence of bile salt. Addition of bile salt significantly enhanced the production of β-galactosidase for strains DSM20016 and MM2-3. The addition of Tween 80 helped bacterial growth and increasing the amount of Tween 80 from 1% to 10% enhanced bile salt tolerance of the microorganisms. Hence, enhanced bile salt tolerance of *L. reuteri* induced by Tween 80 may have the potential for commercial applications and should be considered for useful applications when probiotic strains are selected.

References

Casas, I. A., F. W. Edens, and W. J. Dobrogosz. 1998. *Lactobacillus reuteri*: An effective probiotic for poultry and other animals. In: Salminen, S. and von Wright, A. (Eds.), Lactic Acid Bacteria, Second Ed., Vol. 85, Food Science and Technology Series, Marcel Dekker, Inc. New York, NY 10016-0602, pp. 475–518.

Conway, P. L., S. L. Gorbach, and B. R. Goldin. 1987. Survival of lactic acid bacteria in the human stomach and adhesion to intestinal cells. J. Dairy Sci. 70:1–12.

Dobrogosz, W. J. and S. E. Lindgren. 1988. Antibiotic reuterin. US Patent PCT/US88/01423.

Gilliland, S. E. and H. S. Kim. 1984. Effect of viable starter culture bacteria in yogurt on lactose utilization in humans. J. Dairy Sci. 67:1–6.

Goldin, B. R., and S. L. Gorbach, S. L. 1984. Alterations of the intestinal microflora by diet, oral antibiotics and *Lactobacillus*: decreased production of free amines from aromatic nitro compounds, azo dyes, and glucorinides. J. Nat. Cancer Inst. 73:689–695.

Havenaar, R. and J. H. Huis. 1992. Probiotic. A general view. In: Wood, B. J. B. (Ed.), The Lactic Acid Bacteria, Vol. 1, The lactic acid bacteria in health and disease. Elsevier applied Science, London, pp. 151–171.

Ibrahim, S. A. and D. J. O'Sullivan. 2000. Use of chemical mutagenesis for the isolation of food grade β-galactosidase overproducing mutant of *bifidobacteria*, *lactobacillus* and *Streptococcus thermophillus*. J. Dairy Sci. 83:923–930.

Kaila, M. E. 1992. Enhancement of circulating antibody secreting cell response in human diarrhoea by a human *Lactobacillus* strain. Pediatr. Res. 32:141–144.

Mitsuoka, T. 1992. The human gastrointestinal tract. In Wood, B. J. B. (Ed.), The Lactic Acid Bacteria, Vol. 1, The lactic acid bacteria in health and disease. Elsevier Applied Science, London, pp. 69–114.

Scardovi, V. 1986. In genus *Bifidobacterium*. In: Sneath, P. H. A., Mair, N. S.., Sharpe, M. E. and Holt, J.G. (EDs.) Bergey's manual of Systematic Bacteriology, Vol. 2, Baltimore Wlliams and Wilkins, pp. 1418–1434.

The Effect of Insulin from Jerusalem Artichoke (*Helianthus tuberosus*) Extract on Growth and Viability of *Lactobacillus salivarius* in Fermented Milk

Souod R. Alani, Angela M. Fraser, Mustafa A. Alsharafani, Farouk F. Al-Nouri, and Salam A. Ibrahim

Abstract Inulin, oligosaccharides produced by many plants, has been shown to promote the growth and viability of probiotic cultures in food during cold storage. The objective of this study was to determine the effect of inulin from Jerusalem artichoke extract on the growth and viability of *Lactobacillus salivarius* (SRA01) in fermented milk during cold storage at 5°C. Jerusalem artichoke extract containing 4.8% inulin was added to sterile fresh milk at 2.5% and 5% (v/v, 0.12 and 0.24% inulin), respectively. The milk sample was inoculated with *L. salivarius* and then stored at 5°C. Bacterial growth was monitored during the 21 day incubation period. Milk samples without inulin extract were used as the control. Addition of inulin increased the populations of *L. salivarius* in fermented milk samples by 0.9 log compared to the control after 21 days of storage at 5°C. These results suggest that adding inulin to fermented milk could promote growth and provide protection for *L. salivarius* during cold storage.

Introduction

Inulin has been shown to enhance the growth of beneficial microorganisms, such as *Lactobacilli* and *bifidobacteria,* in food known as probiotics (Gibson et al., 1995; Gallaher and Schneeman, 1996). Probiotics are defined as living microbes added to food to improve the microbial balance in the intestine when consumed (Nyman et al., 1985; Fuller, 1989; Ohr, 2007).

The Fermented Milk and Lactic Acid Bacteria Beverages Association of Japan requires at least 10^7 viable cells per milliliter to be in fresh dairy products, whereas,

S.R. Alani (✉)
Department of 4H Youth Development and Family & Consumer Sciences, North Carolina State University, Raleigh, NC 27695, USA

G.A. Uzochukwu et al. (eds), *Proceedings of the 2007 National Conference on Environmental Science and Technology*, DOI 10.1007/978-0-387-88483-7_4,

the National Yogurt Association (NYA) of the United States specifies 10^8 viable cells/gram be in product at the time of manufacture. In order to describe probiotic fermented milk as having health-promoting properties, a minimum number of viable cells must present at the expiry date of the product (Vuyst, 2000).

Meulen et al. (2004) found *Bifidobacterium animalis* increased to 2.8×10^9 CFU/g in MRS media containing 2% inulin compared to media containing monosaccharide, which bacteria could not consume as a source of energy. While there are literatures reporting the effect of inulin on *L. salivarius* in general, minimal literature is available documenting the behavior of probiotic cultures in milk supplemented with Jerusalem artichoke extract.

Objective. The objective of this study was to determine the effect of inulin from Jerusalem artichoke extract on the growth and viability of *Lactobacillus salivarius* in fermented milk during cold storage at 5°C.

Materials and Methods

Fresh Jerusalem artichoke (*Helianthus tuberosus*) was purchased from a local store in Baghdad. Jerusalem artichoke was refrigerated at 4°C in plastic bags until used the next day. Inulin extract was prepared as follows: an aliquot (100 g) was stir-mixed with 500 ml of distilled water (1:5 w/v), heated at 75°C for 50 min, and cooled to room temperature (24 ± 2°C). The extract was filtered using cotton cloth (Kim and Shin, 1998). Crude extract was clarified by adding lead acetate (30%). The remaining lead was precipitated by adding drops of a saturated solution of sodium oxalate. A 5 ml of 5% HCl was used to hydrolyze inulin to produce free fructose as described by Marchal et al. (2003). Inulin was estimated by the difference between the amount of fructose before and after hydrolysis using a fructose standard curve (Alsharafani, 2006). Titratable acidity and pH were also measured (Alrawi, 1985).

A local isolate of *L. salivarius* was used as the starter to make the fermented milk samples (Alrawi, 1985). *L. salivarius* isolate was maintained on MRS agar and activated by a daily transfer into sterilized skim milk (Lankaputhra and Shah, 1996).

Inulin extract was added to milk at 2.5% (final inulin concentration of 0.12%) and 5.0% (final inulin concentration of 0.24%) and called samples T1 and T2, respectively. Control samples (Tc), without inulin, were included in this experiment as a comparison group.

All samples were autoclaved at 121°C for 5 min and distributed in 100 ml sterilized plastic containers. Active starter culture was added at 5% to each container after cooling to room temperature (24 ± 2°C) and incubated at 37°C until coagulation was achieved. Samples were then stored at 5°C for 21 days. Data was collected at days 1, 7, 14, and 21 of storage at 5°C. The experiment was completed in triplicate and means were calculated for each treatment group at each data collection point.

Results and Discussion

Table 1 shows the number of viable *L. salivarius* (log CFU/g) in milk samples during cold storage at 5°C. Viable *L. salivarius* populations for Tc, T1, and T2 were reduced from 9.8 log CFU/g by 0.9, 0.1 and 0.0 log CFU/g, respectively, after 24 h incubation. The T2 samples had a higher number of viable *L. salivarius* compared to T1 samples after the first day of incubation. These results are related to the presence of inulin extract in both treatment groups. After 7 days of storage at 5°C, the number of viable *L. salivarius* for T1 and T2 samples decreased by 0.3 and 0.2 log, respectively, when compared to the first day. After 14 days of storage, the number of viable *L. salivarius* decreased by 0.8 and 0.7 log in T1 and T2 samples, while control samples did not change much. Although increasing storage time from 14 to 21 days did not affect populations of *L. salivarius*, higher numbers in both treatment groups were recovered than control sample.

The number of viable cells used in this study was within the therapeutic minimum of 10^6–10^7 CFU/g (Akin, 2005) and met the criteria established by the National Yogurt Association (NYA) of the United States, which is 8 log CFU/g. Functional dairy products have to contain a high number of bacteria to function as a probiotic (Vuyst, 2000).

Titratable acidity of T1 and T2 samples increased during cold storage at 5°C for 21 days. Titratable acidity increased in proportion with increasing numbers of viable *L. salivarius*. Increased acidity and decreased pH correlated to the number of viable *L. salivarius*. The higher concentration of inulin would allow *L. salivarius* to produce more acid compared to control sample. These results are consistent

Table 1 Titratable acidity, pH, and populations of viable *L. salivarius* (log CFU/g) in fermented milk supplemented with 0, 0.12 and 0.24% inulin during 21 day incubation at 5°C*

Sample	Storage time (day)	Populations of viable *L. salivarius* (log CFU/g)	pH	Titratable acidity (%)
Control (Tc)	1	8.9	5.15	0.75
	7	8.7	5.12	0.75
	14	8.7	5.10	0.80
	21	8.1	5.10	0.80
Supplemented with 0.12% inulin (T1)	1	9.7	5.00	0.90
	7	9.4	5.00	0.9
	14	9.0	5.00	0.95
	21	9.0	5.00	0.95
Supplemented with 0.24% inulin (T2)	1	9.8	4.97	1.0
	7	9.6	4.97	1.0
	14	9.1	4.90	1.05
	21	9.0	4.90	1.05

*Initial population of *L. salivarius* inoculated at day 0 was 9.8 log CFU per gram of milk.

with the findings of Akin (2005), who studied *Bifidobacterium bifidum* and *Lactobacillus acidophilus* in fermented dairy products and found that higher concentrations of inulin would allow starter culture to produce acid. He also demonstrated that addition of inulin stimulated the growth of probiotic culture in fermented ice cream and improved the viability of these cultures during cold storage at –18°C for 90 days.

Conclusions

Adding inulin at 0.12% and 0.24% to fermented milk increased the number of viable *L. salivarius* during 21 days of storage at 5°C. These results confirm that inulin improved the viability of *L. salivarius* compared to control samples. T2 samples contained twice the concentration of inulin as T1 but the number of viable cells remained the same after 21 days of storage at 5°C, therefore, adding lower concentrations of inulin would be economically feasible. For fermented milk products with added inulin to have a health benefit, the number of viable cells in the product should be at a high level such as 8 log CFU/g.

Acknowledgments We like to express our gratitude to the Institute of International Education, Scholar Rescue Fund for their support, 809 UN Plaza, NY, NY 10017, www.iie.org/SRF.

References

Akin, M. S. 2005. "Effect of inulin and different sugar levels on viability of probiotic bacteria and the physical and sensory characteristics of probiotic fermented ice-cream". *Milchwissenschaft*. 60(3): 297–301.

Alrawi, T. S. 1985. The measurement of inherent acidity in "Practical ways to analyze the milk and its products." Mosul University press, Iraq, pp. 1–84.

Alsharafani, A. M. 2006. "Effects of Inulin extracts from Jerusalem artichoke tubers on cholesterol and mineral availability and as Prebiotic." M.S.Thesis, University of Baghdad, Iraq.

Fuller, R. 1989. "Probiotics in man and Animals." *J. Appl. Bacterial.* 66: 365–378.

Gallaher, D. D. and B. O. Schneeman. 1996. Dietary-fiber. In "present knowledge in nutrition"(Ziegler, E. E & Filer, L. J., eds). International life science institute, Washington, DC.

Gibson, G. R., E. R. Bear, X. Wang, and J. H. Cummings. 1995. "Selective stimulation of Bifidobacteria in the human colon by oligofructose and inulin." *Gastroenterology.* 108: 975–982.

Kim, M. and H. K. Shin. 1998. "The water-soluble extract of chicory influences serum & liver lipid concentrations, fecal short-chain fatty acid concentrations & fecal lipid excretion in rats." *J. Nutr.* 128(10): 1731–1736.

Lankaputhra, W. E. and V. N. P. Shah. 1996. "Survival of *Lactobacillus acidophilus* and Bifidobacterium sp." In presence of acid bile salt. *Cult. Dairy Prod. J.* 30(5).

Marchal, L., M., E. Vorobiev, and J. P. Bonhoure. 2003. Recovery of inulin from Jerusalem artichoke tubers: Development of pressing method assisted pulsed electric field. International Conference on Engineering and Food, France.

Meulen, R.V.D., L. Avonts, and L. D. Vuyst. 2004. "Short fractions of oligofructose are preferentially metabolized by Bifidobacterium animalis DN-173010." *Appl. Environ Microbiol.* 70(4): 1923–1930.

Nyman, M., P. Birthe, and O. E. Bjorn 1985. "Fermentation of dietary fiber in the intestinal tract of rats-a comparison of flours with different extraction rates from six cereals." *J. Cereal Sci.* 3: 207–219.

Ohr, M. L. 2007. "Nutraceuticals: Ingredients for good digestion." *J. Food Technol.* 61(7): 83–88.

Vuyst, L. D. 2000. "Technology aspects related to the application of functional starter cultures." *Food Technol. Biotechnol.* 38(2): 105–112.

Lactic Acid Production from Apple Skin Waste by Immobilized Cells of *Lactobacillus reuteri*

Moussa M.E. Salem, Salam A. Ibrahim, Chyer Kim, Chung W. Seo, Abolghasem Shahbazi, and Amer AbuGhazaleh

Abstract Apple skins generally are not used in food products and often are discarded as waste. Apple skin contains several useful nutrients particularly sugars and more complex carbohydrates. This study was undertaken to determine the feasibility of apple skin waste as a media source for the production of lactic acid (LA) using *Lactobacillus reuteri*. One to nine ratios (10%) of apple skin to water was boiled for 5 min and its filtrate was then added with each of following supplements, respectively; ammonium phosphate, manganese sulfate, peptone, yeast extract, and potassium phosphate monobasic and dibasic. The efficacy of apple skin filtrates with and without supplements was determined by measuring pH, glucose and lactic acid after 2, 4 and 8 h fermentation at 37°C. Moreover, apple skin filtrate was also compared with MRS broth and G-M17 broth for the efficacy of glucose utilization and LA production. At the end of 8 h incubation, pH of apple skin filtrates with peptone and yeast extract reduced from 7 to ~4. The LA concentrations produced from the filtrates with either peptone or yeast extract supplement were the greatest ranging from 0.9 g/L to 1.9 g/L. The efficacy of LA production using apple skin filtrate was about the same as G-M17 broth. Higher amount of LA was yielded with media inoculated with immobilized (encapsulated) cells than free cell. Results also indicated that immobilized *L. reuteri* MM2-3 produced 40% higher amount of LA than free cell. The results obtained from this study indicate that apple skins as a carbohydrate source and immobilization of lactic acid bacteria (LAB) could be a promising method for the production of LA. Further study is necessary to identify ways to increase glucose availability and promote LAB growth and LA production.

M.M.E. Salem (✉)
National Research Center, Dokki, Cairo, Egypt

G.A. Uzochukwu et al. (eds), *Proceedings of the 2007 National Conference on Environmental Science and Technology*, DOI 10.1007/978-0-387-88483-7_5,

Introduction

Lactic acid (LA) is a valuable multi-functional organic acid, widely used as an acidulant, flavorant and preservative in the food, pharmaceutical, leather and textile industries. For example, LA is used in the manufacture of biodegradable and biocompatible polylactate polymers (polylactic acids), an environmentally friendly alternative to non-biodegradable plastics derived from petrochemicals (Magnuson and Lasura, 2004; Litchfield, 1996). In the area of food production, LA has been classified as GRAS as a food additive and is used in a broad range of food-based applications (Naveena et al., 2004).

Large scale production of LA is generally achieved using one of several methods. Conventional batch fermentation is one of the methods commonly used, but it has the limitation of shutting down the growth of lactic acid bacteria (LAB, and therefore LA production) due to inhibitory acid accumulation. Another approach is immobilization of viable LAB, which has attracted considerable interest. Immobilization of LAB by entrapment within a gel bead has been reported to greatly improve the production of LA (Champagne and Cote, 1987).

One potential candidate as a medium for producing large quantities of LA is apple skins. Generally the skins are not used in food products and are discarded. But, apple skin contains various sugars and nutrients and therefore may serve as a viable fermentation medium for the production of LA after pretreatment and supplementation with other nutrients.

Objective. The objective of this study was to determine the feasibility of using apple skins as nutritional supplements for the production of LA via the *Lactobacillus reuteri* immobilization method.

Materials and Methods

Growth Medium Preparation

Fuji apples purchased from a local grocery store (Greensboro, NC) were peeled and the skins were mixed with distilled water at 10% (w/v). The mixture was then boiled at 100°C for 5 min. The suspension fluid only was collected and adjusted to pH 7 using 0.1 N NaOH. The filtrate was then supplemented with either yeast extract (4 g/L), peptone (4 g/L), ammonium phosphate (1 g/L), manganese sulfate (0.2 g/L), or potassium phosphate monobasic (1 g/L) plus potassium phosphate dibasic(1 g/L). The supplemented samples were then heated to 80°C for 10 min in a water bath to eliminate background microflora.

LAB Microorganisms Used

Four strains of *Lactobacillus reuteri* (CF2F, DSM20016, MM7 and MM2-3) obtained from Biogaia Biologics, Inc. (Raleigh, NC) were used in this study. Each strain was cultured individually in 10 ml of MRS broth (Difco^{TM} *Lactobacilli* MRS broth, Becton Dickinson and Co., Sparks, MD) supplemented with 0.5 g/L of L-cysteine at 37°C for 16 h.

Microorganism Immobilization

A 24-h culture of each strain was centrifuged (Model Sorvall RC5B *plus*, Thermo Scientific, Asheville, NC) for 15 min at 4°C (5,000 × g). The pellets were washed with sterile 0.1% peptone water and mixed with 2% (w/v) sodium alginate (Sigma-Aldrich Co., St. Louis, MO). The homogenous mixture of the bacterial cell suspended in alginate solution was then dripped into 0.5 M calcium chloride solution (4°C) using a separatory funnel (ID: 1 cm). The immobilized gel beads were then rinsed with sterile 1% peptone water at 4°C and approximately 1.2×10^6 cells were immobilized in each 1 cm bead.

LA Fermentation by Immobilized L. reuteri

Five milliliters (determined based on previous unpublished work by the authors) of each of the four *L. reuteri* strains were individually inoculated into 100 ml of samples of each of the test filtrates (adjusted to pH 7.00) either in free cell or immobilized form. These samples were then incubated at 37°C up to 24 h. The samples were periodically analyzed for pH, glucose and LA using a pH meter (Accumet® Excel XL15, Thermo Fisher Scientific, Fair Lawn, NJ) and HPLC (Model Waters 717 plus auto sampler, Milford, MA), respectively. Filtrate supplemented with medium, which produced significant ($P<0.05$, Table 1) amount of LA, was chosen and adjusted to pH 7, and compared with commercial MRS and G-M17 broth for the utilization of glucose and the production of LA during 24 h fermentation process.

Data Analysis

Experiments were replicated twice. Data were analyzed by the general linear model procedure of the Statistical Analysis System. Comparisons of means were performed using Duncan's multiple range test.

Results and Discussion

LA Production

Change in the pH values, utilization of glucose and production of LA during 8 h fermentation with *L. reuteri* strains (DSM2016, CF2F, MM7 AND MM2-3) using

Table 1 Effect of supplements on the pH, glucose and LA in apple skin filtrate due to *L. reuteri* fermentation at 37°C*

Growth medium		*L. reuteri* strains	Fermentation time (h) 2	4	8		
Apple skin filtrate	Supplement		pH	pH	pH	Glucose (g/L)	LA (g/L)
	none	CF2F	4.88	4.26	3.90	3.910 b	0.790 a
		DSM20016	5.17	4.74	4.16	3.725 c	0.740 a
		MM2-3	5.23	4.95	4.31	4.015 b	0.720 a
		MM7	5.20	5.02	4.86	4.300 a	0.340 b
	Ammonium phosphate	CF2F	6.91	6.27	4.76	3.450 b	0.870 a
		DSM20016	6.98	6.67	5.27	3.425 b	0.865 a
		MM2-3	7.01	6.66	5.73	3.580 b	0.715 b
		MM7	7.02	6.81	6.51	3.845 a	0.385 c
	Manganese sulfate	CF2F	4.92	4.22	3.81	3.925 b	0.835 a
		DSM20016	5.06	4.73	4.16	3.800 b	0.875 a
		MM2-3	5.19	5.22	4.13	3.855 b	0.915 a
		MM7	4.92	4.22	5.09	4.300 a	0.290 b
	Peptone	CF2F	4.96	4.31	3.87	3.415 b	1.470 a
		DSM20016	5.24	4.65	4.11	3.610 a	1.175 c
		MM2-3	4.83	4.24	3.87	2.175 c	1.400 b
		MM7	4.86	4.30	3.99	2.270 c	1.055 d
	Potassium phosphate monobasic and dibasic	CF2F	6.34	5.51	4.45	3.180 b	0.825 a
		DSM20016	6.35	5.91	4.85	3.125 b	0.825 a
		MM2-3	6.42	6.04	5.03	3.215 b	0.745 b
		MM7	6.57	6.49	6.35	3.470 a	0.410 c
	Yeast extract	CF2F	4.91	4.27	3.74	3.375 b	1.885 a
		DSM20016	5.16	4.55	4.12	3.630 a	1.125 c
		MM2-3	4.97	4.48	3.99	2.310 d	1.475 b
		MM7	5.23	4.73	4.31	2.560 c	0.895 c

*Initial pH and concentrations of glucose and LA in apple skin filtrate were 7.00, 4.5 g/L and 0 g/L, respectively; means with the same letter in the same column of LA concentration within each supplement are not significantly different ($P>0.05$).

different supplements are shown in Table 1. In general, the pH of growth medium decreased with the increase of incubation time. Filtrates inoculated with CF2F obtained relatively lower pH than other strains. Microorganisms subjected to peptone and yeast extract supplements produced significantly ($P<0.05$) higher amount of LA as compared with control samples. Previous work by Chronopoulos et al., (2002) had suggested that organic nitrogen sources from peptone and yeast extract may promote LA production. However, in our study, the rate of glucose utilization shown in Table 1 did not correlate with the rate of LA production. This may have been due to the relatively short fermentation time (8 h).

Results shown in Fig. 1 illustrate increases of LA production with the increase of fermentation time regardless of growth medium for the microorganisms. Figure 1 also indicates higher amount of LA production from MRS broth than apple skin filtrate and G-M 17 broth.

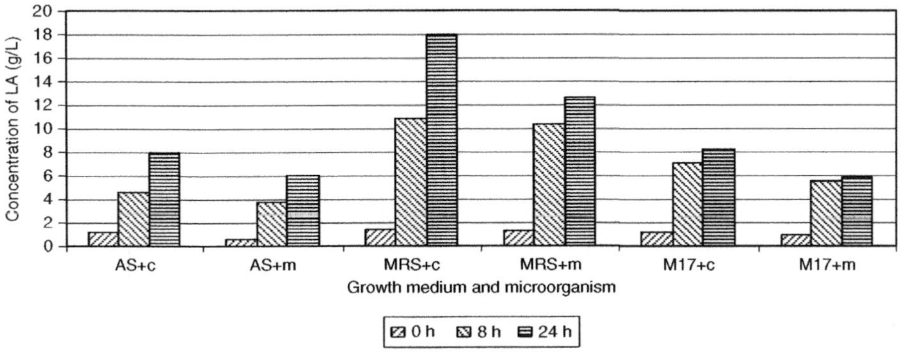

Fig. 1 Effect of growth medium on the production of LA by two strains of *L. reuteri* during 24 h fermentation at 37°C*

*Capital letters preceding symbol "+" represent type of media used; AS: apple skin filtrate, MRS: MRS broth, and M17: G-M17 broth; small letters following symbol "+" represent type of microorganism inoculated; C: *Lactobacillus reuteri* CF2F, and M: *Lactobacillus reuteri* MM7.

Glucose Availability

The effects of the growth medium on the utilization of glucose for two of the strains of *L. reuteri* are shown in Fig. 2. As expected, production of LA was related to the amount of available glucose (Chronopoulos et al., 2002). The results of HPLC analysis (shown in Fig. 2) clearly indicate that there were higher amounts of glucose in MRS broth (15.6 g/L) than in the apple skin filtrate (6.74 g/L) or the G-M17 broth (7.85 g/L). These results are in accordance with the findings of Wee et al. (2006) who also found that the amount of LA produced increased directly with the increase of the quantity of available glucose. Figure 2 also shows that the amount of glucose for MRS, G-M17 and apple skin filtrate decreased from 15.6, 7.85 and 6.74 g/L to 0.17, 2.15 and 3.37 g/L, respectively, after 24 h fermentation at 37°C. Rates of glucose utilization by *L. reuteri*strains were significantly ($P<0.05$) higher for MRS broth than apple skin filtrate and G-M17 broth, while apple skin filtrate was about the same as G-M17. There was no significant difference of glucose utilization due to *L. reuteri* strain.

Immobilized vs. Free Cell Cultures

Comparison of samples of the apple skin filtrate inoculated with each of four strains of *L. reuteri* as immobilized or free cells revealed that lower pH values were obtained from the media inoculated with immobilized cells than free cells (Table 2). Moreover, the amount of LA produced by fermentation of immobilized *L. reuteri* MM2-3 was approximately 40% higher than those inoculated by free cell microroganisms. However, production of LA by other strains was about the same whether the cell was immobilized or not.

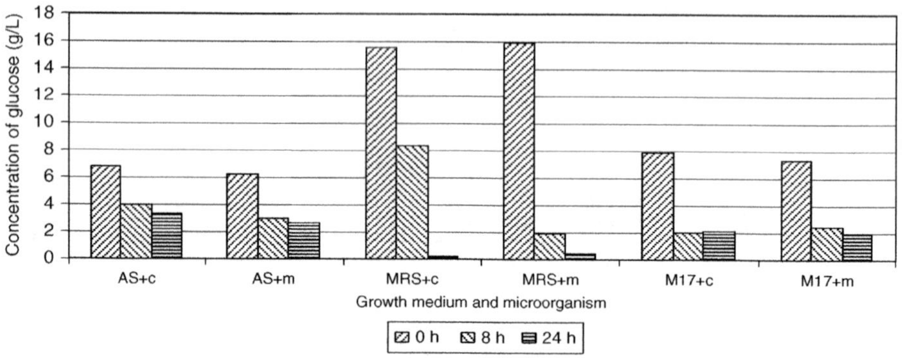

Fig. 2 Effect of growth medium on the utilization of glucose by two strains of *L. reuteri* during 24 h fermentation at 37°C

Table 2 Effect of cell immobilization by four strains of *L. reuteri* on the pH during 24 h fermentation and LA concentration after 24 h fermentation at 37°C, respectively*

		Fermentation time (h)			
		4	8	24	
Strain	Cell type	pH	pH	pH	LA (g/L)
CF2F	Free	4.29	3.79	3.44	9.36
	Immobilized	4.28	3.60	3.28	8.32
DSM20016	Free	6.90	5.32	3.48	9.48
	Immobilized	6.69	5.44	3.31	9.58
MM2-3	Free	4.30	3.85	3.57	5.90
	Immobilized	5.19	3.78	3.24	8.23
MM7	Free	4.48	4.05	3.77	3.97
	Immobilized	6.13	4.65	3.58	4.36

*Initial pH and concentration of LA of apple skin filtrate for all *L. reuteri* strains was 7.00 and 0.7 g/L, respectively.

Moreover, the National Agricultural Statistics Service (World markets and trade, 2007) stated that apple production during marketing year from 2006 to 2007 was 46.1 million tons. Since approximately 10% (5 million tons) of apple total weight are discarded as a waste material, the results obtained from our study (approximately 1 g of LA production from 1 kg apple skin filtrate) indicate possible utilization of apple skin waste for value-added products including production of lactic acid (500 kg).

Conclusions

The higher yield of LA and the lower pH were obtained when the LAB microorganisms (*L. reuteri*) were subjected to fermentation in the growth media studied. Filtrates containing peptone and yeast extract produced a significantly higher amount of LA. The efficacy of LA production by microorganisms using apple skin filtrate as a growth medium was not significantly different from G-M17 broth. MRS broth was the best media for the production of LA. Relatively lower pH values were obtained from the media inoculated with immobilized cells than free cells. The amount of LA produced by fermentation of immobilized *L. reuteri* MM2-3 was approximately 40% higher than its free cell counterpart. The results obtained from this study indicate that apple skins as a carbohydrate source and immobilization of LAB could be a promising method for the production of LA. Further study is necessary to identify ways to increase glucose availability and promote LAB growth and LA production.

Acknowledgment This research was supported in part by a grant from the Egyptian scholarship and funds allocated to North Carolina Agricultural and Technical State University through the Cooperative State Research, Education and Extension Service of the United States Department of Agriculture, Project No. NC.X-173-5-02-170-1. Partial fund was provided by the Nutritional Analysis Service Laboratory at SIUC, Carbondale, IL. The authors also would like to express gratitude to Dr. C.S. Turner, Associate Dean for Research for her support and Dr. K.J. Gruber for his suggestions on this manuscript.

References

Champagne P. C., and B. C. Cote. 1987. Cream fermentation by immobilized lactic bacteria. Biotechnol. Lett. 9, 329–332.

Chronopoulous, G., A. Bekatorou, E. Bezirtzoglou, A. Kaliafas, A. A. Koutinas, R. Marchant, and I. M. Banat. 2002. Lactic acid fermentation by *Lactobacillus casei* in free cell form and immobilized on gluten pellets. Biotechnol. Lett. 24, 1235–1236.

Litchfield, J. H. 1996. Microbiological production of lactic acid. Adv. Appl. Microbiol. 42, 45–95.

Magnuson, J. K., and L. L. Lasure. 2004. "Organic acid production by filamentous fungi." In: J. S. Tkacz and L. Lange (Eds.). Advances in fungal biotechnology for industry, agriculture, and medicine. Kluwer Academic/Plenum Publishers, New York, pp. 207–340.

Naveena, B. J., M. Altaf, K. Bhadrayya, and G. Reddy. 2004. Production of L(+)-lactic acid by *Lactobacillus amylophilus* Gv6 in semi-solid state fermentation using wheat bran. Food Technol. Biotechnol. 42, 147–152.

Wee, Y. J., J. N. Kim, J. S. Yun, D. H. Park, and H. W. Ryu. 2006. Pilot-scale lactic acid production via batch culturing of *Lactobacillus* sp. RKY2 using corn steep liquor as a nitrogen source. Food Technol. Biotechnol. 44, 293–298.

Word markets and trade. 2007. World apple situation. May. http://www.fas.usda.gov/htp/ horticulture/Apples/World_Apple_Situation_053107.pdf

Part II
Bioremediation

Cu and Zn Uptake Inhibition by PAHs as Primary Toxicity in Plants

Zarhelia Carlo-Rojas and Wen-Yee Lee

Abstract Studies of the interference of polycyclic aromatic hydrocarbons (PAHs) on the uptake of copper (Cu) and zinc (Zn) by alfalfa (*Medicago sativa*) were investigated. Alfalfa plants were treated with three PAH compounds individually at 50 $mg{\cdot}kg^{-1}$(soil dry weight) along with Zn and/or Cu. The concentrations of metals were 35 $mg{\cdot}kg^{-1}$(soil dry weight) for Zn and 100 $mg{\cdot}kg^{-1}$(soil dry weight) for Cu which levels were associated with the reported concentrations in soils in the El Paso area. Phenanthrene (Phen) and bezo(a)pyrene (BaP) with combination of Zn or/and Cu were found to reduce the size of alfalfa in 5 days of germination/seedling period. Significant decrease in size in alfalfa was 67–91% in 50 $mg{\cdot}kg^{-1}$BaP treatments, and 30–75% in 50 $mg{\cdot}kg^{-1}$Phen treatments. The uptake of Zn and Cu by alfalfa under the influence of Phen and BaP after 35 days of growth period showed distinctive difference. The Cu uptake was totally inhibited by the two PAHs, while the Zn uptake was inhibited by BaP but enhanced by Phen. Acethylanthracene, however, did not show any effect on the metal uptake. The very different impact of PAH compounds on metal uptake was an indication that there may be various uptake pathways and mechanism of the organics entering into roots.

Introduction

PAHs, which are consist of more than 100 organic compounds containing two or more fused aromatic rings, can be released into the environment during incomplete combustion. They have high melting and boiling points, low vapor pressure, high lipophilicity, and very low solubility in water. Although their acute toxic effect is considered to be moderate to low, their affinity to lipids can cause bioaccumulation and even biomagnifications through the food chain. Thirty PAHs included in the WHO report (UNEP-ILO-WHO, 1998) were identified with genotoxicity to animals

Z. Carlo-Rojas (✉)
Environmental Science and Engineering PhD (ESE) Program UTEP, El Paso, TX, USA

G.A. Uzochukwu et al. (eds), *Proceedings of the 2007 National Conference on Environmental Science and Technology*, DOI 10.1007/978-0-387-88483-7_6,

and 17 were or were suspected to be carcinogenic. Though PAH toxic effects on animals are widely revised, but the direct effect to plants is still understudied (Efroymson et al., 1997). Frequently, some plant species show susceptibility to PAH toxicity. The existing data of toxicity of organics in plants use dose as high as 1000 $mg{\cdot}kg^{-1}$ without visible effect to plants (UNEP-ILO-WHO, 1998; Song et al., 2005), which makes plants ideal candidates for removing contaminants from the environment. In numerous studies, toxicity assessment often does not consider collateral factors such as sorption of the compounds and interaction between contaminants. Plant uptakes of soil-borne heavy metals and organic contaminants were often pursued as two distinct disciplines. This work was therefore aimed to study the interference of PAHs in plant uptake of metals, and to investigate the effect concurrent exposure to Cu/Zn and PAHs on early seedlings of alfalfa.

Materials and Methods

Acute Toxicity Test

To verify the proper concentration levels for this study, acute toxicity test for each single substance was carried out based on the OECD guideline 208 (OECD, 2003) with minor modification. Alfalfa (*Medicago sativa*) was selected as the model plant. The experiment was conducted in laboratory conditions with temperatures between 25°C and 27°C and in total darkness. Twenty alfalfa seeds were placed on a clean filter paper in a petri dish, 5 mL of the corresponding chemical solution were spiked to the system, and the dish was sealed to avoid water evaporation. Seeds germinating in deionized water were the negative control of the assay. Germination rates were recorded after 5 days. Environmental Protection Agency (EPA) analysis Probit software (Probit Analysis Program, Version 1.5) was employed to obtain the Effective Concentration, EC_{50}. As shown in Table 1, four concentration levels were used for each metal and PAH compound. Five additional concentrations of each compound within closer range to the EC_{50} point were then tested to gain better estimate of the EC_{50}.

Table 1 General experimental design to obtain the dose-response curves of each compound

	Levels ($mg{\cdot}L^{-1}$)	
Substance	1st experiment	2nd experiment
Copper (as $CuSO_4$)	10, 100, 1000, 10000	50, 100, 200, 300, 400
Zinc (as $ZnSO_4$)	10, 100, 1000, 10000	100, 200, 400, 600, 800
Benzo(a)pyrene	100, 200, 300	5, 15, 30, 45, 60
Phenantrene	100, 200, 300	5, 15, 30, 45, 60

Effects of PAHs on the Plant Uptake of Zn and Cu

Three PAHs, phenanthrene (Phen), benzo(a)pyrene (BaP), and acetylanthrancene (ACE), and two metals Cu (as in $CuSO_4$) and Zn (as in Zn SO_4) were used in the experiment. The concentrations of the three PAHs were 50 mg·kg^{-1}(soil). The metals levels were referred to regional soil levels which were 35 mg·kg^{-1}for Zn and 100 mg·kg^{-1}for Cu. Three metal treatments were employed: control treatment (neither Zn nor Cu was present); Zn treatment (soil was spiked with 35 mg·kg^{-1}of Zn); and Zn/Cu treatment (soil was spiked with 35 mg·kg^{-1}Zn and 100 mg·kg^{-1}of Cu). The experimental design is summarized in Table 2.

An experimental device modified from Archanvault et al. (2004) was constructed. Briefly, 50 mL HDPE centrifuge tubes were used as the container in a set up designed to avoid analytical interferences and to collect the lechate. Pebbles that previously were washed with nitric acid to assure proper drainage and commercial soil MiracleGro® were used. Laboratory conditions were set to have a light cycle of 12/12 h at 25±1°C.

For each treatment, 4 replicates were prepared. Twenty alfalfa seeds were used for each treatment. The experiment was carried out during a 35 days growth period. Upon harvest, the biomass was composite within each treatment. The parameters evaluated were biomass, size of the plant, and toxicological damages.

Effects of PAHs Combined with Zn and Cu on Alfalfa Seedling

Two PAHs, phenanthrene (Phen) and benzo(a)pyrene (BaP), and two metals Cu (as in $CuSO_4$) and Zn (as in Zn SO_4) were used in the experiment. The concentrations of the PAHs were 50 mg·kg^{-1}(soil). Three metal treatments were employed: control treatment (neither Zn nor Cu was present); Zn treatment (soil was spiked with 35 mg·kg^{-1}of Zn); and Zn/Cu treatment (soil was spiked with 35 mg·kg^{-1}Zn and 100 mg·kg^{-1}of Cu). The experiment was conducted in laboratory conditions with temperatures between 25°C and 27°C and in total darkness. Each Petri dish was sealed to avoid water evaporation. For each treatment, 20 seeds were placed in each petri dish and 4 replicates of each treatment were prepared. The parameters

Table 2 Summary of the 12 treatments of PAHs and Zn/Cu combination in alfalfa seedling experiment

Metals	Organics (mg·kg^{-1})			
Zn/Cu (mg·kg^{-1})	Control	Phenanthrene	Benzo(a)Pyrene	Acetoanthrancene
0/0	0	50	50	50
35/0	0	50	50	50
35/100	0	50	50	50

evaluated were length of hypocotyls and radical, biomass and sign of toxicological damages.

Metal Analysis

0.1 g of oven-dry biomass was placed into Teflon-lined extraction vessels, in which 25 ml of HNO_3 was added to each sample. The extraction was performed at 110°C for 10 min at 100% power (1000 W) in a Microwave Extraction Unit. USEPA method 6010B for Inductively Coupled Plasma–Atomic Emission Spectrometry was used for metal analysis.

Results and Discussion

Acute Toxicity Test

Acute toxicity test for each substance was conducted, and the calculated EC_{50} in alfalfa was 880 mg·L^{-1} for Zn and 290 mg·L^{-1} for Cu ($P < 0.05$). For Phen and BaP, the preliminary result indicated that the EC_{50} values were above 50 mg·L^{-1}. Therefore, the concentrations of metals and PAHs used in the experiment (35 mg·kg^{-1} of Zn, 100 mg·kg^{-1} of Cu, 50 mg·kg^{-1} of PAH) were considered not toxic to the plants.

Effects of PAHs on the Plant Uptake of Zn and Cu

The effects of 3 PAHs, phenanthrene (Phen), benzo(a)pyrene (BaP), and acetylanthrancene (ACE), on the metal uptake by alfalfa revealed that individual PAH compound affected the uptake of Zn and Cu differently. BaP and Phen totally inhibited uptake of Cu while ACE did not show any effect compared to the control

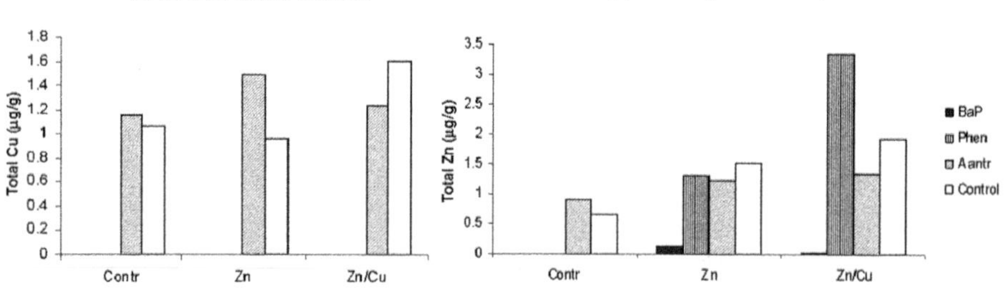

Fig. 1 (**a**) Total uptake of Cu (µg) and (**b**) total uptake of Zn (µg) in lyophilized biomass of alfalfa plants. The values were averages of 4 replicates within treatments. Standard deviations were within 25% of the average values

(Fig. 1a). BaP reduced the uptake of Zn uptake in all treatments, while Phen seemed to enhance the Zn uptake when Cu was present in the soil (Fig. 1b). Again, ACE did not show any effect on Zn uptake in all treatments. Although the plants did not show any toxicological effect by the treatment, they exhibited a high percentage of loss in growth during the germination and establishment phases. Nevertheless, in the post-emergency stage, the surviving plants showed a satisfactory growth, compensating the previous losses. This observation suggests that alfalfa has high susceptibility to metal or organic contaminants in early stages and the toxicological effect was surpassed in the next phase of plant development.

Effects of PAHs Combined with Zn and Cu on Alfalfa Seedling

Based on the findings, Phen and BaP along with metal treatment were tested for their effects on seedling of alfalfa. The size of the seedlings after 120 h treatment showed a significant difference between the organic treatments but no differences were observed between metal treatments (Fig. 2). The growth of the alfalfa seedlings in germination test resulted in a significant inhibition of 67–91% in BaP treatments, and 30–75% in Phen treatments. The more evident alteration was caused by the delay in expansion and inhibition of the hypocotyls elongation. Even thought there was not any visual damage in plants kept over five weeks within the conditions in our laboratory, it was evident that Phen and BaP at the concentration tested affected the germination and development of the seedlings.

BaP clearly affects the uptake of both Cu and Zn. Due its low reactivity and lipophobicity, BaP could affect membrane permeability at high concentrations which resulted in the inhibition of both Cu and Zn uptake. Phen appears to be an interesting case in this study. Phen at the dose tested caused an uptake inhibition of Cu but not Zn in alfalfa. The size of Phen allows easy access into the vegetal cytoplasm (Barhoumi et al., 2000). Research suggested that Phen might alter the

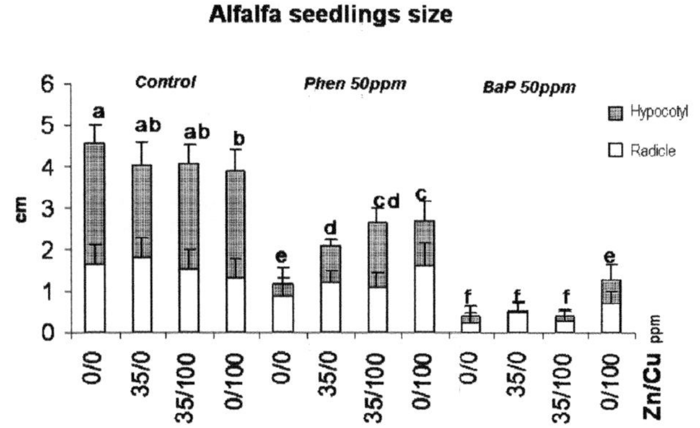

Fig. 2 Size of alfalfa radicles and hypocotyls after 120 h. The effect of the organic treatment result highly significant, meanwhile the metal treatment was not significant

stability and function of lysosomal membranes and therefore change the toxicity of zinc (Shen et al., 2006).

Our results have shown a clear interaction between metals and organics during the germination, seedling, and metal uptake in alfalfa. This implies the need for the characterization of the plant's response to the stress produced by mixtures of organic and inorganic compounds. Future study will test PAH/metal uptake interaction in other plant species and investigate the mechanism involved in such process

Acknowledgments The authors gratefully acknowledge the support of HBCU/MI (Historically Black Colleges and Universities/Minority Institutions) under the grant DE-FC0202EW15254, CERM (Center for Environmental Resource Management) at UTEP, CONACYT (Consejo Nacional de Ciencia y Tecnologia), and SRI (Secretaria de Relaciones Internacionales), Mexico.

References

Archanvault, D.J., J.J. Slaski, and X. Li. 2004. "A rapid, sensitive, seedling-based bioassay for determination of toxicity of solid and liquid substrates and plant tolerance". *Soil and Sediment Contamination* 13:53–63.

Barhoumi, R., Y. Mouneimne, K.S. Ramos, S.H. Safe, T.D. Phillips, V.E. Centonze, C. Ainley, M.S. Gupta, and R.C. Burghardt. 2000. "Analysis of Benzo[a]pyrene Partitioning and Cellular Homeostasis in a Rat Liver Cell Line". *Toxicological Sciences* 53:264–270.

Efroymson, R.A., M.E. Will, E.W. Suter II, and A.C. Wooten. 1997. Toxicological Benchmarks for Screening Contaminants of Potential Concern for Effects on Terrestrial Plants: 1997 Revision. Technical memorandum, US DOE. ES/ER/TM-85/R3, Tennessee.

OECD. 2003. Guideline for testing chemicals, Guideline 208. Terrestrial plant test: Seedling emergence and seedling growth test. Organisation for Economic Co-operation and Development, Paris.

Shen, G., Y. Lu, and J. Hong. 2006. "Combined effect of heavy metals and polycyclic aromatic hydrocarbons on urease activity in soil". *Ecotoxicology and Environmental Safety* 63(3): 474–480.

Song, Y.F., P. Gong, Q.X. Zhou, and T.H. Sun. 2005. "Phytotoxicity assessment of phenanthrene, pyrene and their mixtures by a soil-based seedling emergence test". *Journal of Environmental Science* 17(4):580–583.

UNEP-ILO-WHO. 1998. Selected non-heterocyclic policyclic aromatic hydrocarbons. Environmental health criteria 202. United Nations Environmental Program-International Labour Organisation, Genova.

Metagenomic Environmental Assessment of Aquatic Ecosystems

Parke A. Rublee, Vincent C. Henrich, and Michael M. Marshall

Abstract Environmental assessment may be dramatically improved using a metagenomics approach directed at prokaryotic and eukaryotic microbes. These organisms respond rapidly to environmental change and are sensitive bioindicators, but fundamental knowledge of microbial biogeography is essential for valid environmental inferences to be made. Real-time quantitative PCR was used to determine if representative prokaryotic and eukaryotic taxa are widely dispersed across freshwater sites at narrow scales (regional, within a single state) to broad scales (comparisons among North American and Asian sites). The results demonstrate wide distribution of some taxa, but their abundance varies at individual sites presumably owing to local environmental conditions. Thus, direct assessments of abundance of targeted microbial consortia can be a tool for environmental monitoring and assessment.

Introduction

The use of bioindicators is a well established tenet of ecology and has proven useful as a means of assessment of aquatic systems. It is based on Hutchinson's niche definition: an n-dimensional hypervolume which encompasses the physical, chemical, and biological factors in which the organism can exist (Hutchinson, 1967). For example, trout and mayfly larvae are bioindicators of relatively pristine cold-water systems. Trout and mayfly larvae have overlapping, but not identical niches, since mayfly larvae inhabit the benthic layers, while trout are nektonic. It follows that any community of organisms consists of those whose niche includes tolerance to ambient environmental conditions, and conversely if the conditions change, selection would favor a different community, although it may contain some or even many of the original members. Large, long-lived organisms, such as trout, are integrators of long term environmental condition: unless there is a catastrophic change,

---

P.A. Rublee (✉)
The University of North Carolina at Greensboro, Greensboro, NC, USA

G.A. Uzochukwu et al. (eds), *Proceedings of the 2007 National Conference on Environmental Science and Technology*, DOI 10.1007/978-0-387-88483-7_7,

community transitions will be gradual. Further, while these organisms are generally viewed as indicators of a pristine water environment, such indicators are less useful for the assessment of specific conditions associated with toxins, pollutants, and pathogenic factors which diminish water quality.

In contrast to large bioindicator taxa, microbial taxa respond rapidly to environmental change since they are in intimate contact with their environment (high surface area to volume ratios), and they have high metabolic rates. Accordingly, rather than being integrators of environmental condition over time, they are more likely to be rapid, sensitive responders to even subtle changes, altering their population size as local conditions favor or inhibit their metabolism.

Microbial bioindicators have been used to a limited degree (e.g. coliform bacteria as indicators of fecal pollution), but they have not been used as general ecosystem indicators, because the vast majority of microorganisms remain uncultured and uncharacterized (Sogin et al., 2006). Rapid advancements in molecular biology over the last few decades now give us unprecedented access to this unknown microbial community, which allows a metagenomic approach (Committee on Metagenomics 2007) to community analysis. Further, specific microbes have been associated with the presence of chemicals that are important for water quality assessment (Oremland and Stolz, 2003). Therefore the potential exists for development of single or comprehensive gene probe tests that can serve as quantitative and/or qualitative environmental assessors and monitors.

Materials and Methods

Candidate bioindicator probes (PCR primer pairs) were designed to seven operational taxonomic units (OTUs) based on 97.5% similarity of small subunit ribosomal DNA (SSU rDNA) sequence libraries (n = 50 prokaryotic and 50 eukaryotic sequences) that had been generated by PCR from five lake samples: four from North Carolina reservoirs and one from an arctic Alaskan lake using standard protocols (cf. Marshall et al., 2008 for complete description of sample sites and protocols). Each candidate sequence was compared to GenBank entries for tentative identification using the BLAST program (Altschul, et al. 1990). Samples from additional USA lakes (two from ND, one from SD, 8 from southern CA) and from eight lakes in China (Fig. 1) were then probed by real-time PCR for the presence of the

Fig. 1 Sampling locations in North America and China

candidate bioindicator sequences. Additionally, we used real-time PCR to probe for candidate bioindicators in the genomic DNA from the library source lakes that did not exhibit the indicator of interest in the original sequence library, as a means of testing whether the candidate indicator was present but at low abundance.

Results and Discussion

The seven candidate microbial bioindicators we tested against 24 lake samples represented organisms that were found two or more times among 50 randomly drawn sequences in at least one of our original rDNA libraries. Three were prokaryotic SSU sequences and four were eukaryotic. The prokaryotic sequences represented: an unidentified cyanobacterium, an unidentified bacteria with bioremediation potential for the gasoline additive MTBE, and *Zoogloea ramigera*, a bacterium associated with high levels of dissolved organic matter and often found in nutrient enriched systems. Eukaryotic sequences were identified by BLAST as two widespread algae: the cryptomonad, *Cryptomonas ovata*, and the diatom *Asterionella formosa*, along with two protozoans, a *Dileptus* species and a "*Perkinsus*-like" species. These OTUs were not always found in all of the original rDNA libraries drawn from a given site. For example, the cyanobacterium "LD27" sequence was only found in one of the three libraries derived from three separate microbial collections. One collection had been taken at a different lake near the lake in which LD27 had been found. Another collection was taken at the same location where LD27 was found but during a different season. Primer pairs from each OTU to be tested were used to determine if the LD27 sequence in our original lake samples was actually present, but at abundances too low to be represented in the small sequence libraries that we had generated (that is, at a frequency below about 1 in 50; Fig. 2). With real-time PCR, the cyanobacteria LD27 target sequence was detected in the other two lake samples. Its presence, while detectable, was an order of magnitude less abundant ($0.071\times$) in the sample taken from a different and nearby lake. LD27 was also found in even lower abundance ($0.029\times$) in a sample taken during a different season in the original source lake.

We probed additional lakes in North America and China for presence of the candidate bioindicators (Fig. 3). Algal OTUs were widespread: *Asterionella formosa* was detected in all samples, *Cryptomonas ovata was* detected in 19 of 24 samples, and cyanobacteria LD27 was detected in 18 of 23 samples. Patterns for other OTUs were more variable. For example, *Zoogloea ramigera,* characteristic of high dissolved organic carbon, was originally recovered as an rDNA library sequence in the most eutrophic North Carolina lake sample and found in lower abundance using real-time PCR in the other NC lakes. It was not detected in the highly oligotrophic arctic lake or in any of the generally nutrient poor California lakes. In contrast, it was commonly detected in the samples from lakes in China which are generally considered to be highly eutrophic and which are often polluted. The lakes in China were also more likely to contain the "MTBE degrading" bacterium than the lakes in North America (75% incidence in vs. 25%).

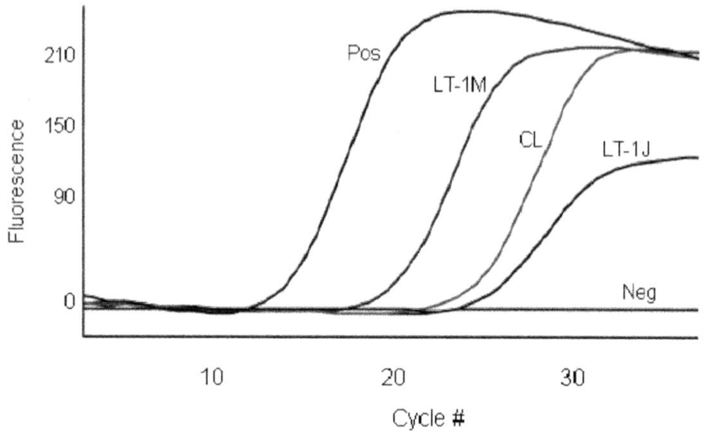

Fig. 2 Example of real-time qPCR output probing for OTU cyanobacterium LD27. Pos indicates a positive control clone, LT-1M genomic DNA from the sample where the sequence was originally identified in the library, and CL and LT-1J are samples where the OTU was not found in the original library. Abundance of the target is proportional to the cycle # where the curve passes a standard fluorescence value

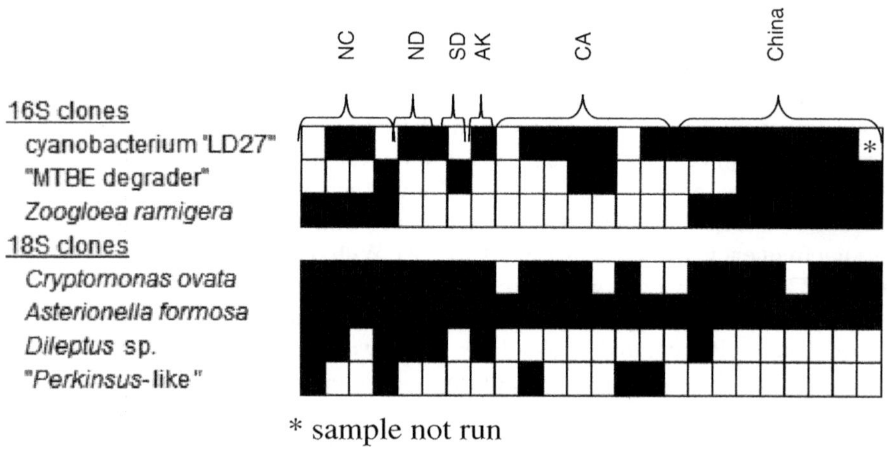

Fig. 3 Results of PCR probing of samples for candidate bioindicators from representative lakes in North Carolina (n = 4), North Dakota (n = 2), South Dakota (n = 1), Alaska (n = 1), California (n = 8), and China (n = 8)

Overall, the results are consistent with the potential to use a metagenomic approach for developing comprehensive environmental assessment and monitoring tests. Further progress will require identification and validation of target sequences indicative of physical, chemical, and biological characteristics. It will also require basic research to understand the levels of sequence variation among target taxa and potential complications in detection protocols. Once a sufficient number of target sequences are identified and combined with engineering advances in microfluidics and nanotechnology, monitoring and assessment can be conducted *in situ* in real-time. The availability of real-time data will improve management strategies including proactive efforts to address pollutant, disease, and even bioterrorist threats before they become acute.

References

Altschul, S.F., W. Gish, W. Miller, E.W. Myers, and D.J. Lipman. 1990. "Basic local alignment search tool." *J. Mol. Biol.* 215(2):403–410.

Committee on Metagenomics. 2007. *The New Science of Metagenomics: Revealing the Secrets of Our Microbial Planet.* National Academies Press. Washington, D.C. 171p.

Hutchinson, G.E. 1967. *A Treatise on Limnology, Vol. II.* J. Wiley & Sons, Inc., New York, 1115p.

Marshall, M.M., R. Amos, V.C. Henrich, and P.A. Rublee. 2008. "Metagenomic profiling of aquatic microbial communities using SSU rDNA sequences: Implications for environmental assessments." *Ecological Indicators*, 8(5):442–453. (doi:10.1016/j.ecolind.2007.04.007)

Oremland, R.S. and J.F. Stolz. 2003. "The ecology of arsenic." *Science* 300(5621):939–944.

Sogin, M.L., H.G. Morrison, J.A. Huber, D.M. Welch, S.M. Huse, P.R. Neal, J.M. Arrieta, and G.J. Herndl. 2006. "Microbial diversity in the deep sea and the underexplored rare biosphere." *PNAS* 103(32):12115–12120.

Part III
Environmental Justice and Ethics

Water Quality in an Environmental Justice Community in Durham, NC

Blakely Adair, John J. Bang, Yolanda B. Anderson, Saundra F. DeLauder, Marcia Bradshaw, Marion Lamberth, Faustina Meheux, Rakesh Malhotra, Roy Fortmann, Peter Egeghy, Ronald Williams, and Donald Whitaker

Abstract Environmental Justice Communities are usually minority communities of low socio-economic status with a concern of increased risk from point source pollution not present in other communities. A priority of the U.S. EPA is to empower these communities to advocate for themselves. Toward that end, EPA has entered into a cooperative agreement with the Department of Environmental Earth and Geospatial Sciences at North Carolina Central University, Durham, NC, to perform preliminary community based participatory studies and to establish the infrastructure for long-term interactions with environmental justice communities. In central Durham, housing and commercial properties are intermingled creating a concern for increased pollution and poor water quality in area creeks. To determine if current or past commercial practices impacted creeks, organochlorines (pesticides and polychlorinated biphenyls), polycyclic aromatic hydrocarbons and heavy metals were quantified in water and sediment samples collected during the summer of 2007 from nine creeks located in six parks in the study area. Water quality measurements and contaminant concentrations were compared to regulatory values. To date, few water quality and no contaminant concentrations were above regulatory levels. The results are being shared with members of community organizations at community functions.

Introduction

The concept of Environmental Justice (EJ) gained great awareness about two decades ago when the disproportionate distribution of environmental pollution was recognized primarily in communities of low socio-economic status and of color

Although this work was reviewed by EPA and approved for publication, it may not necessarily reflect official Agency policy.

B. Adair (✉)
North Carolina Central University, Durham, NC, USA

G.A. Uzochukwu et al. (eds), *Proceedings of the 2007 National Conference on Environmental Science and Technology*, DOI 10.1007/978-0-387-88483-7_8,

throughout the world. Studies have shown that these communities face a higher level of exposure to environmental hazards (Brown et al., 2003; Sexton, 2000), often living close to major streets, hazardous waste sites, traffic stations, heavy industrial facilities, power plants, and incinerators (Krieg, 2004; Perlin et al., 1995). Maantay et al. (2007) used proximity analysis to determine that in Bronx, New York, people living within specified distance buffers of noxious land use areas were up to 66% more likely to be hospitalized for asthma, were 30% more likely to be poor, and were 13% more likely to be a minority than those outside the buffers.

In 2001, North Carolina Central University (NCCU) received Congressional funding through a cooperative agreement with the National Exposure Research Laboratory at U.S. EPA. The primary objectives were to establish a research and training initiative for assessing environmental exposures in EJ communities impacted by environmental hazards and to develop the research infrastructure in the Environmental Science Program (ESP) at NCCU for long-term sustainability to better serve local communities. The long-range goal of this project is to develop community-level tools that can be adapted for use in other communities in order to better understand environmental exposures and their impacts in the context of EJ.

East Central Durham in Durham County, North Carolina, was selected for study because it is an EJ community where members had expressed concerns previously and where organizations were prepared to participate in multiple aspects of the study. In addition, there were existing environmental data from the area that could be used to broaden the scope. Meetings were held with community organizations to identify and prioritize environmental issues of concern. The environmental concerns of the community included contaminant exposure from runoff or standing water around auto salvage shops, metal scrap yards, or other industrial sources. A technical approach to assess the environmental hazards began with site screening at eight junkyards. However, a more inclusive approach that examined creek health and impact from pollution sources including junkyards was deemed more effective for addressing community concerns.

Materials and Methods

Study Design. A study was designed to determine water quality and contaminant presence in creeks impacted by the identified areas of concern. The area for study is located in East Central Durham where community demographics have been identified to fit environmental justice parameters. Historical and current human activity has possibly resulted in an increased level of pollution in the community. Table 1 lists industrial sites and locations along with possible pollutants from the different sources. A map of the study area (Fig. 1) demonstrates the location of these industrial sites.

Study Area and Sampling Sites. Creek health was the basic target for impact from contaminants in this community because multiple contaminants with short and long environmental half lives can exist in creek water and sediment. Site locations (Fig. 1) were selected to represent areas up and downstream of possible contaminant

Table 1 Industries and possible pollutants distributed from these sites

Industries	Target pollutant
3 Cement Plants	Particulate matter, VOCs, and SVOCs
2 Junk Yards	Metals, ethylene glycol
3 Auto Repair Shops	Metals, VOCs, and SVOCs
2 Auto Paint Shops	Metals, VOCs and SVOCs
4 Gas Stations and Oil Housings	Metals, Ethylene glycol, acids,
1 Light Industrial Plant	Solvents: Methylene chloride and n-propyl bromide carrier with latex, latex/synthetic blends, hot melt (100% solids with no solvent or carrier)
1 Dry Cleaners	Trichloroethylene (TCE), VOCs, SVOCs
1 Metals Fabrication Plant	Acid/alkaline waste, paint overspray, solvents: halogenated and non-halogenated
1 Chemical Company	VOCs, SVOCs

VOCs, volatile organic compounds.
SVOCs, semi volatile organic compounds.

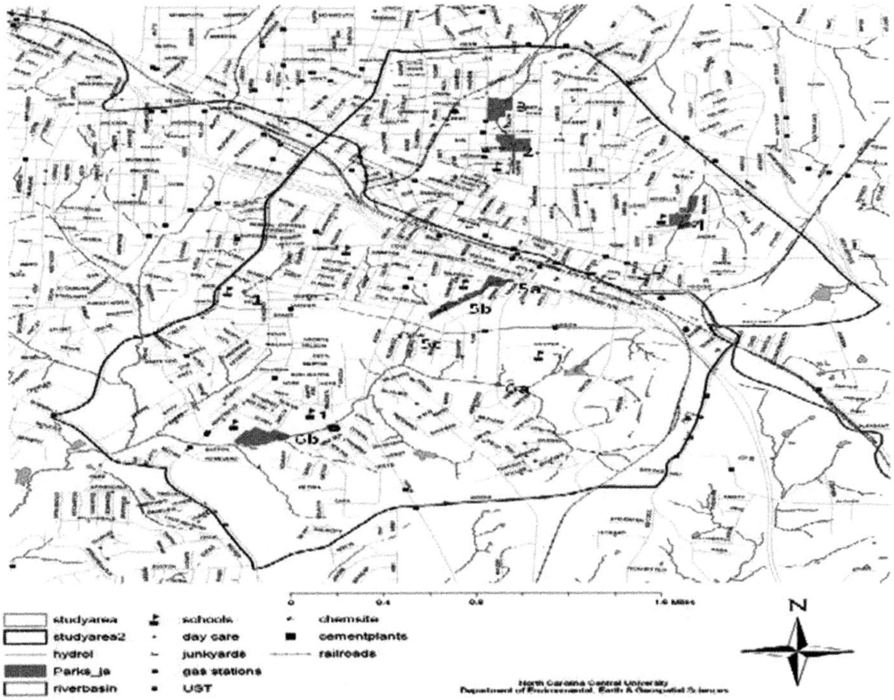

Fig. 1 Map of study area with creek sites *(1-6b)*, possible pollution sources, and community concern areas (e.g., schools and daycares)

sources in the Neuse and Cape Fear River Basins. Therefore, a range of pollution and water sources was examined by the surface water and sediment samples collected at the sites.

Sample Collection and Analysis. Sediment and water samples were collected using methods recommended by USGS and EPA (USGS, web 2007; EPA, 1982). Water samples were collected using a dip sampler. The top two inches of sediment were collected directly in the sample containers to examine the sediment that may reenter the water column. The samples were shipped to an accredited analytical laboratory for analysis of metals, polycyclic aromatic hydrocarbons (PAHs), pesticides/polychlorinated biphenyls (PCBs), and total organic carbon (TOC) using EPA methods 6010B, 8270, 8081/8082, and 415.1 respectively.

Results and Discussion

The water quality measurements were similar among sites and were within regulatory recommended values of pH 7.3 $\pm$ 0.3, temperature 21.9 $\pm$ 2, and salinity 0.129 $\pm$ 0.05 (http://www.epa.gov/waterscience/criteria/wqcriteria.html, August 14, 2007). Detection frequencies for all contaminants were below 100% and several contaminants were detected in only a few samples (Table 2). Zn and Mn were the only contaminants consistently detected in water for comparison with water quality regulations.

The overall water quality is fair if the outliers associated with known discharge events are removed. Dissolved oxygen and turbidity varied greatly among sites depending on water level and flow rates. Dissolved oxygen was below the regulatory minimum levels at several sites; however, plants and animals were present at those sites which indicate a temporary condition of poor oxygen levels. Turbidity was only above regulatory maximum levels during a discharge or rain event. Preliminary analytical results revealed that few contaminants were present in water samples collected from multiple sites. Mn was the only contaminant detected regularly and at levels above drinking water standards, but there is no standard for wildlife protection. Therefore, we can report to the community that contamination of creeks in the study area is limited to small areas and does not appear to reach regulatory levels.

One major community concern that we were able to address was an intermittent smell that can be detected at least a mile along the creek in central Durham (Fig. 1, Sites 5a-5c). The creek is fed by storm water and storm water discharge pipes from an industrial source at the headwaters. Anecdotal data suggests monthly discharge events that result in increased water levels and may contribute to the smell. During daily monitoring, we observed a discharge and collected samples during the event that contained semi volatile organic compounds not seen in previous samples. The discharge water temperature was 10°C higher than creek water temperature minutes before the discharge and the average water temperature. The pH went from neutral to strongly basic, and the chemical smell was overpowering, which indicates the

Table 2 Representative data collected from creeks in central Durham

Contaminant		Water			Sediment		
Group	Quantified	N	Concentration range (μg/L)	detect (%)	N	Concentration range (μg/kg)	detect (%)
6[1] Metals		41			34		
	As		BQL[2]	0		BQL – 2780	38
	Cd		BQL	0		BQL – 591	12
	Pb		BQL – 285	29		BQL – 99900	97
	Mn		BQL – 896	98		BQL – 242000	97
	Hg		BQL – 0.081	2		BQL – 2010	71
	Zn		BQL – 231	71		BQL – 760000	97
18 PAHs		49			38		
	Chrysene[3]		BQL	0		BQL – 771	34
	Fluroanthene					BQL – 1580	50
	Phenanthrene					BQL – 1280	26
	Pyrene					BQL – 1330	39
7 PCBs							
	Aroclors	1	BQL	0	39	BQL	0
22 OCs		1	BQL	0	36		
	Dieldrin[4]					BQL – 18.8	22
	α-Chlordane					BQL – 4.52	30
	γ-Chlordane					BQL – 23.8	11

[1] The number of contaminants in the group that were analyzed.
[2] BQL-below quantitation limit.
[3] PAHs with >25% detects are presented.
[4] OCs with >10% detects are presented.

presence of volatile compounds that were beyond the scope and abilities of this study. Although the contamination was not detected in water collected at two points less than one mile down stream of the discharge site, pH and temperature were still impacted. These results provide scientific evidence to support the community concerns, and we are working with the local division of water quality to determine if the discharge is legal.

Acknowledgments We would like to thank Kai Ngegba, Aiqun Yu, Janay Jones, and Kimberly Outen for their hard work. We would also like to thank Wayne Randolph for his assistance in planning and implementing the sample collection. This study was supported by U.S. EPA cooperative agreement R-82946901.

References

Brown, P., Mayer, B., Zavestoski, S., Luebke, T., Mandelbaum, J., McCormick, S. The health politics of asthma: environmental justice and collective illness experience in the United States. Social Science & Medicine. 2003, 57: 453–464.

Krieg, E.J., Faber, D.R. Not so black and white: environmental justice and cumulative impact assessments. Environmental Impact Assessment Review. 2004, 24: 667–694.

Maantay, J. Asthma and air pollution in the Bronx: methodological and data considerations in using GIS for environmental justice and health research. Health and Place. 2007, 13: 32–56.

Perlin, K., Setzer, R.W., Creason, J., Sexton, K. The distribution of industrial air emissions by income and race in the United States: an approach using the Toxic Release Inventory. Environmental Science and Technology. 1995, 29: 68–80.

Sexton, K. Socioeconomic and racial disparities in environmental health: Is risk assessment part of the problem or part of the solution? Human and Ecological Risk Assessment. 2000, 6(4): 561–574.

The Impact of Social Capital on Environmental Risk Reduction in Moncure

John J. Bang, Yolanda B. Anderson, Saundra F. DeLauder, Marcia Bradshaw, Faustina Meheux, Rakesh Malhotra, Roy Fortmann, Peter Egeghy, Ronald Williams, and Donald Whitaker

Abstract Despite rigorous efforts to understand and resolve environmental justice (EJ) issues, the complexity of the issues associated with many of these EJ communities continues to bring challenges to community residents and environmental researchers. Moncure, NC, located in northern Chatham County, is a rural community proximal to a cluster of several major industries. While the presence of these industries has helped the local residents maintain higher than average household incomes, it has also created environmental concerns (air quality and potential respiratory illness) for the community. In this study, the authors examine the impact of social capital in the form of a community coalition called the Southeastern Chatham Citizens Advisory Council (SCCAC) on the resolution of EJ issues in Moncure. EJ issues in Moncure have been investigated by examining trends in archived data from the U.S. EPA's Toxic Release Inventory (TRI) and by addressing residents' pollution concerns, identified in Town Hall meetings, by designing a field monitoring study. Monitoring for the field study was conducted during one week periods in January and June of 2007. Preliminary results from the January sampling was compared with regulatory standard values, TRI data, and with results from a modeling study conducted by NC DENR Division of Air Quality (DAQ). Data indicate that the levels of released pollutants from local industries have been reduced during the overlapping time period when the SECCA has been active in trying to resolve the pollution issues, illustrating the potential impact of social capital on EJ risk reduction.

Although this work was reviewed by EPA and approved for publication, it may not necessarily reflect official Agency policy.

J.J. Bang (✉)
North Carolina Central University, Durham, NC, USA

G.A. Uzochukwu et al. (eds), *Proceedings of the 2007 National Conference on Environmental Science and Technology*, DOI 10.1007/978-0-387-88483-7_9,

Introduction

The Moncure Air Quality Screening Study (MAQSS) is one of three independent study components under the Environmental Risks and Impacts in Communities of Color (ERICC) project conducted between January 2006 and August 2007. The ERICC project will provide guidelines, as a case study, that other EJ communities in similar settings may use to design studies to assess environmental risks known to be present in EJ communities (Krieg and Faber, 2004; Perlin et al., 1995; Sexton and Anderson, 1993; Mantay, 2007). The MAQSS was designed in response to concerns expressed by the Southeast Chatham Citizens Advisory Council (SCCAC). The primary purpose of the study was to inform residents about the air quality in the community by providing measured ambient levels of air pollutants that have potential respiratory health relevance. Moncure, NC, is located in northern Chatham County, about 25 miles southwest of Raleigh, the capital of North Carolina. The demographic composition of the community is a little different than that of traditional Environmental Justice (EJ) communities as defined by Mbamalu (2002). There are fewer minority residents and the overall socioeconomic status is higher compared to typical EJ communities. Potentially major pollution sources are located in a clustered pattern near residential areas. According to Scorecard (www.scorecard.org), a popular internet resource that utilizes publicly available data collected by federal regulatory agencies to provide information about local community pollution problems, the levels of some Hazardous Air Pollutants (HAPs) released by the local industries reached well above national averages during the late 1990s and early 2000s. The strong economic ties between the residents and the industries have presented challenges in dealing with these complex issues. Social capital, vested in the SCCAC, has been instrumental in addressing potentially detrimental health impacts from these possible environmental risks in Moncure. Empirical data are presented in order to better understand the impact of a consistent voice from residents on the reduction of environmental risks present in Moncure.

Materials and Methods

Study Design. Trends in industrial releases of chemicals in Moncure were examined using archived data from the U.S. EPA's annual Toxic Release Inventory (TRI). Hazardous Air Pollutants (HAPs), including Volatile Organic Compounds (VOCs), and Criteria Air Pollutants (CAPs) were assessed. Results of modeling performed by the North Carolina Department of Environmental and Natural Resources (NC DENR) Division of Air Quality of air concentrations of SO_2, formaldehyde, and particulate matter (PM_{10}) were compared to National Ambient Air Quality Standards (NAAQS). A field screening study was designed by NCCU in consultation with U.S. EPA after a review of the TRI trends and DENR modeling results and following community meetings to determine the residents' environmental concerns. A ranking order for the selection of pollutants for the field screening survey was

Table 1 List of air pollutants measured in the MAQSS

Criteria	Names of pollutants
CAPs:	CO
	PM (2.5 and 10 μm)
	SO_2
	O_3
	$NO/NO_2/NO_X$
HAPs:	Formaldehyde/Acetaldehyde
	VOCs (total of 25):
	1,3-Butadine; 1,2-Dichloro-1,1,2,2-tetrafluoroethane; Trichlorofuoromethane; 1,1-Dichoroethane; 1,1,2-Trichloro-1,2,2-trifluoroethane; 1,1-Dichloroethane; cis-1,2-Dichloroethene; 1,2-Dichloroethene; 1,1,1-Trichloroehane; Benzene; Carbon tetrachloride; 1,2-Dichloropropane; Trichloroethene; Toluene; Tetrachloroethene; Chlorobenzene; Ethylbenzene; m,p-Xylene; Styrene; o-Xylene; 4-Ethyltoluene; 1,3,5-Trimethylbenzene; m-Dichlorobenzene; p-Dichlorobenzene; o-Dichlorobenzene
Others:	Heavy metals and other elements (total of 46 elements)

assigned using EPA's Risk-Screening Environmental Indicator (RSEI Version 2.1.2, www.epa.gov/opptintr/rsei). RSEI is a tool that assesses the potential impact of local industrial releases based on quantity (pounds released), hazard (pounds multiplied by toxicity), and risk (hazard multiplied by surrogate dose) requiring the following information: amount of chemical released, chemical toxicity, fate and transport through the environment, the route and extent of human exposure, and the number of people affected. The final list of pollutants for the field survey, reflecting assigned rankings and feasibility of measurement, is presented in Table 1. The sampling methods were chosen with consideration of the availability of equipment and other sampling and analysis requirements as shown in the Table 2.

Results and Discussion

The yearly average amounts of area relevant HAPs released into air over five years (2001–2005) are presented in Fig. 1. The Southeastern Chatham Citizens Advisory Council (SCCAC) focused their attention on three specific pollutants: SO_2, formaldehyde, and PM_{10}. A decreasing trend in formaldehyde and sulfuric acid as a surrogate measurement for SO_2 emissions over the four years is observable.

The field measurement data from the January 2007 survey performed by NCCU are shown in Table 3. The data shown in the Table 3 do not include results from integrated (Ogawa) sampling due to space limitations, but the results are similar to those obtained with the continuous emissions monitoring (CEM) methods. In the

Table 2 Sampling parameters and equipment for the MAQSS

Measurement parameters	Equipment	Sampling methods	Notes
Carbon Monoxide	-API 300 (CEM)	Continuous	Vehicle mounted
$NO/NO_2/NO_X$	-API 200 (CEM)	Continuous	Vehicle mounted
	-Ogawa	24-h integrated	IC analysis
Sulfur Dioxide	-TECO 43-A	Continuous	Vehicle mounted
	-Ogawa	24-h integrated	IC analysis
Ozone	-TECO 49	Continuous	Vehicle mounted
	-Ogawa	24-h integrated	IC analysis
$PM_{2.5}$	-PEM	24-h integrated	gravimetric
PM_{10}	-PEM	24-h integrated	gravimetric
Heavy Metals	-PEM filters	24-h integrated	XRF analysis
Aldehydes	-2,4-DNPH cartridges, Ozone scrubbers, HPLC	24-h integrated	HPLC analysis
VOCs	-Carbopack-X loaded Perkin Elmer tubes	24-h integrated	Thermal Desorption
Weather Conditions	-Weather Oracle WS-2000, HOBO	Continuous	

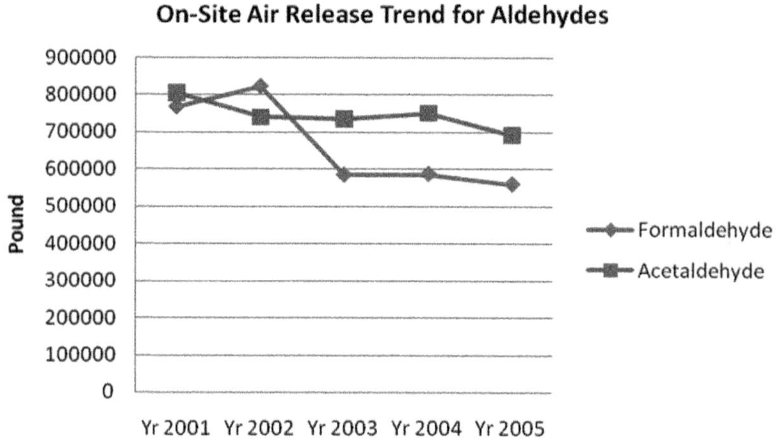

Fig. 1 Five-year release trend for selected HAPs in chatham county

case of NO_2, the average of the weekly data was used as a surrogate for annual average value. All measured levels, with the exception of $PM_{2.5}$ and PM_{10}, were below regulatory levels. For both $PM_{2.5}$ and PM_{10}, 24-h average values exceeded the regulatory levels on only one day and at only one site (out of five locations).

The SCCAC was created before the year 2000. However, the active involvement of the Council for the reduction of environmental risks began in early 2000 according to the members of SCCAC and news articles. Public awareness about environmental quality issues was further heightened by a consistent media presence lead

Table 3 Measured levels of pollutants from the January 2007 screening study

Measurement parameters	Regulatory levels	Measured levels
Carbon Monoxide (CEM)	1 h: 35 ppm	1.4 ppm (range 1.1–1.6)
	8 h: 9 ppm	1.1 ppm (range 1.0–1.5)
NO_2 (CEM)	50 ppb (annual avg.)	7.1 ppb (weekly average)
Sulfur Dioxide (CEM)	3 h: 0.5 ppm	
	24 h: 0.14 ppm	0.82 ppb (weekly average)
	Annual: 0.03 ppm	Daily Avg. Range (0.31–1.5 ppb)
Ozone (CEM)	1 h: 0.12 ppm	Below 0.055 ppm
	8 h: 0.08 ppm	Below 0.039 ppm
PM_{10}	24 h: 150 $\mu g/m^3$	Daily Avg. Range (9.9–170 $\mu g/m^3$)
$PM_{2.5}$	24 h: 35 $\mu g/m^3$	Daily Avg. Range (1.5–84 $\mu g/m^3$)
	Annual: 15 $\mu g/m^3$	Wkly Avg. Range (12–26 $\mu g/m^3$)
Aldehydes (formaldehyde)	NC TAP AAL	
	150 $\mu g/m^3$ (1 h)	0.02–3.1 $\mu g/m^3$ (24 h average)
	OSHA 8 h PEL	
	0.75 ppm (900 $\mu g/m^3$)	0.01–3.1 $\mu g/m^3$
Aldehydes (acetaldehyde)	n/a	up to 0.90 $\mu g/m^3$
VOCs including BTEX	n/a	Toluene spike (up to 13 ppb)

NC TAP ALL, North Carolina Toxic Air Pollutant Acceptable Ambient Level. OSHA PEL, Occupational Safety and Health Administration Permissible Exposure Limit.

by a few community residents. As a consequence, voluntary reduction of pollutant release and installation of equipment for pollution abatement and similar actions from area industries followed (The Raleigh News & Observer, December 8, 2005, 1B). According to a modeling study results made available on July 14, 2006 by North Carolina Department of Environmental and Natural Resources, Division of Air Quality, early year 2000 levels of pollutant release were lower (34–80% of the regulatory levels) than the NAAQS in the categories of SO_2, formaldehyde, and particulate matter (PM_{10}).

Conclusions

There is a high level of uncertainty inherent in the interpretation of the positive role of social capital on the reduction of environmental risks in a community like Moncure. More empirical data from additional sampling surveys and/or modeling exercises may be needed to fully understand the dynamics. However, three sets of independent data (Toxic Release Inventories, the NC DENR DAQ modeling study, and the NCCU field measurements) in combination supports the conclusion of a positive role of the community coalition effort. At the very least, the persistence of the SCCAC members resulted in a comprehensive air quality screening survey that may help alleviate some of the environmental concerns of the community. Considering the limitations presented in voluntary emissions reports and modeling studies, a periodical monitoring effort is recommended for future EJ studies.

Acknowledgments This study was supported by EPA cooperative agreement R-82946901. We are grateful to other community members, a group of volunteers from NCCU and Moncure residents as well as technical support from other departments at NCCU and EPA.

References

Krieg, E.J., Faber, D.R. 2004. "Not so Black and White: Environmental Justice and Cumulative Impact Assessments". *Environmental Impact Assessment Review.* 24: 667–694

Mantay, J. 2007. "Asthma and air pollution in the Bronx: Methodological and data considerations in using GIS for environmental justice and health research". *Health and Place.* 13: 32–56

Mbamalu, G.E. 2002. "Normative Concerns and Practical Solutions" In G.A. Uzochukwu, K. Schimmel, G.B. Reddy, S. Chang, and V. Kabadi (Eds.), Proceedings of the 2002 National Conference on Environmental Science and Technology. Battelle, Columbus, OH, pp. 113–121

North Carolina Division of Air Quality. 2006 "Moncure Area Criteria and Toxic Pollutant Impact Study Dispersion Modeling Report" North Carolina Division of Air Quality, Air Quality Analysis Branch, Raleigh, NC (available: http://daq.state.nc.us/permits/mets/moncure˙asof07072006.pdf)

Perlin, K., Setzer, R.W., Creason, J., Sexton, K. 1995 "The distribution of industrial air emissions by income and race in the United States: an approach using the Toxic Release Inventory". *Environmental Science and Technology*. 29: 68–80

Sexton, K.; Anderson, Y.B. 1993 "Equity in Environmental Health: Research Issues and Needs". *Toxicology and Industrial Health*.9(5): 679–959

Air Quality Issues for an Aging Population

Richard V. Crume, Kathleen E. Sykes, and Yoko S. Crume

Abstract The number of older adults in the U.S. population is increasing as the baby-boomer generation reaches retirement age. Older adults are more sensitive to air pollution, and many have existing medical conditions that are worsened by air pollution exposure. Additionally, a majority lives in metropolitan areas where air contaminant concentrations are higher. For these reasons, there is concern over future adverse health effects as the percentage of older adults grows substantially larger. A related problem for older adults is excessive heat, made worse by summertime air pollution episodes. Indoor air contaminants can also be a problem because older adults spend more time indoors than they did at a younger age. Steps can be taken now to plan for the aging of America's population, including programs aimed at education, notification, and research involving air pollution exposure and excess heat events.

Introduction

The population of adults age 65 and older in America is growing at a dramatic rate (Fig. 1), increasing from 12% of the U.S. population in the year 2004 to 20% by 2030 (65+ in the United States, 2005; A Profile of Older Americans, 2005). Because older adults are more susceptible to the adverse effects of air pollution than people who are younger, greater demands will be placed on the health care system as the population ages. State and local officials may be hard pressed to provide the health and long-term care services required by increasing numbers of older adults, many with chronic conditions made worse by air pollution.

Several factors contribute to the increasing older population in the U.S. Most important is the aging of the 70 million strong baby-boomer generation, individuals born between 1946 and 1964 who will begin turning 65 in the year 2011. Also

R.V. Crume (✉)
U.S. EPA, Research Triangle Park, NC, USA

G.A. Uzochukwu et al. (eds), *Proceedings of the 2007 National Conference on Environmental Science and Technology*, DOI 10.1007/978-0-387-88483-7_10,

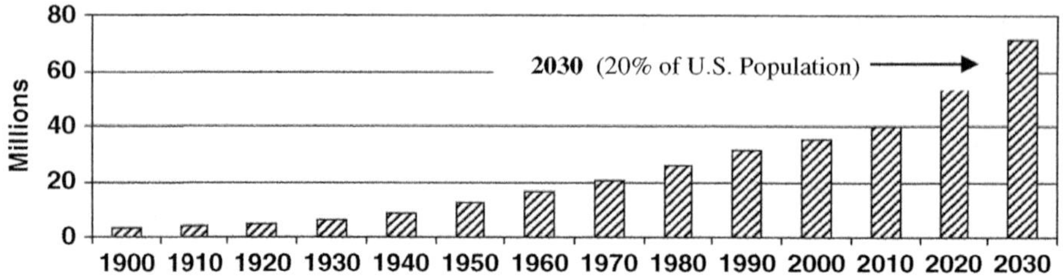

Fig. 1 Population of Americans age 65 and older

important is the trend towards a longer life expectancy at birth, reaching a new high of 77.8 years for the total population in 2004, up from 75.4 years in 1990 (Health, United States, 2006). Improved health care is another key factor today enabling older Americans to live a longer life, as is the decline in cigarette smoking among adults.

Air pollution emissions have dropped dramatically under the Clean Air Act. For example, between 1970, when the Act came into existence, and 2006, aggregate emissions of the six principal air pollutants tracked by the U.S. Environmental Protection Agency (EPA) dropped by 54% (Fig. 2). This is remarkable given that during the same timeframe, gross domestic product increased 203%, vehicle miles traveled was up 177%, energy consumption rose by 49%, and the U.S. population grew by 46% (Air Trends, 2007). Even so, over 75% of older adults live in metropolitan areas (about 50% in suburbs and 27% in central cities) where air pollution concentrations can be elevated (A Profile of Older Americans, 2005), and as many as 23% live in counties having air pollution concentrations exceeding federal standards (Older Americans Update, 2006).

There are several reasons why many older adults choose to live in metropolitan areas. It is reported that upwards of 90% want to continue living in their own homes (Mahoney, 2007), which are often in urban areas near their family and pre-retirement employer. (This is sometimes referred to as "aging in place.")

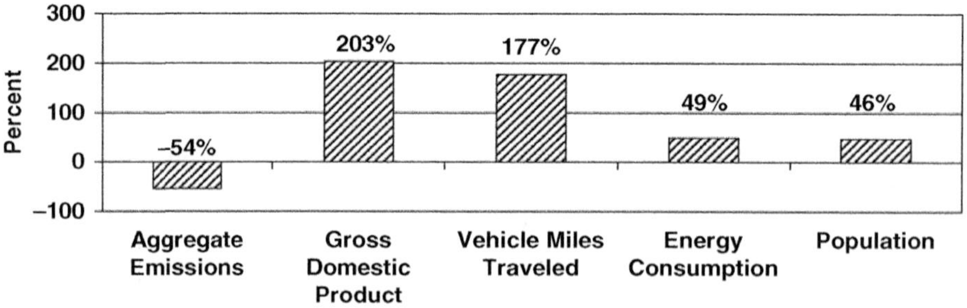

Fig. 2 Aggregate emissions of six criteria pollutants have dropped under the Clean Air Act (1970–2006)

Furthermore, older adults want to live in close proximity to friends, shopping, transportation, entertainment, and medical care. Also, it is not uncommon for retirement communities, assisted living facilities, and nursing homes to be located in metropolitan areas near freeways, busy intersections, and commercial areas, where air pollution levels may be high. (Such areas often provide greater convenience and visibility and sometimes lower land costs.) Poor older Americans may have little choice but to remain where they are, which for many is in polluted urban areas. Thus, despite the impressive advances made in cleaning up the nation's air, an increasing number of older Americans may find themselves living in polluted urban areas as greater numbers of adults reach retirement age.

The Effects of Air Pollution Exposure

As people age, their bodies are less able to compensate for the effects of exposure to pollutants. In particular, air pollution is known to aggravate heart and lung disease, leading to greater medication use, more visits to health providers, and increased admissions to emergency rooms (Older Americans Update, 2006). Generally, we know that older adults "*... may be more vulnerable to chemical exposure because of a decreased capacity to repair DNA damage caused by mutagens. Decreased immunologic defenses may also increase the vulnerability of the elderly to chemical carcinogens. Also, the elderly have a decreased ability to detoxify free radicals and other reactive metabolites ...*" (Aging and Toxic Response, 2005).

Linkages between air pollution and the health of older adults are well established. Of particular concern for older adults is the association of fine particulate pollution with cardiac arrhythmias and heart attacks, asthma attacks, development of chronic bronchitis, and premature death (Older Americans Update, 2006). Additionally, ozone pollution can exacerbate respiratory diseases, including chronic obstructive pulmonary disease (COPD) and asthma, leading to increased hospital admissions and mortality, and exposure to fine particles, nitrogen dioxide, and carbon monoxide can increase the risk of pneumonia among the older population (Aging and Toxic Response, 2005). Recognizing the air pollution threats to older adults, the EPA's Air Quality Index program has established the "Unhealthy for Sensitive Groups" (index color "orange") concern level for particulate matter specifically for older adults as well as children and people with heart and lung disease (Air Quality Index, 2007).

Several recent studies shed new light on the air pollution hazards for older adults. Based on a four-year study of 11.5 million Medicare enrollees, one report finds that small increases in fine particulate air pollution resulted in increased hospital admissions (1.28% increase for each 10 $\mu g/m^3$ of fine particulates) for heart and vascular disease, heart failure, COPD, and respiratory infection (Dominici et al., 2006). A study of 65,893 postmenopausal women (median age 63 when first enrolled in the study) without previous cardiovascular disease found a 24% increase in the risk of a cardiovascular event and a 76% increase in the risk of death from cardiovascular disease associated with each 10 $\mu g/m^3$ increase in fine particulate concentration

(Miller et al., 2007). (This study also found an association between fine particulate concentrations and the risk of cerebrovascular events.) Another study hypothesizes that adults with higher cholesterol levels when exposed to urban air pollution may be more vulnerable to heart disease, finding that "*[d]iesel exhaust particles and oxidized phospholipids synergistically affect the expression profile of several gene modules that correspond to pathways relevant to vascular inflammatory processes such as atherosclerosis*" (Gong et al., 2007).

Three other studies indicate that: (1) background exposure to certain persistent organic pollutants may be involved in the pathogenesis of arthritis in women (Lee et al., 2007), (2) an association exists between particulate exposure and adverse post-myocardial infarction (MI) in persons (mean age 76.1) who survived an MI (Zanobetti and Schwartz, 2007), and (3) people with diabetes are at increased risk of heart attack and stroke when exposed to particulate air pollution (Diabetes and Environmental Hazards, 2007). Regarding the later study, 14% of Whites, 23% of African Americans, and 24% of Hispanics age 65 and older in the U.S. have diabetes.

It is reasonable to speculate that older adults living in facilities located close to busy roadways may suffer even more from the effects of air pollution. For example, one study of more than 3,600 children found that living within 500 yards of a freeway adversely affects lung function (Sandström and Brunekreef, 2007), a result that might also have implications for older adults under similar circumstances. (Although older adults spend more time indoors than children, indoor and outdoor concentrations of the principal air pollutants may be similar.)

Older adults are believed to have increased susceptibility to environmental exposure for various reasons (Aging and Toxic Response, 2005), such as:

- Declining physiological processes,
- Age associated diseases, conditions, and nutritional status,
- Lifestyle choices such as smoking and consumer product use,
- Exposure to a variety of toxicants that may accumulate over a lifetime and impact health conditions later in life, and
- Interaction with prescription drugs. (Even healthy older adults filled an average of 10 prescriptions in 2000, while some with multiple medical conditions filled many more.)

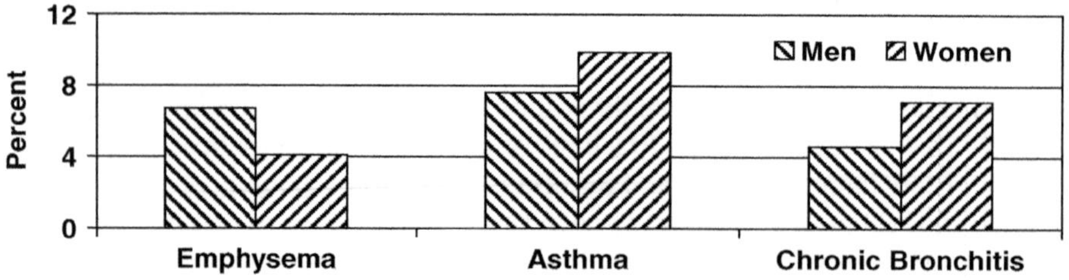

Fig. 3 Adults age 65 and older with selected chronic conditions

Whether caused by air pollution, occupational exposure, years of cigarette smoking, or some combination, a significant percentage of older adults suffer from chronic conditions involving the lungs. For example, see Fig. 3 (Older Americans Update, 2006).

Climate Change and Excess Heat Events

A related concern for older adults is climate change, caused by global warming gases (principally, carbon dioxide) emitted into the atmosphere by industrial and commercial facilities, electric utility plants burning coal, and automobiles. While debate continues about the causes and consequences of climate change, there is little doubt that global warming gases like carbon dioxide (Fig. 4) are increasing in atmospheric concentration (Carbon Dioxide Concentration Measurements, 2006) and that the climate is warming. The EPA reports that some climate change effects are already occurring, including "*... sea level rise, shrinking glaciers, changes in the range and distribution of plants and animals, trees blooming earlier, lengthening of growing seasons, ice on rivers and lakes freezing later and breaking up earlier, and thawing of permafrost*" (Climate Change, 2007). The World Health Organization reports that the Earth's warming climate may currently contribute to more than 150,000 deaths and 5 million illnesses each year, mainly among the poor (Patz et al., 2005).

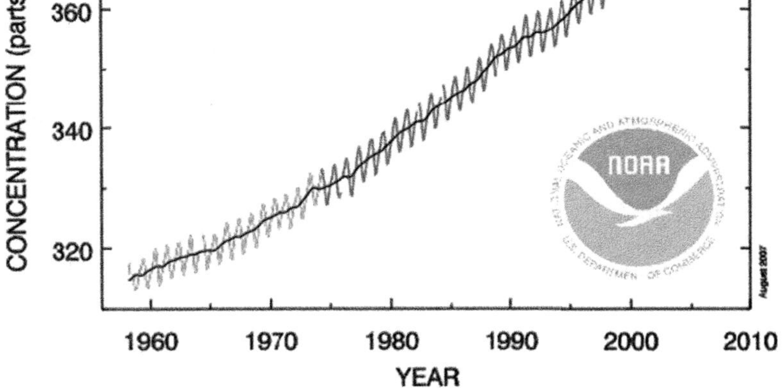

Fig. 4 Atmospheric carbon dioxide measurements at the Mauna Loa Observatory

Aside from the global environmental changes destined to affect all age groups, older adults are especially vulnerable to rising temperatures. It is estimated that as many as 1,700–1,800 heat-attributable deaths occur in the U.S. each summer, plus an unknown number of nonfatal adverse health outcomes among affected populations (i.e., older adults, the very young, the poor, and those with physical challenges or mental impairments) (Excessive Heat Events Guidebook, 2006). Perhaps the most dramatic excess heat event in recent years occurred in France during the summer of 2003, when over 14,000 deaths, many among elderly citizens, were attributed to high temperatures. The French Parliament blamed these deaths on a complex health system, widespread failure among agencies and health services to coordinate efforts, and notably, chronically insufficient care for the elderly (France Heat Wave, 2003). Excess heat events are usually worse in metropolitan areas due to the heat island effect, which describes the higher summertime temperatures common in cities due to the absorption of heat by buildings and roadways.

It is reported that present-day European and North American heat waves coincide with an atmospheric circulation pattern that is intensified by ongoing increases in greenhouse gases, indicating that more severe heat waves can be expected in the future (Meehl and Tebaldi, 2004). Regardless of the cause, global temperatures have been rising – 2006 was the second warmest year on record in the contiguous U.S., and the past nine years have been among the 25 warmest years on record (Climate of 2006, 2007). This trend is unprecedented in the historical record. Climate change models differ in their predictions of the magnitude and timeframe of global temperature rises, but any increase in the frequency and duration of excessive heat events would put older adults at even greater risk for heat exhaustion and stroke.

Indoor Exposure

Because older adults spend most of their time indoor, the quality of indoor air is critically important. Inhalation hazards can result from exposure to various air contaminants arising from the use of wood stoves, cleaning products, and pesticides. Dust mites, cockroaches, pet dander, mold, and dust can also contribute to indoor air contaminant levels. Carbon monoxide, which limits the blood's ability to carry oxygen, can emanate from furnaces, water heaters, gas ranges, fireplaces, wood stoves, and an automobile idling in a garage (Williams, 2005). Indoor air contamination can be made worse by the poor air circulation found in some modern, well insulated, energy efficient structures. For some seniors, indoor air pollution may be more of a hazard than breathing the outside air.

Discussion

As adults age and become more susceptible to air pollution and heat episodes, many unfortunately have difficulty meeting their medical and long-term care needs because of high out-of-pocket costs, even with Medicare coverage. This is especially a problem for the 28% of "near poor" (100–199% of the poverty threshold) adults age 65 and older, most of whom do not qualify for Medicaid, the health and long-term care assistance program for the lowest income population.

For some, racial/ethnicity disparities within major metropolitan areas with respect to air pollution exposure make matters worse. The American Lung Association (ALA) has published lists of the worst metropolitan statistical areas (combined) for two major air pollutants: particulate matter less than 2.5 μm in diameter (separate lists published for 24-h and annual average concentrations) and ozone (State of the Air, 2007). (Worst metropolitan areas are ranked by counting poor air quality days using the EPA's Air Quality System data and Air Quality Index rankings.) An examination of the age 65 and older adult populations living in the ALA's worst five metropolitan areas with populations greater than one million (Table 1) reveals that most have a substantially greater percentage of non-White older adults than the U.S. population average. In other words, a greater proportion of older adults in metropolitan areas with poor air quality, as characterized by the ALA, are minorities.

While pollution control can impart substantial costs to society, the benefits to older Americans are clear: reduced respiratory symptoms and serious, debilitating diseases like cancer; less medication use; fewer hospital and emergency room admissions; and longer life. The U.S. Office of Management and Budget examined the economics of EPA air pollution control programs for the period 1992–2002, concluding that benefits outweigh costs by a factor of five to seven (Informing

Table 1 Percent non-white among population age 65 and older

Metropolitan statistical areas (Combined)*	Percent** (U.S. Average = 18.9)
Top Five for Particulate (24-h PM2.5)	
Los Angeles–Long Beach–Riverside, CA	41.5
Pittsburgh–New Castle, PA	6.5
Birmingham–Hoover–Cullman, AL	20.5
Salt Lake City–Ogden–Clearfield, UT	7.9
Detroit–Warren–Flint, MI	20.1
Top Five for Particulate (Annual PM2.5)	
Los Angeles–Long Beach–Riverside, CA	41.5
Pittsburgh–New Castle, PA	6.5
Birmingham–Hoover–Cullman, AL	20.5
Detroit–Warren–Flint, MI	20.1
Cleveland–Akron–Elyria, OH	15.3
Top Five for Ozone	
Los Angeles–Long Beach–Riverside, CA	41.5
Houston–Baytown–Huntsville, TX	34.8
Dallas–Fort Worth, TX	22.4
Sacramento–Arden-Arcade–Truckee, CA–NV	22.2
New York–Newark–Bridgeport, NY–NJ–CT–PA	28.6

* Populations greater than one million.
** Includes Hispanic and Latino origin.

Regulatory Decisions, 2003). This suggests that for all Americans, including older adults, federal air pollution regulatory programs are justifiable on a cost/benefit analysis basis.

Recommendations

By the year 2030, over 71 million baby-boomers in the U.S. will be age 65 and older. Yet, only 46% of towns, counties, and other municipalities are developing strategies to address their aging populations (The Maturing of America, 2006). More effort is needed among the nation's municipalities to plan for the influx of older adults and to incorporate environmental considerations into the planning process.

While moving out of municipal areas to escape the pollution is an option for many older adults, most want to remain in their communities after retirement. Some communities offer a high quality of life for retirees and have established programs to help baby-boomers become an asset rather than a drain on resources (Mahoney, 2007). For this trend to continue, there are specific steps local and state governments can take to reduce the effects of air pollution exposure and excess heat events, including:

- Locating senior housing and care facilities away from freeways and commercial/industrial areas,
- Minimizing sources of indoor air pollution, including second-hand smoke, and encouraging out-of-door activities on good air quality days,
- Ensuring adequate air circulation in modern, energy-efficient homes,
- Reaching out to at-risk individuals for additional support, including frail older adults and those with asthma, COPD, heart failure, coronary artery disease, and arrhythmias,
- Issuing heat event warnings and ensuring older adults are aware of programs offering fans, air conditioners, and transportation to air conditioned facilities,
- Educating older adults, local officials, and the community about the hazards associated with air pollution and excessive heat episodes and the need to limit strenuous, outdoor activities,
- Increasing awareness about air quality and heat indicators such as the Air Quality Index and the National Weather Service's heat advisories and warnings, and
- Encouraging public/private collaboration aimed at implementing programs to reduce air pollution exposure, such as:
  - Promoting public transit options,
  - Establishing smoke-free policies,
  - Locating recreational areas away from roadways,
  - Encouraging energy efficient building construction,
  - Providing flexible working hours and locations, and
  - Developing heat warning/response systems.

Valuable information on aging and the environment can be found at the EPA's Aging Initiative web site (www.epa.gov/aging).

References

"A Profile of Older Americans: 2005." 2005. Administration on Aging, U.S. Department of Health and Human Services, www.aoa.gov/prof/Statistics/profile/2005/ profiles2005.asp, Washington, DC.

"Aging and Toxic Response: Issues Relevant to Risk Assessment." 2005. National Center for Environmental Assessment, U.S. Environmental Protection Agency, EPA-600/P-03-004A, www.epa.gov/ncea, Research Triangle Park, NC.

"Air Quality Index (AQI) – A Guide to Air Quality and Your Health." 2007. Office of Air Quality Planning and Standards, U.S. Environmental Protection Agency, http://airnow.gov/index.cfm?action=static.aqi, Research Triangle Park, NC.

"Air Trends." 1997. Office of Air Quality Planning and Standards, U.S. Environmental Protection Agency, www.epa.gov/airtrends/, Research Triangle Park, NC.

"Carbon Dioxide Concentration Measurements at Mauna Loa Observatory." 2006. Earth System Research Laboratory, National Oceanic and Atmospheric Administration, www.esrl.noaa.gov/gmd/ccgg/trends/co2_data_mlo.html, Boulder, CO (Figure reproduced with permission).

"Climate Change." 2007. U.S. Environmental Protection Agency, www.epa.gov/ climatechange, Washington, DC.

"Climate of 2006 – in Historical Perspective, Annual Report." 2007. National Climatic Data Center, National Oceanic and Atmospheric Administration, www.ncdc.noaa.gov/oa/climate/research/2006/ann/ann06.html, Asheville, NC.

"Diabetes and Environmental Hazards – Fact Sheet." 2007. Aging Initiative, U.S. Environmental Protection Agency, EPA-100-F-07-020, www.epa.gov/aging, Washington, DC.

Dominici, F, et al. 2006. "Fine Particulate Air Pollution and Hospital Admission for Cardiovascular and Respiratory Diseases." The Journal of the American Medical Association 295(10): 1127–1134.

"Excessive Heat Events Guidebook." 2006. Office of Atmospheric Programs, U.S. Environmental Protection Agency, EPA 430-B-06-005, www.epa.gov/hiri/about/ heatguidebook.html, Washington, DC.

"France Heat Wave Death Toll Set at 14,802." 2003. News story in USA Today, September 25, 2003, McLean, VA.

Gong, K.W., et al. 2007. "Air-Pollutant Chemicals and Oxidized Lipids Exhibit Genome-Wide Synergistic Effects on Endothelial Cells (Provisional)." Genome Biology 8(7): 1–47.

"Health, United States, 2006, with Chartbook on Trends in the Health of Americans." 2006. Centers for Disease Control and Prevention, U.S. Department of Health and Human Services, DHHS No. 2006-1232, www.cdc.gov/nchs/data/hus/hus06.pdf, Atlanta, GA.

"Informing Regulatory Decisions: 2003 Report to Congress on the Costs and Benefits of Federal Regulations and Unfunded Mandates on State, Local, and Tribal Entities." 2003. Office of Management and Budget, www.whitehouse.gov/omb/inforeg/2003_cost-ben_final_rpt.pdf, Washington, DC.

Lee, D.-H., M. Steffes, and D.R. Jacobs Jr. 2007. "Positive Associations of Serum Concentration of Polychlorinated Biphenyls or Organochlorine Pesticides with Self-Reported Arthritis, Especially Rheumatoid Type, in Women." Environmental Health Perspectives 115(6): 883–888.

Mahoney, S. 2007. "Great Places to Live." ARRP The Magazine, September/October 2007: 56–63.

Meehl, G.A., and C. Tebaldi. 2004. "More Intense, More Frequent, and Longer Lasting Heat Waves in the 21st Century." Science 305(5686): 994–997.

Miller, K.A., et al. 2007. "Long-Term Exposure to Air Pollution and Incidence of Cardio-vascular Events in Women." The New England Journal of Medicine 356(5): 447–458.

"Older Americans Update. 2006: Key Indicators of Well-Being." 2006. Federal Interagency Forum on Aging Related Statistics, www.agingstats.gov, Washington, DC.
Patz, J.A., et al. 2005. "Impact of Regional Climate Change on Human Health." Nature 438(7066): 310–317.
Sandström, T., and B. Brunekreef. 2007. "Traffic-Related Pollution and Lung Development in Children." The Lancet. 369(9561): 535–537.
"State of the Air: 2007." 2007. American Lung Association, http://lungaction.org/reports/stateoftheair2007.html, New York, NY.
"The Maturing of America: Getting Communities on Track for an Aging Population." 2006. National Association of Area Agencies on Aging, www.n4a.org/pdf/ MOAFinalReport.pdf, Washington, DC.
Williams, R. 2005. "Exposure Assessment of Older Adults to Particulate Matter." Conference on the Effects of Air Pollution on Older Adults, Arlington, VA.
Zanobetti, A., and J. Schwartz. 2007. "Particulate Air Pollution, Progression, and Survival After Myocardial Infarction." Environmental Health Perspectives 115(5): 769–775.
"65+ in the United States: 2005, Current Population Reports." 2005. National Institute on Aging and U.S. Census Bureau, www.census.gov/prod/2006pubs/p23-209.pdf, Washington, DC.

Environmental Pollution, Race and Place: Research and Policy Implications

Ibipo Johnston-Anumonwo

Abstract A key theme in geographic research is to analyze changes over time and across places. Studies on trends about racial differences in settlement patterns can inform analysts about whether the thesis of differential racial proximity to toxic sites or exposure to hazardous wastes is still tenable or not. The purpose of this study is to provide a locational and longitudinal comparative analysis of the role of race and place on environmental pollution based on the conceptual and empirical literature of research on racial disparities in exposure to pollutants for varied locations in the US. The findings reveal racial inequalities in environmental pollution that are consistent with a systematic and widespread pattern of defacto racial residential segregation in contemporary US. These findings are synthesized to inform future research agenda and the policy options for alleviating adverse impacts of environmental pollution on racial minorities. With its emphasis on place-specific analyses, a geographic perspective offers a comprehensive approach to understanding and addressing the multiple burdens of environmental racism.

Introduction: Race and Place

The paper starts with a summary of US residential geography at different geographic scales that underscores the home-employment disadvantages of African Americans. Next is a review of findings from case studies on environmental pollution, focusing on locational analyses followed by longitudinal analyses. The paper concludes with a discussion of research and policy implications.

A very striking aspect of US ethnic geography is the regional concentration of African Americans in the South. Before 1910, black settlement was relatively fixed

---

I. Johnston-Anumonwo (✉)
State University of New York, Cortland, New York, USA

G.A. Uzochukwu et al. (eds), *Proceedings of the 2007 National Conference on Environmental Science and Technology*, DOI 10.1007/978-0-387-88483-7_11,

and concentrated mostly in the South. With the post 1910 Great migration and the post 1970 reverse migration as significant dimensions of change in the national distribution of African Americans, the persistence of a large Black American presence in the Southern states remains in the early 2000s.

In urban areas, racial residential segregation also remains high with minorities still concentrated in central cities. There has been little change in the residential segregation of Blacks (Denton, 1994; Massey and Hajnal, 1995). Because of their relative concentration in central cities, the growth of employment in the suburbs impacts black workers' access to jobs, and increasing numbers of African American workers are commuting outward to suburban workplaces. In light of the continuing residential differences between Whites and Blacks, perhaps the most important dimension in locational access to employment is differences among workers who reverse commute from inner city residences to suburban workplaces. Few recent studies have examined racial differences among reverse commuters. But some researchers note that if employment opportunities continue to expand in suburban locations and not in central city locations, African American workers face the penalty of longer commutes to suburban workplaces than do white counterparts (Johnston-Anumonwo, 2001; Johnston-Anumonwo and Sultana, 2006). Conversely, if more jobs are available in inner city locations, there will be less need for suburban commutes.

Alternatively, if African Americans face less discrimination for suburban housing, the racial disparity in access to suburban jobs may reduce. Yet, even for those who reside in suburban locations of US metropolitan areas, African Americans still experience segregation (e.g., Darden, 1990; Darden and Kamel, 2000). Because of the combination of constrained access to suburban housing and suburban employment for African Americans, the reality of residential immobility is linked with employment outcomes as widespread research and overwhelming evidence of spatial mismatch (e.g., Johnston-Anumonwo, 2001), lead many researchers to conclude that persistent racial disparities in access to employment continue to limit the full economic participation of black workers in US metropolitan areas (e.g., Dickerson, 2007).

A variety of reasons underlie these geographies of home and work for African Americans. They range from preference, poverty, and prejudice to a host of discriminatory exclusionary policies. In spite of the almost 40 year old Fair Housing Act, a combination of subtle and blatant unfair practices including white flight continue to make US residential location patterns largely differentiated along racial lines. Research on racial residential segregation has sometimes circumvented the connections with environmental pollution, but these well observed locational patterns of the residential geographies of African Americans often correlate with observations on adverse environmental impacts for minorities. Thus research findings on environmental pollution can best be understood in cognizance of the structural nature of residential segregation in the US. There is indeed a compelling body of research literature showing that minorities and the poor are disproportionately affected by environmental pollution (Bullard, 1990a and 1990b).

Environmental Pollution: Findings from Case Studies

Locational Analyses. The classic work by Robert Bullard, Dumping in Dixie, (revisited by Mitchell et al., 1999), is applicable to the regional case of the US South, the region touted as a "sacrifice zone" for the rest of the country's toxic waste, and the prevalence of locally unwanted land uses (LULUs). Allen (2001), using data for over 2000 counties nation-wide, found that in 1995, as the percentage of the Black population in American counties increased, so did the level of toxic releases. But the study goes further in corroborating Bullard's conclusions for the US South by documenting that the increase in level of toxic releases as a function of the Black population is stronger in counties of the Sunbelt. Counties within this region are home to large percentages of Blacks, and large volumes of toxic releases are found in the Sunbelt. The result is that Black citizens are subjected to higher levels of toxic releases in this region than is the case elsewhere in the country (Allen, 2001). It is in this light that studies on environmental justice such as Cutter and Solecki's (1996) findings on acute and chronic airborne toxic releases in the Southeastern US, and Checker's (2005) account of "environmental racism and the search for justice in a southern town" remain salient. Other county-level case studies including of Georgetown County in South Carolina (Cutter et al., 2000) allude to this regional prevalence of the vulnerability of people and places in the US South.

As previously noted, there is metropolitan and urban concentration of non Whites, and much empirical investigations of environmental pollution have been conducted at the metropolitan scale. At the county, metropolitan and urban level, there are ample studies that have examined the role of class and race as factors in patterns of residential segregation (e.g., Phelan and Schneider, 1996; Darden and Kamel, 2000), and of job access (e.g., Mouw, 2000; Dickerson, 2007), but there is far less work documenting the connections between segregated urban neighborhoods and the geography of environmental pollution within cities. Cutter et al. (2001) is a notable exception. Focusing on occupants of public housing (more than two thirds of whom are nonwhite), the researchers found that families residing in Section 8 HUD housing had greater potential risk exposure as a result of their proximity to the locations of hazardous facilities. The study of eight medium-sized metropolitan areas found that minority populations (defined as percentage non-White) had significantly greater locational exposure than non minority populations in all but one of the US metropolitan areas that were studied.

Other metropolitan level studies of Detroit, Buffalo, and Los Angeles point to a significant relationship between race and toxic releases. Boer et al.'s (1997) results suggest that communities most affected by hazardous waste treatment, storage, and disposal facilities are working class communities of color located near industrial areas. Among studies on intra urban variation among ethnic groups, Cutter et al. (2001) documented intra-urban environmental pollution inequities. Pulido's (2000 empirical study of Southern California, reports that blacks' exposure to environmental hazards is largely a function of severe spatial containment, and the historic practice of locating hazardous land uses in black neighborhoods. She states that in the region as a whole, it is clear that people of color are disproportionately exposed to a particular set of environmental hazards (Pulido, 2000:32). But a significant contribution is her insight about examining the reality of environmental racism through the expanded lens of White privilege. According to Pulido, White privilege allows us to see how environmental racism has been produced. She points out that explanations of environmental inequality must include careful consideration of residential patterns, land use, and industrial development, since past patterns continue to inform present patterns. Therefore, apart from locational analyses, longitudinal analyses also inform our knowledge base on institutional racism and environmental pollution.

Longitudinal Analyses. Pulido shows the value of integrating conceptual and empirical research in the study of environmental racism. It follows then that a temporal framework aimed at identifying the role of place and race on environmental pollution will increase our understanding of how environmental injustice is perpetuated. In this regards, the changing geographical context is also important since there are likely to be different implications for a geographically extensive location compared to a more geographically compact setting, or even an area experiencing large scale demographic shifts within a short period of time. In addition to paying appropriate attention to the geographic unit of analysis and scale concerns, more recent studies show that geographic information systems GIS techniques and tools lend themselves to deeper analytical research as is well demonstrated in the study of South Carolina by Cutter et al. (2000) and the study of Buffalo by Margai (2003).

Cutter et al. (2001) emphasize the critical need for longitudinal studies when attempting to unveil any relationship between race and environmental pollution. Specifically, they cite the provision of a historical time series of sociodemographic change that addresses temporal changes in the social and industrial geography that create the landscape of risk. To the extent that long term monitoring data remains a critical need, studies that jointly examine temporal patterns of residential segregation and environmental pollution remain scarce overall. Research initiatives with a process-oriented focus can answer questions such as whether there is any increase in industrial facilities in minority neighborhoods over time; or whether industrial zones become disproportionately the residential locations of more low income minority populations. Synthesizing these sets of analyses raises a pertinent research question: "who bears the burden of environmental pollution?" (Pollock and Vittas, 1995).

Unfair Burden

Cutter (1994) explicitly addresses the question of whether the burdens of toxic risks are fair. Reports from Atlanta to San Francisco Bay Area verify that minorities still face a disproportionate burden of exposure to environmental hazards. It is reported that three out of every five African Americans and Latinos live in communities with toxic waste sites; and 71 percent of African Americans and Latinos live in communities where air pollution violates federal clean air standards (compared with

58 percent of Whites). Based on findings in Indiana that suggest a pattern of race and class discrimination in the sites of toxic facilities, Maher (1998) presents the unfair burden of environmental pollution as environmental oppression. Such a viewpoint bears merit and could be backed by conceptual research that establishes the relevance of theoretical frameworks of multiple forms of oppression. The interplay of marginalization and powerlessness as two faces of oppression operating in the context of environmental pollution is clear, and the role of cultural imperialism as a face of oppression could be examined as a specific instance of institutional racism. Indeed, Maher concluded that "regardless of whether racism or classism are at work in the toxic polluting of communities, public policy should focus on relieving the burden placed on already distressed communities" (Maher, 1998:364). Lastly, Cutter et al. (2001) also report that minority populations had higher risk burdens than non minority residents. Thus much of the empirical research strongly suggests that when it comes to environmental pollution, non Whites bear a race tax burden.

Drawing on data, case studies, and the growing body of research that compare environmental pollution of whites with non-white counterparts, and taking into account key socioeconomic, locational, and accessibility factors, this paper has reported continuing empirical evidence of persistent adverse environmental health hazards for non Whites ranging from industrial toxic dumping to solid waste facility siting. The findings on geographic and demographic differences reveal racial inequalities in environmental pollution that are consistent with a systematic and widespread pattern of defacto racial residential segregation in contemporary US. The remainder of the paper discusses some implications.

Implications for Research

Some of the literature on the employment disadvantages of US minorities as captured in the spatial mismatch hypothesis links increased commuting, and reduced employment access of African Americans to their segregation in inner city residences that are distant or isolated from suburban centers of job growth. But this literature by social and urban geographers on place and work and on intra urban spatial accessibility constraints has not been fully utilized by environmental geographers in the study of environmental pollution (see Cutter, 1995). Furthermore, compared to studies on racial residential segregation, there is as yet far less documentation of the degree of workplace segregation of ethnic groups and the possible environmental pollution differentials.

In light of emerging geographical patterns of new immigrants, the findings documented for these study areas could be used as an indicator for other US locations with high and persisting levels of residential segregation and recent large-scale immigration. The possibility that nonwhites who are employed work outside the home also calls for more thorough examination of workplace environmental pollution in appraising racial differential pollution exposure. Finally, gender is known to be a significant factor in employment (e.g., Hanson and Pratt, 1995). In future, research inquiries should investigate comparisons between men and women. For instance, is there a gendered connection between residential and workplace environmental pollution? Geographical and accessibility inequalities in the race and sex segmented labor market (e.g., Preston and McLafferty, 1998) raise the possibility that foreign born workers, particularly women of color usually with comparatively limited private automobile access, and who are thus more dependent on public transit may operate within more circumscribed local geographies and daily activity spaces with different dimensions of pollution exposure. This is an area of inquiry that should feature prominently in future research agenda. Case studies in Miami that link race, ethnicity, gender and immigrant status with labor force characteristics (Johnston-Anumonwo, 2003), as well as in New York City that document the interacting effects of race, foreign born status, and residential segregation on the incidence of low birth weight among mothers (Grady, 2003) strongly suggest that more workplace environmental pollution research is necessary for immigrant women of color.

Combining locational (geographical) and longitudinal (historical) research analyses offers a more comprehensive assessment of the nature of environmental racism. Such research could provide non equivocal evidence about the relationships between race and place and environmental pollution. Furthermore, such research inquiries will expose continuing patterns of racially differential environmental pollution across the country. Pulido's (2000) case study of Los Angeles clearly demonstrates that this approach is analytically robust.

Conclusion and Policy Implications

This paper has attempted to show that with its emphasis on place-specific analyses over time, a geographic perspective offers a comprehensive approach to understanding and addressing the multiple burdens of environmental racism. Long term studies are needed to better characterize the nature of the problem of differential environmental pollution. The synthesis of past and more recent findings in different geographical settings is used in this paper to inform future research agenda and the policy options that can alleviate adverse impacts of environmental pollution on racial minorities.

Instead of focusing on individuals' behaviors, the significance of locational context is emphasized. Thus from a public policy perspective, steps for addressing environmental pollution disparities among ethnic and racial groups are likely to be more effective when place-based methodologies are used as a first approach to identifying locations that are prone to environmental pollutants. Policy initiatives and efforts that emphasize housing and transportation are needed across the board from large industrial northern cities to rural Southern counties. But simply improving the locational access and transportation of inner city black workers for suburban homes or jobs is not a sufficient policy goal if they are being relocated or transported to polluted environments.

In conclusion, as Margai (2003) points out, while efforts at expanding resources and improving the people's well being in high risk locations are important, such efforts must be accompanied by the remediation and clean up of pre-existing brownfields and other environmental hazards, particularly in urban communities with a long history of pollution problems (Margai, 2003:273). Pulido (2000) and even more emphatically Bullard (1996 and 2007) stress the same message of bridging longitudinal analyses with locational analyses when they note that it is not simply circumstances of the placement of proposed or new hazardous waste sites that are important, but that the conditions in existing sites are probably more paramount precisely because these locations remain the residential spaces of a disproportionate number of non-Whites who find themselves living in and unable to relocate from these endangered environs.

References

Allen, D. W. 2001. "Social Class, Race and Toxic Releases in American Counties, 1995." Social Science Journal, 38: 13–25.

Boer, J. T., M. Pastor, J. L. Sadd, and L. D. Snyder. 1997. "Is There Environmental Racism? The Demographics of Hazardous Waste in Los Angeles County." Social Science Quarterly, 78(4): 793–810.

Bullard, R. 1990a. "Dumping in Dixie." Westview Press, Boulder, CO.

Bullard, R. D. 1990b. "Housing Trends in the Nation's Fourth-Largest City." Journal of Black Studies, 21: 4–14.

Bullard, R. 1996. "Environmental Justice: It's More than Waste Facility Siting." Social Science Quarterly, 77: 493–499.

Bullard, R. 2007. "Toxic Waste and Race at Twenty 1987–2007". The United Church of Christ, Cleveland, OH.

Cutter, S. 1994. "The Burdens of Toxic Risks: Are They Fair?" B & E Review. October-December, 3–7.

Cutter, S. 1995. "Race, Class and Environmental Justice." Progress in Human Geography, 19: 107–118.

Cutter, S. and W. Solecki. 1996. "Setting Environmental Justice in Space and Place: Acute and Chronic Airborne Toxic Releases in the Southeastern US." Urban Geography, 17: 380–399.

Cutter, S. L., J. T. Mitchell, and M. S. Scott. 2000. "Revealing the Vulnerability of People and Places: A Case Study of Georgetown County, South Carolina." Annals of the Association of American Geographers, 90(4): 713–737.

Cutter, S. L., M. E. Hodgson, and K. Dow. 2001. "Subsidized Inequities: The Spatial Patterning of Environmental Risks and Federally Assisted Housing." Urban Geography, 22: 29–53.

Darden, J. T. 1990. "Differential Access to Housing in the Suburbs." Journal of Black Studies, 21: 15–22.

Darden, J. T. and S. Kamel. 2000. "Black Residential Segregation in the City and Suburbs of Detroit: Does Socioeconomic Status Matter?" Journal of Urban Affairs, 22: 1–13.

Denton, N.A. 1994. "Are African Americans Still Hypersegregated?" In R.D. Bullard, J. E. Grigsby III and C. Lee (Eds.), Residential Apartheid: The American Legacy. Center for Afro-American Studies, University of California, Los Angeles, pp. 49–81.

Dickerson, N.T. 2007. "Black Employment, Segregation, and the Social Organization of Metropolitan Labor Markets." Economic Geography, 83: 283.

Grady, S. 2003. "Low Birth Weight and the Contribution of Residential Segregation, New York City, 2000." In J. W. Frazier and F. M. Margai (Eds.), Multicultural Geographies: The Changing

Racial/Ethnic Patterns of the United States. Global Academic Publishing, Binghamton University, Binghamton, NY, pp. 65–75.
Hanson, S., and G. Pratt. 1995. Gender, Work, and Space. Routledge, London.
Johnston-Anumonwo, I. 2001. "Persistent Racial Differences in the Commutes of Kansas City Workers." Journal of Black Studies, 31(5): 651–670.
Johnston-Anumonwo, I. 2003. "Ethnicity, Immigration, and the Gender Division of Labor in Miami." The Florida Geographer, 34: 6–21.
Johnston-Anumonwo, I. and S. Sultana. 2006. "Race, Location, and Access to Employment in Buffalo, NY." In J. W. Frazier and E. L. Tettey-Fio (Eds.), Race, Ethnicity and Place in a Changing America. Global Academic Publishing, Binghamton University, Binghamton, NY, pp. 115–130.
Maher, T. 1998. "Environmental Oppression: Who is Targeted for Toxic Exposure?" Journal of Black Studies, 28(3): 357–367.
Margai, F. M. 2003. "Using Geo-Data Techniques to Analyze Environmental Health Inequities in Minority Neighborhoods: The Case of Toxic Exposures and Low Birth Weights." In J. W. Frazier and F. M. Margai (Eds.), Multicultural Geographies: The Changing Racial/Ethnic Patterns of the United States. Global Academic Publishing, Binghamton University, Binghamton, NY, pp. 263–274.
Massey, D. S. and Z. L. Hajnal. 1995. "The Changing Geographic Structure of Black-White Segregation in the United States." Social Science Quarterly, 76(3): 527–542.
Mitchell, J. T., D. S. K. Thomas, and S. L. Cutter. 1999. "Dumping in Dixie Revisited: The Evolution of Environmental Injustices in South Carolina," Social Science Quarterly, 80(2): 229–243.
Mouw, T. 2000. "Job Relocation and the Racial Gap in Unemployment in Detroit and Chicago 1980–1990." American Sociological Review, 65: 730–753.
Phelan, T. and M. Schneider. 1996. "Race, Ethnicity and Class in American Suburbs." Urban Affairs Review, 31: 659–80.
Pollock, P. and Vittas, M. 1995. "Who Bears the Burden of Environmental Pollution? Race, Ethnicity and Environmental Equity in Florida." Social Science Quarterly, 76: 294–310.
Preston, V., S. McLafferty, and X. F. Liu. 1998. "Geographical Barriers to Employment for American-Born and Immigrant Workers." Urban Studies, 35(3): 529–545.
Pulido, L. 2000. "Rethinking Environmental Racism: White Privilege and Urban Development in Southern California." Annals of the Association of American Geographers, 90(1): 12–40.

Financial Reporting of Environmental Risks and Liabilities

Gwendolyn McFadden-Wade, Jean T. Wells, and Diana R. Robinson

Abstract Financial accounting and environmental issues are not strange bedfellows. A standard for financial statement reporting and disclosure of environmental risks and liabilities existed 30 years ago but this standard proved inadequate in the post-Sarbanes Oxley financial reporting environment. This paper, by discussing the evolution of accounting reporting standards related to environmental risks, seeks to provide a brief summary of the accounting profession's contribution to the business discussion of environmental issues. Accounting faculty may use this paper to discuss, not only the evolution of standards related to the financial reporting liabilities associated with environment risks, but also the limitations and shortcomings inherent in current reporting and disclosure standards.

The Relevancy of Financial Reporting of Environmental Risks and Liabilities

Risks and liabilities resulting from the enactment of environmental laws such as the Clean Air Act, and the Toxic Substance Control Act in the 1970s and the Congressional Comprehensive Environmental Response, Compensation and Liability Act ("CERCLA" or "Superfund") of the 1980s, have long been a concern for regulatory agencies, standards setters, investors, and creditors. Superfund legislation, which empowers the Environmental Protection Agency (EPA) to clean up waste sites and charge the clean-up costs to potentially responsible parties, exposes all entities and individuals responsible for contamination. In most cases, these costs typically result in a liability that significantly impacts the financial statements.

Accounting standards, promulgated during the mid 1970s, can be traced to Statements of Financial Accounting Standards (SFAS) No. 5, Accounting for Contingencies. This standard required the accrual of a liability if two conditions were met. First, the occurrence of a future event must be probable. Second, the amount must

---

G. McFadden-Wade (✉)
North Carolina A & T State University, Greensboro, NC, USA

G.A. Uzochukwu et al. (eds), *Proceedings of the 2007 National Conference on Environmental Science and Technology*, DOI 10.1007/978-0-387-88483-7_12,

be subject to reasonable estimation. However, reasonably estimating environmental liabilities proved difficult to apply because clean-up costs could vary depending on the information available at the time of the disaster or incident.

Take the case of the disasters which occurred in the 1970s and 1989. During the late 1970s, toxic waste was discovered at a landfill site near the Love Canal in upstate New York. In 1989, eleven million gallons of crude oil leaked into Alaska's Prince William Sound after the Exxon Valdez struck a reef. Not only did these catastrophes effect the environment, but they also resulted in substantial economic risks and liabilities. The liability amounted to roughly $129 million to Occidental Chemical, the company charged with the repaying the EPA for cleanup cost associated with the Love Canal and over $850 million for Exxon.

At the time of the Love Canal and Exxon disasters, neither Generally Accepted Accounting Principles nor any other regulations or standards mandated disclosure of and/or reporting of the potential environmental risks and liability before the incidents occurred. Even after the incidents occurred, accountants had only the guidance of SFAS No. 5 regarding the amount of liability that should be recorded. As a result, accountants did not record in the financial statements the potential liability for these risks which, had they been reported, could have alerted financial statement users to existing or potential financial risks to the company.

Evolution of Financial Reporting Guidelines

Other factors contributed to the inability to make a reasonable estimate of the liability associated with environmental risks. Some companies simply failed to record the Superfund liability by foregoing activities that would trigger the EPA's scrutiny of known or suspected contaminated facilities. Other companies were faced with the inability to establish the appropriate factual circumstances that would trigger recognition of a liability. The most significant of which was the liability associated with asset retirement obligations (ARO) related to environmental remediation.

The AICPA attempted to address the ARO issue in 1996 when it issued Statement of Position (SOP) No. 96-1, Environmental Remediation Liabilities. SOP 96-1 provided specific guidelines on when companies should recognize, measure, and disclose environmental remediation liabilities in the financial statements. Because of the measurement difficulties associated with the "probable and reasonably estimable approach" required under SFAS No. 5, additional guidance was necessary. In 2001, roughly thirty years after SFAS No. 5 was released, SFAS No. 143, Accounting for Asset Retirement Obligations (ARO), was issued. SFAS No. 143 moved to a "fair value" measurement approach, which did not recommend, but mandated that a company must measure and recognize liability for ARO associated with environmental remediation issues. The standard also provided companies with several options by which they could measure the liability. Companies could estimate fair value based on market prices of similar liabilities, present value methods, or the best obtainable information.

Nonetheless, this standard did not deter some companies that only recorded liabilities for sites that posed an imminent threat or faced imminent closure. The practice resulted in an inconsistency in recognizing, measuring, and reporting of

ARO in the financial statements. To resolve this inconsistency, FASB Interpretation Number (FIN) 47, Accounting for Conditional Asset Retirement Obligations, was issued effective December 15, 2005. FIN 47 provides that companies should record liabilities for conditional ARO (CARO) in the financial statements. A CARO is a legal obligation conditioned on future events that may or may not be within the company's control, e.g., future asbestos remediation. Thus, FIN 47 tightened the financial reporting rules for environmental liabilities forcing companies to record future environmental clean-up costs even in instances when the company had no immediate plans to retire the contaminated assets. Some companies also delayed their Securities and Exchange Commission (SEC) filings to have more time to ensure compliance with FIN 47.

The financial impact of FIN 47 was immediate. The Table 1 below shows the financial effect of Fin 47 on manufacturers, energy generating companies, and on service companies as well.

Table 1 FIN 47: Financial statement impact during first year of implementation

Company	Expense	Environmental remediation	Industry
Ford Motor Company	$251 million	Asbestos abatement and PCB removal	Automobile Manufacturing
ConocoPhillips	$88 million	Asbestos abatement at refineries and removal and disposal of offshore oil and gas platforms pipelines	Energy
Citigroup Inc	$49 million	Real estate restoration activities in branches and office space obligations	Financial Services
FirstEnergy	$57 million	Asbestos remediation	Energy Generation
E. I. DuPont de Nemours & Co	$29 million	Closure, reclamation and removal costs for mining operations related to production of titanium dioxide	Chemical Manufacturing
Dow Chemical	$20 million	Asbestos encapsulation as a result of planned demolition and remediation activities at various sites	Chemical Manufacturing
Marathon Oil	$19 million	Removal and disposal of fire-retardant materials from refining facilities	Energy Generation
Kimberly Clark	$12 million	Retirement of long-lived assets	Health/Hygiene Products Manufacturer
USG Corp	$11 million	Reclamation requirements related to mines, quarries, landfills, pools and wells	Mining and raw materials

Source: Form 10-Ks filed with the SEC, available at http://www.sec.gov/edgar/searchedgar/webusers.htm

Current disclosure guidelines have been significantly more useful in providing information to financial statement users than in past years. Still, there is concern about company compliance with these guidelines. As indicated in the Securities and Exchange Commission's (SEC's) Summary by the Division of Corporation Finance of Significant Issues Addressed in the Review of the Periodic Reports of the Fortune 500 Companies, many companies do not provide sufficient disclosure pertinent to environmental liabilities. The SEC also mentioned that companies with material contingent liabilities need to improve their disclosures by providing all required information and explanations supporting their calculation of the liabilities. The SEC, various congressional committees, and the Justice Department have conducted investigations and have addressed the role of management in providing more meaningful protection to investors. These investigations, coupled with numerous recent corporate debacles prompted the issuance of the Sarbanes-Oxley Act of 2002 (SOX) which highlights the need for management to establish and maintain a sufficient internal control system over financial reporting.

Internal Control Weaknesses

Management must attest to the effectiveness of the company's internal control system over financial reporting. Under SOX, corporate officers are also responsible for certifying that their company's financial statements "fairly present" the financial condition and results of operations of the company. Corporate officers face civil and criminal penalties for misleading financial statements. As a result, companies must ensure that their internal controls are sufficient to identify, assess, measure and report environmental risks and liabilities in the financial statements. Otherwise, companies must report to the SEC material weaknesses in their internal control system. The financial impact of a material weakness in a company's internal control system can be significant.

SOX's requirements resulted in the restatement of the financial statements in several companies and the reporting of internal control weaknesses in others. For example, in 2005, Millennium Chemicals Inc. restated its financial statements for the three-year period ending December 13, 2003 and the first three quarters in 2004 to recognize an increase of $52 million in recorded liabilities for environmental remediation. Other companies, such as Ultra Petroleum, Westmoreland Coal and Clean Harbors, Inc. also reported internal control weaknesses in their SEC filings.

Internal control weaknesses are not unique to public companies, but can be found in the federal government as well. A 2006 U.S. Government Accountability Office report concluded that the Department of Defense's "policies and procedures for determining, reporting, and documenting environmental liability estimates were not consistently followed" and that "none of the military services had adequate controls in place to help ensure that all identified contaminated sites were included in their environmental liability cost estimates."

Conclusion

While, financial reporting of environmental risk has evolved over the past thirty years, the most significant changes have occurred during the past eight years. Accounting standards, promulgated in the early 1970s during the enactment of the environmental laws, proved insufficient in the post SOX disclosure requirement era. Current standards, established by accounting regulators and mandated by SOX, although limited, seek to tighten the reporting and disclosure requirements which should lead to more transparency regarding environmental risks and liabilities in the financial statements.

Appendix

Generally accepted accounting principles (GAAP) govern the financial accounting practices in the United States. This common set of standards and procedures is developed by four organizations within the public and private sectors. The Financial Accounting Standards Board (FASB), a private standard-setting body, develops and issues Statements of Financial Accounting Standards (SFAS) and FASB Interpretation Numbers (FIN). Other private groups, such as the American Institute of Certified Public Accountants (AICPA), establish guidelines for the competent and ethical practice of accountants entrusted with preparing and auditing financial statements. The AICPA issues Statements of Position (SOP) on relevant financial reporting issues before the FASB sets the final standard. The Governmental Accounting Standards Board (GASB) sets the standards for state and local government reporting. The Securities and Exchange Commission (SEC) oversees the accounting standards-setting process.

GAAP is designed to ensure that financial information contained in financial statements is relevant, reliable, and comparable. The goal is to ensure that the users can trust the accuracy of the information, determine its relevance to decision making and compare financial information between and among companies and industries.

Despite GAAP, major misstatements of financial information at Enron, WorldCom, HealthSouth, and other companies have resulted in a loss of confidence in the reliability and accuracy of financial statements. These accounting scandals prompted Congress to pass the Sarbanes-Oxley Act of 2002 (SOX) which forces companies to establish more stringent internal controls and oversight mechanisms that would promote transparency, accountability and truthfulness in financial reporting.

References

American Institute of Certified Public Accountants. 1996. Statement of Position 96-1: Environmental Remediation Liabilities (Including Audit Guide). New York, NY: Environmental Accounting Task Force.

Financial Accounting Standards Board. 1975. Accounting for Contingencies, SFAS No. 5, Stamford, CT: FASB.

Financial Accounting Standards Board. 2001. Asset Retirement Obligations, SFAS No. 143, Stamford, CT: FASB.

Financial Accounting Standards Board. 2005. Conditional Asset Retirement Obligations, FIN 47, Stamford, CT: FASB.

Government Accountability Office. 2006. Environmental Liabilities: Long-Term Fiscal Planning Hampered Control Weaknesses and Uncertainties in the Federal Government's Estimates, GAO-06-427. Washington, DC.

Sarbanes-Oxley Act of 2002. P.L. 107–204.

Securities and Exchange Commission. Edgar Database. Retrieved August 23, 2007 from http://www.sec.gov/edgar/searchedgar/webusers.htm

Summary by the Division of Corporation Finance of Significant Issues Addressed in the Review of the Periodic Reports of the Fortune 500 Companies. Retrieved August 23, 2007 from http://www.sec.gov/divisions/corpfin/fortune500rep.htm.

Oak Biodiesel: State Building and Fire Codes Affecting Biodiesel Manufacturing

Myra M. Shird and Regina M. Williams

Abstract Long chain mono alkyl esters or Biodiesel is a renewable energy efficient fuel manufactured from basic vegetable oils and recycled cooking greases. For the environmentally conscious home manufacturer or home brewer, the *Discovery Channel*, *The Learning Channel*, and Ebay encourage biodiesel usage and manufacturing by offering how-to-tips as well as do-it yourself kits. This paper focuses on the Oak Biodiesel company formerly located in High Point, North Carolina that was shut down because the "biodiesel business did not comply with the state fire and building codes" (Jarboe, 2007). This research surveys news reports about four biodiesel fires as a basis for query into biodiesel fire and building code regulations, biodiesel fire safety, and the associated ethical dilemmas facing the home and small-scale manufacturers. It also stimulates discussions about how local, state and federal policies may have to be adjusted to accommodate the increased homemade biodiesel trend. The current canon of scholarly biodiesel research does not address how emissions and fire and building codes affect the in-home manufacturer, the small less than 100,000 gallon per year producing companies, as well as the communities surrounding those facilities.

Introduction

Biodiesel, unlike fossil fuels, can be produced from vegetable oils. It can be used in its most pure form B100 or as an additive to petroleum products in popular B20, B5 and B2 proportions. "Making biodiesel is not complicated," explains an article in the *Carolina Country* magazine (Cruze, 2005). The article further outlines, " If you take 10 gallons of any vegetable oil, then add one gallon of methyl alcohol and some sodium hydroxide—commonly known as boxed lye—you create 10 gallons of biodiesel fuel and one gallon of glycerin. The lye acts as the catalyst for the

M.M. Shird (✉)
North Carolina Agricultural and Technical State University, Greensboro, NC, USA

G.A. Uzochukwu et al. (eds), *Proceedings of the 2007 National Conference on Environmental Science and Technology*, DOI 10.1007/978-0-387-88483-7_13,

chemical process called transesterification, which replaces the mixture's glycerin molecule with an alcohol molecule (p. 12). This description invites the novice to enter the field of biodiesel production. For the environmentally conscious want-to-be home brewer, the *Discovery Channel*, *The Learning Channel*, and Ebay encourage biodiesel usage and manufacturing by offering how-to-tips as well as do-it yourself kits.

The state of North Carolina is aggressively promoting and strategically planning for biodiesel expansion in the state. According to House Representative Brad Miller, "North Carolina is among the top consumers of biodiesel fuel in the United States, using more than 1.5 million gallons in 2005. In addition 25 percent of North Carolina farm land is planted with soybeans, making a state ripe for developing biodiesel fuels" (Miller, n.d.)

Legislation is in place to promote individual manufacturing as well as larger multi-million gallon per year producers. In July 2006, The North Carolina General Assembly ratified Senate Bill 2051, the Energy Independence Act. The following month, Governor Mike Easley signed the bill into law. The Energy Independence Act promotes the conservation of energy and promotes the use of alternative fuels in state-owned fleets and the development of a plan to expand the biofuels industry in North Carolina. Biodiesel is one of the recommended alternative fuels covered in this law. Senate Bill 1741 and North Carolina General Statues 105-129.16F provides a tax credit to biodiesel producers, who produce at least 100,000 gallons of biodiesel in the taxable year, a tax credit equal to the per gallon excise tax the producer paid in accordance with the motor fuel excise tax rate. North Carolina General Statutes 105-164.13 makes the retail sale, use, storage, and consumption of alternative fuels exempt from the retail sales and use tax.

The state of North Carolina has been fairly definitive on biodiesel tax laws and incentives for the 100,000 gallons per year producer. The state is not as definitive on the health hazards that may result from the increased biodiesel production nor is it clear on regulations for the below 100,000 gallons per year producers, specifically the home brewer.

Objective. The current canon of scholarly biodiesel research does not address how safety regulations affect the home brewer as well as the small less than 100,000 gallon per year producing companies. This research uses news reports of four biodiesel fires as a basis for query into North Carolina state biodiesel fire and building code regulations and the associated ethical dilemmas facing the home and small-scale manufacturers.

In July 2007, High Point, North Carolina city officials ordered operations at the Oak Biodiesel facility to cease because the "biodiesel business did not comply with state fire and building codes" (Jarboe, 2007). Background information reveals that the 100,000 gallons per year maximum capacity location did not have the proper permits to stay in business. The owner of the Oak facility, which also houses a Volkswagen repair shop, says of the permit problem, "It was a comedy of errors" (Jarboe, 2007).

There is no comedy in the results of improperly built biodiesel processing and storage facilities. The results of poor biodiesel management have been disastrous, as noted in the cases below. The long-range environmental implications for those disasters are not yet known. The aggressive push for biodiesel production and the socio-eco push to be more environmentally conscious require that scientists investigate hazards associated with biodiesel production on multiple levels.

Cases. Many states do not have specific rules regulating the storage of pure biodiesel either above or below ground, because biodiesel is not considered a petroleum product. Biodiesel does, however, require the use of methanol or ethanol, which are both highly flammable agents. Biodiesel plant fires are a source of information regarding how the agents forming biodiesel react in processing, spill, and storage situations. Biodiesel fire research can further the understanding of the environmental impact of increased small-scale operations.

In the past two years, there have been several biodiesel plant fires in the United States. Early news accounts and official statements about these fires are often ambiguous about the toxicity of the fumes and the harm to the surrounding communities. Some of these fires blazed in areas were residents lived in close proximity of the facility.

"Multiple explosions took place in February 2006 at a biodiesel storage facility located on the Hondo Chemical Plant property in Bakersfield, CA, after an accident ignited hundreds of gallons of dangerous ingredients" (Burger and Williams, 2006). A 250-gallon container of methanol accidentally spilled as workers moved the container. The methanol, which is combustible, spread quickly. The director of American Biofuels apologized to the Bakersfield community and commented, "This is a really freak-type fire. We don't know exactly what happened. It could have been static electricity" (Burger and Williams, 2006).

In June 2006, a fire destroyed a startup biodiesel production company being operated out of a barn in rural Canby, Oregon. "Firefighters responding to the fire did not know the barn contained biodiesel and had a hard time suppressing the fire because of the fuel stored there" (Redden, 2006). After the Canby incident, the Oregon Department of Environmental Quality determined that producing biodiesel involves hazardous materials and must be done in areas zoned for industrial uses.

In July 2006, a New Plymouth, Idaho biodiesel plant explosion claimed the life of a 25-year-old man. The man was welding a tank that contained biodiesel when the explosion occurred. KTVB.COM & Associated Press reported, "It is not known if the fumes from the fire are toxic. They don't appear to be posing a danger to humans" (KTVB.COM). According to another report from the Idaho Press-Tribune, "Officials were concerned that the thick, black smoke might carry poisons" (Rush, 2006). The same report quotes an Idaho State Police spokesman as saying, "The smoke was not believed dangerous." OSHA levied citations against the company alleging "the company failed to provide adequate training in the physical and health hazards of the chemicals in the work area, failed to clean, ventilate or otherwise inert the metal tank before cutting it with a torch and allowed cutting to take place in the presence of flammable compounds" (Grainnet, 2006).

In August 2007 in Lenoir, North Carolina, a fire damaged the Foothills Bio-Energies plant. The co-owner of the land and building attributed the fire to poultry fat and soybean oil stored outside of the processing plant. "Although firefighters were able to prevent most of the liquids from going into the stream at the back of the property, some of them did get into the stream" (Newell, 2007). On the Lenoir case, the Lenoir Battalion Chief said, "We don't know the source [of the fire]. I'm not sure if we'll ever know that" (Teague, 2007). The chief also expressed concerns about the fire spreading to methanol storage tanks located near the fire. Carswell said, "We did not want the fire to reach those." For every two shipments of biodiesel, the Foothills plant takes in one 700-gallon shipment of methanol (Fuller, 2007).

In each of the preceding cases, the level of uncertainty regarding biodiesel-manufacturing fires is alarming. This ambiguity is especially alarming if we consider that these biodiesel fires could be replicated on small residential levels, meaning that mismanaged home brewing facilities have the potential to present enormous environmental concerns.

Discussion

Ethical Concerns. The manufacturing of biodiesel fuel unmistakably is an environmental and economical benefit to our society in that it will allow the United States to be significantly less dependent on imported oil. It will require minimal fossil fuel consumption and relieve the citizens of the expense of the offshore drilling process, thus eliminating the disturbance of sea life. With all of the benefits to the environment and the economy, biodiesel manufacturing comes with its' own ills.

Evidently, biodiesel manufacturing requires the use of methanol or ethanol, both have been identified by the National Institute of Safety and Health as violent reactants with oxidants causing fire and explosion hazards. Essentially these chemicals may not be exposed to open flames, sparks, and cigarette smoke or have contact with oxidants. They require a closed system, ventilation, explosion-proof electrical equipment and lighting. Therefore due to the severity of potential harm, regulations and guidelines for material handling, building and fire codes are crucial as outlined previously.

Given the cost of gasoline and the capitalistic nature of our society, individual home-owners are choosing to use vegetable oil, chicken fat, soybean oil and other known flammables for their home brewing of this fuel. Home brewers produce for personal use and re-sale, resembling the days of prohibition. Unfortunately, these individual producers are not yet required to adhere to the building and fire codes. Consequently, their actions can cause havoc.

America is a nation addicted to oil. Obsessed with the need for fuel-operated machinery, many individuals have become oblivious to the fact that their actions affect the balance of nature and the environmental concerns of the many. Yet, most moral and ethical standards demonstrate concern for the needs of the many, which generally will outweigh the needs of the few. Ethical frames are complicated because of the complexities of values and the fact that we live in a pluralistic society. Nevertheless, many identify with virtue ethics, which is often associated with a teleological ethical philosophy. Taking a teleological position suggests that people should not be just reactive in accomplishing their goals, but be more often proactive.

To be proactive is to be deliberate and plan in advance; consider the outcomes before acting. Researching, testing and generating the necessary regulations, guidelines, and codes for safety concerns of biodiesel companies, and more specifically, home brewers will be the ethical safety net for prevention of harm to these companies and the surrounding homes and communities.

To be proactive is a way of practicing virtue ethics, assessing consequences before acting, which does not limit individuals completely to one set of rules or guides. Because it is generally dynamic in nature, the uniqueness of every situation is considered.

Ensuring extensive testing, research and on-going investigations of best practices protects the utilitarian view, (virtue ethics) which suggests the most ethical human actions are those that produce the best consequences for the greatest number of people. Therefore, consuming fossil fuels faster than they can be produced would be considered ethically wrong. This concern is the catalyst for seeking other natural resources (i.e., biodiesel) to be used to satisfy the requirement for oil. Because biodiesel is not manufactured in mass quantity, small businesses or home brewing plants are emerging. Launching and growing of these home brewing facilities without the needed occupational safety standards on personally-owned farmlands and compounds violates the utilitarian view, thus it is also morally and ethically incorrect.

Conclusion

Therefore, if biodiesel manufacturing or home brewing facilities are an environmental benefit for reasons such as fueling diesel engines thus making the US less dependent on fossil fuels and imported oil, then the appropriate ethical actions should be taken to launch these small manufacturers and home brewing businesses as well as assist in their growth and sustainability.

More scientific research is needed on the hazards of biodiesel fires and the consequences of manufacturing the product in residential and other non-industrially zoned areas. Building and fire regulations must accommodate the growing trend in home brewers and less than 100,000 gallon per year producers.

References

Burger, D. and M. Williams (17 February, 2006). Explosions tear through plant. *The Bakersfield Californian*. Retrieved 8/10/2007 from http://www.bakersfield.com/local/v-pring/story/5920492p-5920492p-5934453c.html

Cruze, S. (June, 2005) Soybean Energy: North Carolina Soy Biodiesel is catching on as a locally produced fuel. *Carolina Country* Retrieved on (8/27/07) from http://www.carolinacountry.com/StoryPages/howtos/biodiesel/SoybeanEnergy6.05.pdf

Fuller, B. (25 August, 2007). Town, Residents embroiled in fuel fight. Record & Landmark. Retrieved (8/29/07) from http://www.statesville.com/servlet/Satellite?pagename=Common%2FMGAArticle%2FPrint

Grainnet (4 October, 2006). OSHA Alleges Safety Violations at Blue Sky Biodiesel Plant in New Plymouth, ID. Retrieved (8/29/07) from http://www.grainnet.com/articles/OSHA_Alleges_Safety_Violations_at_Blue_Sky_Biodiesel_Plant_in_New_Plymouth__ID-37899.html

KTVB.COM & Associated Press. (9 July, 2006). Explosion rocks Biodiesel Plant, Meridian Man Killed. Retrieved 8/10/2007 from http://www.ktvb.com/new/localnews/stories/ktvbn-jul0706-new_plymouth.3504db03.html

Jarboe, M. (19 July, 2007). Biodiesel Business Shuts Down, Plans Move. *News and Record.* Retrieved 7/19/2007 from http://www.news-record.com/apps/pbcs.dll/article?AID=/20070719/NEWSREC0101/70719025/-1/NEWSRECRSSARKIVE

Miller, B. (n.d). Biodiesel Fuels. Retrieved (8/27/07) from http://www.house.gov/bradmiller/legissuesag.html

Newell, S. (26 August, 2007). Tanks catch fire: Chick fat, Soybean Oil Burn at Biodiesel Plant. *Hickory Daily Record.* Retrieved 8/10/2007 from http://www.hickoryrecord.com/servlet/Satellite?pagename=Common%2FMGArticle%2FP

Redden, J. (11 September, 2006). State: Biodiesel Production is Potentially Hazardous. Portland Tribune. Retrieved (8/15/07) from http://www.portlandtribune.com/news/story.php?story_id=1158001175411056600

Rush A. (8 July, 2006). Blast rocks local town; 1 Man Killed; 2 Injured.

The Idaho Press-Tribune. Retrieved (8/22/07) from http://www.idahopress.com/articles/2006/07/08/news/news1.txt

Teague, P. (26 August, 2007). Fire Damages Bio-Diesel Plant. Retrieved (8/29/07) http://www.newstopic.net/articles/2007/08/26/news/22fireatbiodiesel.prt.

The Ecological Effects of Salt Water Intrusion on the Agriculture Industry After Hurricane Katrina

Vereda Johnson Williams

Abstract Hurricane Katrina made landfall August 29, 2005 becoming the costliest hurricanes in U.S. history. The economic and ecological consequences were considerable. This research will (1) provide updates of the ecological impact of hurricane Katrina (2) review the literature relating to salt water intrusion, and (3) analyze the ecological impact on the agriculture industry attributed to salt water intrusion after hurricane Katrina.

Introduction

Hurricane Katrina caused widespread loss of life, with over 700 bodies recovered in New Orleans by October 23, 2005. Before Hurricane Katrina, the region supported approximately one million non-farm jobs, with 600,000 of them in New Orleans. As such, the total economic impact to Louisiana and Mississippi has exceeded $150 billion. The economic effects of Hurricane Katrina, which hit Louisiana and Mississippi in late August 2005, are far-reaching.

The ecological impacts of hurricane Katrina are insurmountable. These impacts include the following: (1) storm surge floods into coastal areas (2) destroyed havens for wildlife and migratory birds, (3) felled forests in south Louisiana and Mississippi, (4) saltwater intrusion and (5) polluted floodwaters from New Orleans which surged into Lake Pontchartrain. Now, almost three years after Hurricane Katrina hammered the U.S. Gulf Coast and spurred massive flooding in New Orleans, the ecological impacts are still being felt throughout the region. In particular, human-driven coastal erosion and saltwater intrusion which are both issues that have long

V.J. Williams (✉)
Economics, Transportation and Logistics Department, North Carolina A&T State University, Greensboro, NC 27411, USA
e-mail: kingvj@ncat.edu

G.A. Uzochukwu et al. (eds), *Proceedings of the 2007 National Conference on Environmental Science and Technology*, DOI 10.1007/978-0-387-88483-7_14,

been damaging the region's natural storm buffers. These issues were made worse by hurricane Katrina.

A Review of Salt Water Intrusion

When fresh water is withdrawn at a faster rate than it can be replenished, a draw down of the water table occurs with a resulting decrease in the overall hydrostatic pressure. When this happens near an ocean coastal area, salt water from the ocean intrudes into the fresh water aquifer. The result is that fresh water supplies become contaminated with salt water.

More specifically, saltwater intrusion is the movement of salt water into a non-salt water environment, such as a freshwater marsh. This intrusion may occur as the result of a natural process like a storm surge from a hurricane or from human activities such as construction of navigation channels or oil field canals. These channels and canals provide conduits for salt water from the Gulf of Mexico to reach deep into interior marshes. Saltwater intrusion is a natural process that occurs in virtually all coastal aquifers. This behavior is caused by the fact that sea water has a higher density (which is because it carries more solutes) than freshwater.

Saltwater intrusion can be detrimental because water with high salt concentrations can adversely affect vegetation. For instance, when highly saline water enters a low-saline or non-saline area, most or all of the native plant life will be destroyed. Because plant root systems are essential in holding the marsh soil together, loss of plant life eventually leads to rapid erosion.

The flow of saltwater inland is limited to coastal areas. Although saltwater intrusion is a natural process, it becomes an environmental problem when excessive pumping of fresh water from an aquifer reduces the water pressure and intensifies the effect, drawing salt water into new areas.

The first physical formulations of saltwater intrusion were made by W. Baydon-Ghyben (1888, 1889) and A. Herzberg (1901), thus called the Ghyben-Herzberg formulation. They derived analytical solutions to approximate the intrusion behavior, which are based on a number of assumptions that do not hold in all field cases. The Ghyben-Herzberg ratio states, for every foot of fresh water in an unconfined aquifer above sea level, there will be forty feet of fresh water in the aquifer below sea level (Daniel et al. 2004).

Over the years, many laboratory and field experiments have been conducted to estimate decreases in crop yield as a function of increasing levels of salinity in the soil root zone. Knowledge of the ability of crops to tolerate salinity and still produce a measurable agricultural product is critical to optimal management of salt-affected lands. In 2004, Harold Steppuhn, Van Genuchten and Grieve made a detailed assessment of available literature and devised a general relationship that accurately estimates crop yield in response to soil root- zone salinity. Their data indicated that product yields of agricultural crops grown subject to increasing levels of root-zone salinity decrease according to a sigmoidal-shaped response curve (Steppuhn et al. 2005).

Hurricane Katrina's Impact on the Ecology

A major economic problem created by disasters such as hurricane Katrina is the calculation of ecological damages. The ecological consequences from Katrina were considerable: storm surges flooded coastal areas, powerful winds felled forests in south Louisiana and Mississippi which were havens for wildlife and migratory birds, saltwater and polluted floodwaters from New Orleans surged into Lake Pontchartrain, and wetland restoration closed or restricted the Mississippi River Gulf Outlet.

The continuing effects of saltwater intrusion driven by Katrina can be still seen in the wilting trees and plants far from the coast. The inland saltwater intrusion caused the state's rice crop to decrease by 20 percent last year. The canals that connect the city to the coast allow storm surges to travel inland, bringing salt water that damages the land. An example of such a canal is the Mississippi River Gulf Outlet, which was built in the mid-1960s to be a 76-mile (122-kilometer) shortcut between the Gulf of Mexico and New Orleans. Before the record hurricane season of 2005, salt water brought inland by the canal was fingered as the culprit in the death of thousands of acres of cypress swamp, a natural buffer against storms. When Katrina hit, levee failures on the canal allowed water to pour into St. Bernard Parish and New Orleans East.

Some fields could take up to two years to recover. The rice acreage was down in southwest Louisiana by about 80,000–90,000 acres [32,400–36,400 hectares] because of the storm surge. Many plants were not planted because of soil contamination or contaminated waterways that would be used to flood the rice fields. Lake Pontchartrain is one of the few examples of quick ecological recovery following the record storm season.

In the weeks after Katrina, polluted floodwaters called "the toxic stew" were pumped directly from the streets of New Orleans into the lake. Approximately 10 billion gallons (38 billion liters) of contaminated water were dumped into Pontchartrain. One saving grace was that it was less than 10 percent of the lake's volume, therefore the bacteria counts dropped almost immediately after the pumping stopped in October.

The barrier islands that extend from Texas to Florida have historically served as the Gulf Coast's major natural protection from hurricanes. These long thin islands just off shore buffer the mainland from strong winds and ocean waves. The constant battering from intense tempests over the years has taken its toll on these islands. Hurricane winds cut Petit Bois Island from Dauphine Island 150 years ago, and in 1969 Hurricane Camille sliced Ship Island in half. Additionally, canals and levees along the Mississippi River prevent fresh sediment deposits from building the islands back up.

Louisiana's Chandeleur Islands—which have shrunk considerably in the past decade—were almost wiped off the map because of Katrina. Katrina's towering storm surge completely washed over most of the barrier islands along the Mississippi coast, scattering hefty amounts of debris. Thick layers of sediments deposited over the coast can help create wetland sustainability, but it will become more difficult to maintain in coming years because of more frequent and more intense hurricanes. Raising public and political awareness of the threats to Louisiana's ecology is the most critical path of action.

Katrina's Ecological Impact on the Agriculture Industry

Economic impact studies measure direct and indirect effects on an area economy. Direct impact refers to the area's spending on goods and services, its various sources of income, and employment levels. Indirect impact is the effect of the area's spending and employment on other sectors of the economy through the multiplied impact of spending and job creation. The indirect impact is an attempt to measure the impact on various businesses and industries that depend on the re-spending of income and other expenditures generated by an area. The analysis in this research begins the creation for estimates to be used in explaining the direct and indirect effects of salt water intrusion as a result of hurricane Katrina.

Fresh water is a scarce resource. Only 2.5 percent of the total water volume on the Earth is fresh water and the remainder is saline. The largest available source of fresh water still lies underground and the availability of surface water has become sparse in many areas in the world. Groundwater is used extensively to supplement available surface water to meet ever increasing water demand in the world. Hurricanes Katrina caused tidal surges that left concerns about salt deposition in soil. The unprecedented flooding inundated almost 40,000 acres of sugarcane in the coastal parishes of Louisiana. While a considerable acreage of sugarcane was damaged or killed by the direct effects of the flood waters, the extent of the effects on sugarcane from the salt content of sea water is not completely known. The level of salinity across tested sites was not predictable and did not appear to be associated with texture or any other variable.

Published reports from the U.S. Department of Agriculture suggest that sugarcane is moderately sensitive to salt, with a saturated-extract electrical conductivity (EC) threshold for yield reduction at 1.7 dS m-1 (U.S. Department of Agriculture, Cooperative Extension Service).

Flooding of agricultural land by hurricane storm surges can have both short-term and long-term effects on both crops and soil structure. While most of the "salt" in seawater is sodium chloride (table salt), it also contains appreciable amounts of magnesium sulfate (Epson's salts) and other elements. After heavy rains, sodium and chloride will be preferentially lost in runoff and leachate. Therefore, within the next few years, much of the agricultural land flooded by previous season's storm surges should naturally recover and return to previous levels of productivity.

Monitoring of soil salt levels will continue until sugarcane harvest, at which time a yield impact assessment will be made in an attempt to confirm the applicability of the salinity damage threshold for sugarcane in Louisiana.

Table 1 contains preliminary estimates of the economic impact from hurricane Katrina to Louisiana agriculture were calculated as well over $1 billion dollars (Burton et al. 2005). The major contributing factors which led to these impacts

Table 1 Louisiana Agriculture Preliminary Economic Damages: the direct effects of Salt Water Intrusion ($Million)

Commodity	Est. economic damages
Timber	$611
Christmas Trees	2
Sugar Cane	145
Cotton	10.4
Rice	0.483
Vegetables	2.57
Usuries	19
Citrus	11
Alligators	3.8
Turtles	5.4
Fisheries	142.2
Total	$953

Source: Hurricane Katrina: Preliminary Estimates of Commercial and Public Sector Damages. Marshall University: Center for Business and Economic Research. September, 2005.

were reduced revenue, lower market prices and increased production cost. The three largest commodities which were impacted were forestry ($612.9 M), agronomic crops (sugarcane, cotton and rice) ($156.1 M) and fisheries ($142.2 M). Damage estimates for forestry were developed based on forestry industry personnel's assessment of the percentage of timber affected in impacted regions and the expected Christmas tree loss. Damage estimates for sugarcane were based on personnel's assessment of production losses, increased planting and harvesting costs and revenue loss and increased costs associated with flooded cane acres. Experience with past storms has indicated that sugarcane production could be affected by 10–15 percent in subsequent future years because of carry-over effects including saltwater intrusion. These estimates include an estimate for the direct economic effects of salt water intrusion.

Summary

Groundwater systems in coastal areas will continue to come into contact with saline water. Saltwater intrusion will continue to replace the fresh water in coastal aquifers by saltwater due to the motion of a saltwater body into the freshwater aquifer. Saltwater intrusion will reduce the available fresh groundwater resources in coastal aquifers. At present, many coastal aquifers in the world, especially shallow ones, experience an intensive saltwater intrusion caused by both natural and human-induced processes.

Changes in climatic variables, such as precipitation and temperature, and the land-use change have significantly altered groundwater recharge rates for major aquifer systems. Under such circumstances it is important to study the problem of saltwater intrusion and to accurately estimate the saltwater intrusion due to changes in recharge or discharge of groundwater and related activities. Therefore, information on the local or regional impacts of climate change on hydrological processes and coastal water resources is becoming more important. Inclusion of these activities as a part of the management planning stage is paramount for coastal areas. A large number of land-use impact studies on water resources have been carried out for watersheds with a focus on water scarcity, flood, erosion, and water management.

The accumulation of economic and ecological damages from hurricane Katrina is still increasing. Many effects still need to be evaluated, and only time will determine the devastation, including salt water intrusion. Future studies, which will account for the immediate and the long- term effects and as well the indirect economic effects of salt water intrusion, will become increasing more important as the dollar value for Katrina continues to mount. Modeling these economic impacts will increase our knowledge and assist planners in preparing for future variable weather patterns.

References

Daniel W. Urish, Thomas E. McKenna. 2004. Tidal effects on ground water discharge through a sandy marine beach. Ground Water 42(7): 971–982, doi:10.1111/j.1745-6584.2004.tb02636.x

Steppuhn, H., Van Genuchten, M.T., Grieve, C.M. 2005. Root-zone salinity: I. Selecting a product-yield index and response function for crop tolerance. Crop Science 45: 209–220.

U.S. Department of Agriculture, Cooperative Extension Service, University of Florida, IFAS, Florida A. & M. University Cooperative Extension Program, and Boards of County Commissioners Cooperating. Larry Arrington, Dean. Economics and finance department, North Carolina A and T State University, Greensboro, NC.

Burton, Mark L., Hicks, Michael J. September, 2005. Hurricane Katrina: Preliminary Estimates of Commercial and Public Sector Damages. Marshall University: Center for Business and Economic Research.

Municipal Wastewater Concentrations of Pharmaceutical and Xeno-Estrogens: Wildlife and Human Health Implications

Maxine Wright-Walters and Conrad Volz

Abstract Most pharmaceutical estrogens and xenoestrogens are introduced into the environment through municipal waste water treatment plant (WWTP) effluent sources. These effluents contain synthetic compounds; surfactants, flame retardants and halogenated hydrocarbons that can mimic estrogens; and are discharged directly into rivers and lakes. As rivers and lakes are used for water and food supply, and recreation, and wastewater effluent usage increases, the presence and concentration of xenoestrogens in surface water becomes a valid public health concern. Additionally, many USA cities have significant combined sewer overflows releasing untreated sewage directly into surface waters, thus increasing the amounts of xenoestrogens finding their way into drinking water supplies and commercial and subsistence fishing habitat.

In the United States, humans are exposed daily to both pharmaceutical and xenoestrogens which have been implicated in various human health outcomes, such as testicular dysgenesis syndrome including testicular cancer and breast cancer in women. Also, they can have adverse reproductive effects in aquatic wildlife through sex reversals, production of intersex individuals, alterations in mating, and prevention of gonadal maturation. Combinations of estrogenic compounds are present in municipal WWTP effluents but, the natural estrogens, 17β-estradiol (E2) and estrone (E1), and the synthetic E2 derivate 17α-ethinylestradiol (EE2) are most responsible for in vitro estrogenic activity. Each xenoestrogen exhibits its own wildlife or human health risk, but synergistic effects could occur with xenoestrogen mixtures. Less than 1 ng/L EE2 can cause feminization of male fishes, 4 ng/L caused abnormal reproductive development (male fathead minnows). E2 has been detected at concentrations from 1 ng/L to 80 ng/L. Total estrogenicity (E2 equivalents) of 147 ng/L has been measured in WWTP effluent. Nonylphenol, a surfactant and brominated biphenyls, a flame retardant have been detected between 0.1–3.7 μg/L and 0.3–4.6 mg/kg (on suspended particles) respectively.

M. Wright-Walters (✉)
Department of Environmental and Occupational Health, Graduate School of Public Health, University of Pittsburgh, Pittsburgh, PA, USA

G.A. Uzochukwu et al. (eds), *Proceedings of the 2007 National Conference on Environmental Science and Technology*, DOI 10.1007/978-0-387-88483-7_15,

Understanding the species and xenoestrogen concentrations in surface water is imperative for environmental public health tracking of associated disease states. Such research will determine the necessity for utilizing limited and competing public financial resources to invest in technology to remove xenoestrogens from surface waters and, in regulation of fish or wildlife consumption from our rivers and lakes.

Introduction

Most pharmaceutical estrogens and xenoestrogens are introduced directly into surface waters through municipal wastewater treatment plant (WWTP) effluent sources, also called sewage treatment works (STW) (Daughton 1999; Norris 2007). The low concentrations of individual pharmaceutical estrogens (possibly exceeding the catabolic enzyme affinities of sewage microbiota), coupled with their metabolic "novelty," leads to incomplete removal from STWs (Daughton 1999). Municipal wastewaters are a complex mixture containing estrogens and estrogen mimics called xenoestrogens, (Kidd et al. 2007) natural and synthetic xenobiotics, household and agricultural chemicals, pharmaceuticals, hormones, and other compounds, many of which remain unidentified (Stevens et al. 2003). The majority of natural and pharmaceutical estrogens excreted by humans as well as xenoestrogens from numerous domestic and municipal sources (e.g., detergents, plastics, cosmetics,) enter WWTPs (Norris 2007). STWs receiving domestic and pharmaceutical waste release a complex (and ill-defined) mixture of natural and synthetic chemicals into the aquatic environment, due to their partial or complete resistance to biodegradation during the treatment process (Desbrow et al. 1998). Most of these compounds are retained in biosolids and a smaller portion typically appears in the wastewater effluent depending on the chemical and the type of treatment and retention times. Currently, more than half of the biosolids produced by municipal wastewater treatment systems is applied to land as a soil conditioner or fertilizer and the remaining solids are incinerated or landfilled (USEPA 2004; King et al. 2006). These disposal practices provide numerous routes for xenoestrogen reentry into environmental media and ultimately surface water. The use of biosolids as a soil conditioner and fertilizer allows for pharmaceutical estrogens and xenoestrogens exposure through the food supply chain and also reentry into surface water systems through run off, and contaminated groundwater outflow. Through incineration, compounds such as dioxins and furans are released into the air and may be deposited in watersheds through wet and dry deposition. Thus, there are many routes of reentry of xenoestrogens in and attached to the surface of biosolids from WWTPs, thus possibly, increasing their environmental concentration and exposure routes for both humans and animals. There are over 16,000 municipal WWTPs nationwide and over 75% of the nation's population is being served by centralized wastewater collection and treatment systems. The remaining population uses septic or other onsite systems.

(USEPA 2004), which have not been adequately studied for xenoestrogens release but, due to their high failure rate and lack of maintenance, could be considered potential non-point releasers of estrogenic compounds. Therefore, there can be an extremely varied mixture of pharmaceutical estrogens and xenoestrogens reentering surface waters possibly contaminating municipal drinking water supplies. But, what are the environmental concentrations of these compounds and are these concentrations significant enough to cause harm to human and wildlife health. Is human pharmaceutical estrogen and xenoestrogen exposure a valid public health concern?

The Problem

In the United States humans are exposed daily to xenoestrogens in food (e.g., phytoestrogens, various pesticides) and from contact with detergents [e.g., nonylphenols (NP)] and ingestion of plastic additives from plastic bottles, metal beverage can linings and food packaging (e.g., phthalates, bisphenol A (BPA), NP). Health care contributes estrogenic chemicals through pharmaceuticals and plastics used for surgical tubing and gloves. In addition, many personal care products (e.g., shampoos, cosmetics, aftershave lotions) contain xenoestrogens such as phthalates, NP and BPA. Most of these pharmaceutical and xenoestrogens are introduced into the environment via municipal WWTPs. Treated WWTP effluents are directly discharged in rivers and lakes. A recent publication by the U.S. Geological Survey reported that reproductive hormones and estrogenic alkylphenols were present in 40% and 70%, respectively, of the surveyed U.S. surface waters (USEPA 2001). Thus, as rivers and lakes are used for municipal water sources, to help produce our food supply and for recreation, and as wastewater effluent water reuse increases, the presence and concentration of xenoestrogens in surface water becomes a valid public health concern. Advances in civilization coupled with rising population levels have resulted in an increasing need to treat and recycle available water resources. In the United States surface water provides for 62% of the public water drinking supply (University of Michigan 2005). Irrigation remains the largest use of freshwater in the United States and totaled 137 Bgal/d for 2000. Since 1950, irrigation has accounted for about 65% of total water withdrawals, excluding those for thermoelectric power. Historically, more surface water than ground water has been used for irrigation (Hutson et al. 2004). Following use, water is returned to the aquatic environment, usually via STWs of varying processes and performance, which improves its quality, but it has a high probability of being withdrawn downstream for municipal or industrial reuse. In US cities with a high population density, the volume of effluent discharged from STWs can be considerable, sometimes contributing up to 50% of the flow of a river, a figure that can rise as high as 90% in periods of low rainfall (Routledge et al. 1998).

STWs permanently receive a complex mixture of industrial, domestic, and agricultural wastewater containing a load of synthetic and natural chemical compounds.

It has been demonstrated that, because of incomplete removal or conversion to an active form during the process of sewage treatment, pharmaceutical estrogens and xenoestrogens are released into surface water like rivers, lakes, and seas or adsorbed to sewage sludge or sediment (Liney et al. 2006). These chemicals are found in low parts per trillion in the aquatic environment (Ternes et al. 1999).

Impact on the Population

Both pharmaceutical estrogens and xenoestrogens have been implicated in various human health outcomes, such as testicular dysgenesis syndrome including testicular cancer and breast cancer in women (Giwercman et al. 1993; Carlsen et al. 1995; Toppari et al. 1996). Also, they can have adverse reproductive effects in aquatic wildlife through sex reversals, production of intersex individuals, alterations in mating, and prevention of gonadal maturation.

Human Effects. Xenoestrogens have been implicated in a variety of medical problems. Foremost is the concern that xenoestrogens as false messengers disrupt the process of reproduction. Reproductive issues, which are of concerns in humans, are fetal exposure (perhaps leading to hypospadias) and decreased reproductive ability in men (i.e. decrease in sperm numbers and abnormal sperm shapes) and testicular carcinoma in situ. Another issue is the potential effect of xenoestrogens on oncogenes, specifically, it is implicated in breast cancer in women (Giwercman et al. 1993; Carlsen et al. 1995; Toppari et al. 1996; Körner et al. 2001), endometriosis (Adlercreutz 1995), heart disease (Meyer 2001), osteoporosis (Meyer 2001) and Alzheimer's disease (Meyer 2001). It is important to note that a recent comprehensive literature survey of 48 endocrine disrupting chemicals (EDCs) revealed that 79% of these EDCs were also carcinogenic or mutagenic, 52% were also immunotoxic, and 50% were also neurotoxic (Choi et al. 2004). Both 4-*tert*-nonylphenol (4-NP) and BPA, for example, are contaminants found at appreciable concentrations in the aquatic environment that can cause endocrine disruption by interacting with both the estrogen receptor as agonists (Gaido et al. 1997) and the androgen receptor (Sohoni and Sumpter 1998). In addition, 4-NP can disrupt steroidogenesis in the liver and can interfere with the dynamic control of follicle-stimulating hormone release from the pituitary (Harris et al. 2001).

Wildlife Effects. Assessing whether any pharmaceutical estrogens or xenoestrogens pose a threat to the natural environment requires balancing information on its potency against observed environmental concentrations (Colborn et al. 1993). In many instances it is difficult to assign causality because of the complexity of environmental contaminants and the lack of analytical data that document contaminant levels during critical windows of exposure (Safe 2000). Nevertheless, there have been several incidents in wildlife populations that strongly correlate with exposure to specific industrial chemicals; this includes altered sex determination in alligators in Lake Apopka, Florida, exposed to a spill of organochlorine pesticides from a chemical waste site (Guillette et al. 1994, 1995, 1996). Several studies on wildlife populations have documented adverse effects that correlate with exposure to one or more putative endocrine-disrupting chemicals (Giesy et al. 1994; Sumpter and Jobling 1995; Jobling and Tyler 2003). Reproductive abnormalities have been observed in several wildlife populations living in polluted areas. (Guillette et al. 1994, 1995, 1996; Jobling et al. 1998).

There is considerable evidence that fishes inhabiting waters that receive untreated municipal wastewaters or effluents from municipal WWTPs are exposed to chemicals that affect reproductive endocrine function (Kidd et al. 2007). Estrogenic effects of treated wastewater, released into the aquatic environment, were first verified by Purdom et al. (1994). Purdom and colleagues reported that STW effluent was estrogenic to fish, causing feminization. The STW effluents tested were mainly domestic (rather than industrial) in source, indicating that the estrogenic component(s) were likely to be domestic in origin and were probably common to most STWs (Desbrow et al. 1998). Male fishes downstream of some wastewater outfalls produce vitellogenin (VTG) (a protein normally synthesized by females during oocyte maturation) and early-stage eggs in their testes, and this feminization has been attributed to the presence of estrogenic substances such as natural estrogens [17β-estradiol (E2), the synthetic estrogen used in birth-control pills 17α-ethinylestradiol (EE2), or weaker estrogen mimics such as NP in the water (Kidd et al. 2007). Recent studies have also shown that concentrations of 4-NP and BPA that inhibit gonadal development and reproductive function in fish can also cause damage to the kidneys (as a consequence of VTG induction), and decreased body weight and induce stressed behavior (Magliulo et al. 2002). DNA damage in barnacles has also been reported (Atienzar et al. 2002). Similarly, steroid estrogens that are known to be present in WWTP effluents (Desbrow et al. 1998) and to cause feminizing effects in fish have been reported to be genotoxic in mammals both in cell lines and *in vivo* (Nutter et al. 1991, 1994; Banerjee et al. 1994; Han and Liehr 1994). In laboratory studies, it has been confirmed that environmental contaminants with endocrine disrupting properties (EDCs) can disturb the development and expression of sexual characteristics in fish (Gimeno et al. 1996; Gray and Metcalfe 1997), amphibians (Hayes et al. 2002), reptiles (Crain et al. 1999; Willingham and Crews 1999), birds (Feyk and Giesy 1998), and mammals (Gray et al. 1994; Sharpe et al. 1995). However, the extent to which the sexual characteristics and reproductive capabilities of natural populations are impacted by these EDCs is still not well understood.

Source of the Problem

Municipal Wastewater Treatment Plants and the Clean Water Act. Potable water utilities select a treatment train that is most appropriate for the contaminants found in their source water. The most commonly used processes include flocculation, sedimentation, filtration, and disinfection for surface water. Some treatment trains also include ion exchange and adsorption techniques (USEPA 2001). These conventional processes, according to the EPA are inefficient for substantially reducing certain pesticide concentrations and other EDCs including pharmaceutical and xenoestrogens from source water (USEPA 2001). Additionally, the Clean Water Act, amended in 1972 which addresses WWTP releases and the Safe Drinking Water Act (SDWA), first enacted in 1974 and amended in 1986 and 1996, which regulates contaminants in public water supplies does not have any provisions for removal or testing of xenoestrogens or pharmaceutical estrogens. Furthermore, many USA cities such as Pittsburgh and Los Angeles have significant combined sewer overflows releasing untreated sewage directly into surface waters, thus increasing the amounts and concentrations of xenoestrogens finding their way into drinking water supplies and commercial and subsistence fishing habitat. Many of the earliest sewer systems were combined sewers, designed to collect both sanitary wastewater and storm water runoff in a single system. These combined sewer systems were designed to provide storm drainage from streets and roofs to prevent flooding in cities. Later, lines were added to carry domestic wastewater away from homes and businesses. Early sanitarians thought that these combined systems provided adequate health protection. We now know that the overflows designed to release excess flow during rains also release pathogens and other pollutants (USEPA 2004).

Concentrations in the Environment

Steroid estrogens have the potential to exert estrogenic effects in the low ng/L level, whereas alkylphenolic compounds are estrogenic at μg/L concentrations (Routledge et al. 1998). Natural and synthetic hormones are frequently detected in STW effluents and receiving surface waters with concentrations ranging from pg/L to ng/L (Belfroid et al. 1999; Baronti et al. 2000; Kuch and Ballschmiter 2001), whereas alkylphenolic compounds are found in concentrations up to μg/L (Bolz et al. 2001; Stachel et al. 2003; Jin et al. 2004).

Ternes et al. (2007) were able to show in aerobic batch experiments that steroid conjugates such as glucuronides of E2 are rapidly cleaved in contact with activated sludge, and thus the active form of the estrogen is released. To date, estrogenic effects on aquatic wildlife have not been conclusively linked to only one particular compound, but some chemicals are mainly responsible for higher estrogenicity indexes. Among them, the natural estrogens estrone (E1) and E2, and the exogenous, EE2, the active ingredient in oral contraceptive pills, possess the highest estrogenicity indexes. Apart from these steroids, alkylphenols such as 4-tert-octylphenol and the technical isomer mixture of 4-NP, both breakdown products of nonionic surfactants (Desbrow et al. 1998), and BPA, a widely used monomer for epoxy resins, show estrogenic potentials of approximately 4 orders of magnitude lower than E2 (Ternes et al. 1999). Also representatives of the groups of PCBs, dioxins, phytoestrogens, pesticides, preservatives, antioxidants, or phthalic esters contribute to the daily exogenous burden of humans and wildlife with hormonally active agents (Belfroid et al. 1999; Larsson et al. 1999; Thorpe et al. 2003; Zhou et al. 2007).

Due to their incomplete breakdown in current municipal WWTP processes (Ternes et al. 1999), natural and synthetic estrogens can be found in the aquatic environment at low parts per trillion concentrations, typically at less than 5 ng/L (Belfroid et al. 1999; Larsson et al. 1999; Zhou et al. 2007). WWTP effluents contain mixtures of individual estrogens and their mimics that differ in their ability to elicit estrogenic responses (Thorpe et al. 2003). Combinations of estrogenic compounds are present in municipal WWTP effluents but the natural estrogens, E1 and E2, and the synthetic E2 derivate EE2 are most responsible for in vitro estrogenic activity. Each xenoestrogen exhibits its own wildlife or human health risk, but synergistic effects could occur with xenoestrogen mixtures. Wildlife species are rarely exposed to single chemicals but instead are exposed to complex, fluctuating mixtures of contaminants that may act in various ways (Thorpe et al. 2001; Silva et al. 2002; Sumpter 2003; Thorpe et al. 2003) and that may induce combination effects (Rajapakse et al. 2002) via the same or different mechanisms. Less than 1 ng/L EE2 can cause feminization of male fishes, 4 ng/L caused abnormal reproductive development in male fathead minnows. E2 has been detected at concentrations from 1 ng/L to 80 ng/L in surface water. Total estrogenicity (E2 equivalents) of 147 ng/L and 17 ng/L has been measured in final WWTP effluent and surface water respectively. NP and brominated biphenyls, a flame retardant have been detected between 0.1–3.7 μg/L and 0.3–4.6 mg/kg (on suspended particles) respectively and at 6–135 ng/L in river water and 2–15 ng/L in drinking water (Kuch and Ballschmiter 2001). Diethyl hexyl phthalate has been detected in sewage sludge between 15–50 ng/g (Petrovic and Barcelo 2000) and at 5–10 ng/L in surface and drinking water (Hyötyläinen et al. 1997). Di butyl phthalate has been reported at a concentration of 5–10 ng/L in surface water and drinking water (Hyötyläinen et al. 1997). Bisphenol A concentrations have been reported in drinking water from 300 pg/L to 2 ng/L and in river water from 500 pg/L up to 16 ng/L (Kuch and Ballschmiter 2001).

Conclusion

Understanding the species and concentrations of pharmaceutical and xenoestrogen in WWTP effluent, septic leachate, groundwater and surface water is imperative for environmental public health tracking of associated disease states. The current environmental concentrations of both pharmaceutical and xenoestrogens seem to be adequate to cause harmful health effects for human and wildlife and thus, human exposure to these compounds is a very valid public health concern.

Further research is needed to determine specific associations between disease states and pharmaceutical estrogen and or xenoestrogen exposure. Through the Toxic Substance Control Act (TSCA) the USEPA provides a very limited and inadequate framework to evaluate the estrogenic potential of new and existing chemical substances, intermediates and products. The TSCA program does this, as do all modern chemical evaluation programs, by evaluating each chemical singly. It does not allow for testing the natural combinations of environmentally relevant concentrations of pharmaceutical and xenoestrogens and resulting synergies that may take place. Through modification of the CWA discharges such as WWTP effluents must be regulated due to their estrogenic potential. Additionally, through the SDWA the concentrations and number of potential estrogenic compounds present in our drinking water can be regulated. Most studies that have quantitatively measured pharmaceutical estrogens or xenoestrogens in WWTP effluents or surface water use a detection limit (DL) of 0.02 ng/L thus concentrations below this level are not measured. Yet, it is evident from the physical and reproductive malformations of fishes in these waters that there are possible health effects from concentrations below these DLs. Thus, quantitative methodologies must be established that can measure both xenoestrogen and pharmaceutical estrogenic concentration well below these levels. Estrogenic assays such as proliferation of MCF-7 human breast cancer cells (an estrogen receptor positive cell line), can be used to evaluate combined chemical estrogenicity. Naturalistic wildlife and holistic ecosystem effects studies also provide valuable risk information concerning cumulative estrogenicity at the level of a system or population, but these studies are few. Additionally, an important aspect of the total risk to humans and the environment from ingestion of xenoestrogens is the profusion of substances with estrogenic activity that are being introduced into water through municipal (household) sources or industrial processes. The EPA must incorporate appropriate methodologies to evaluate total estrogenic and other risk from environmental mixtures of the many commonly found pharmaceutical estrogens and xenoestrogens. We propose that categories of estrogens and their mimics be created for testing purposes that take into account their bioavailability, their probable uses and likely disposal paths. Ecological and other methods must be developed to assess their likely combined impacts on living receptors in the aquatic environment through the use of sentinel species. Such research will determine the necessity for utilizing limited and competing public financial resources to invest in technology to remove pharmaceutical estrogens and xenoestrogens from surface waters and, in regulation of fish or wildlife consumption from our rivers and lakes.

This work is made possible through a grant from the Centers for Disease Control and Prevention's Environmental Public Health Tracking (EPHT) Network to the University of Pittsburgh, Academic Center for Excellence (UPACE) in EPHT. Support was also provided by the Heinz Endowments and the Highmark Foundation through the University of Pittsburgh Cancer Institute, Center for Environmental Oncology.

References

Adlercreutz H. (1995). "Phytoestrogens: epidemiology and a possible role in cancer protection." *Environ Health Perspect* **7** (103): 103–112.

Atienzar FA, Billinghurst Z, Depledge MH. (2002). "4-n-Nonylphenol and 17-beta estradiol may induce common DNA effects in developing barnacle larvae." *Environ Pollut* **3** (120): 735–738.

Banerjee SK, Banerjee S, Li SA, Li JJ. (1994). "Induction of chromosome-aberrations in Syrian-hamster renal cortical-cells by various estrogens." *Mutat Res Fundam Mol Mech Mutagen* **2** (311): 191–197.

Baronti C, Curini R, D'Ascenzo G, Di Corcia A, Gentili A, Saperi R. (2000). "Monitoring natural and synthetic estrogens at activated sludge sewage treatment plants and in a receiving river water." *Environ Sci Technol* **34**: 5059.

Belfroid A, Van der Horst A, Vethaak AD, Schäfer AJ, Rijs GBJ, Wegener J, Confino WP. (1999). *Sci Total Environ* **225**: 101–108.

Bolz U, Hagenmaier H, et al. (2001). "Phenolic xenoestrogens in surface water, sediments, and sewage sludge from Baden-Wurttemberg, south-west Germany." *Environ Pollut* **115** (2): 291–301.

Carlsen E, Giwercman A, Keiding N, Skakkebaek NK. (1995). "Declining semen quality and increasing incidence of testicular cancer: Is there a common cause?" *Environ Health Perspect* **103**: 137–139.

Choi SM, Yoo SD, Lee BM. (2004). "Toxicological characteristics of endocrine-disrupting chemicals: developmental toxicity, carcinogenicity, and mutagenicity." *J Toxicol Environ Health B Crit Rev* **1** (7): 1–32.

Colborn T, vom Saal SF, Soto AM. (1993). "Developmental effects of endocrine-disrupting chemicals in wildlife and humans." *Environ Health Perspect* **101**: 378–384.

Crain DA, Spiteri I, Guillette LJ Jr. (1999). "The functional and structural observations of the neonatal reproductive system of alligators exposed in ovo to atrazine, 2,4-D or estradiol." *Toxicol Ind Health* **15**: 180–185.

Daughton CG, Ternes TA. (1999). "Pharmaceuticals and Personal Care Products in the Environment: Agents of Subtle Change?" *Environ Health Perspect* **107** (Suppl 6): 907–938.

Desbrow C, Routledge EJ, Brighty GC, Sumpter JP, Waldock M. (1998). "Identification of estrogenic chemicals in STW effluent. 1. Chemical fractionation and in vitro biological screening." *Environ Sci Technol* **11** (32): 1549–1558.

Feyk LA, Giesy JP. (1998). "Xenobiotic modulation of endocrine function in birds. In: Principles and Processes for Evaluating Endocrine Disruption in Wildlife (Kendall R, Dickerson R, Giesy J, Suk W, eds)." Pensacola, FL, SETAC Press, pp. 121–140.

Gaido KW, Leonar L, Lovell S, Gould JC, Babai D, Portier CJ, et al. (1997). "Evaluation of chemicals with endocrine modulating activity in a yeast-based steroid hormone receptor gene transcription assay." *Toxicol Appl Pharmacol* **1** (143): 205–212.

Giesy JP, Ludwig JP, Tillitt DE. (1994). "Deformities of birds in the Great Lakes region: assigning causality." *Environ Sci Technol* **28**: 128A–135A.

Gimeno, S. G., A. Bowmer, T. Komen, H. (1996). "Feminization of male carp" *Nature* **384**(6606): 221–222

Giwercman A, Carlsen E, Keiding N, Skakkebaek NE. (1993). "Evidence for increasing incidence of abnormalities of the human testis: A review." *Environ Health Perspect* **101**: 65–72.

Gray LE, Ostby JS, Kelce WR. (1994). "Developmental effects of an environmental antiandrogen: the fungicide vinclozolin alters differentiation of the male rat." *Toxicol Appl Pharmacol* **129**: 46–52.

Gray, M.A., and Metcalfe, Chris D. (1997). "Induction of Testis–Ova in Japanese Medaka (Oryzias Latipes) Exposed to p-Nonylphenol." *Environmental Toxicology and Chemistry* 1082–1086.

Guillette LJ Jr, Crain DA, Rooney AA, Pickford DB. (1995). "Organization versus activation: the role of endocrine-disrupting contaminants (EDCs) during embryonic development in wildlife." *Environ Health Perspect* **103** (Suppl 7): 157–164.

Guillette LJ Jr, Gross TS, Masson GR, Matter JM, Percival HF, Woodward AR. (1994). "Developmental abnormalities of the gonad and abnormal sex hormone concentrations in juvenile alligators from contaminated and control lakes in Florida." *Environ Health Perspect* **102**: 680–688

Guillette LJ Jr, Pickford DB, Crain DA, Rooney AA, Percival HF. (1996). "Reduction in penis size and plasma testosterone concentrations in juvenile alligators living in a contaminated environment." *Gen Comp Endocrinol* **101**: 32–42.

Han XL, Liehr JG. (1994). "8-Hydroxylation of guanine bases in kidney and liver DNA of hamsters treated with estradiol-role of free radicals in estrogen-induced carcinogenesis." *Cancer Res* **21** (54): 5515–5517.

Harris CA, Santos EM, Janbakhsh A, Pottinger TG, Tyler CR, Sumpter JP. (2001). "Nonylphenol affects gonadotropin levels in the pituitary gland and plasma of female rainbow trout." *Environ Sci Technol* **14** (35): 2909–2916.

Hayes TB, Collins A, Lee M, Mendoza M, Noriega N, Stuart AA, et al. (2002). "Hermaphroditic, demasculinized frogs after exposure to the herbicide atrazine at low ecologically relevant doses." *Proc Natl Acad Sci USA* **99**: 5476–5480.

Hutson SS, Barber NL, Kenny JF, Linsey KS, Lumia DS, Maupin MA. (2004). Estimated Use of Water in the United States in 2000. USGS. U.S. Department of Interior, U.S. Geological Survey. U.S. Geological Survey Circular 1268.

Hyötyläinen T, Grob K, Biedermann M, Riekkola ML. (1997). "Reversed phase HPLC coupled on-line to GC by the vaporizer/precolumn solvent split/gas discharge interface; analysis of phthalates in water." *J High Resolut Chromatogr* **20**: 410.

Jin XL, Huang GL, et al. (2004). "Simultaneous determination of 4-tert-octylphenol, 4-nonylphenol and bisphenol a in Guanting Reservoir using gas chromatography-mass spectrometry with selected ion monitoring." *J Environ Sci (China)* **16** (5): 825–828.

Jobling S, Tyler CR. (2003). "Endocrine Disruption in Wild Freshwater Fish." *Pure Appl Chem* **75** (11–12): 2219–2234.

Jobling S, Nolan M, Tyler CR, Brighty G, Sumpter JP. (1998). "Widespread sexual disruption in wild fish." *Environ Sci Technol* **17** (32): 2498–2506.

Kidd KA, Blanchfield PJ, et al. (2007). "Collapse of a fish population after exposure to a synthetic estrogen." *Proc Natl Acad Sci USA* **104** (21): 8897–8901.

King TE, Ballereau SJ, et al. (2006). "Genetic signatures of coancestry within surnames." *Curr Biol* **16** (4): 384–388.

Körner W, Bolz U, Rita T, Schwaiger J, Rolf-Dieter N, Hagenmaier AMaH. (2001). "Steroid Analysis and Xenosteroid Potentials in the Small Streams in Southwest Germany." *J Aquat Ecosyst Stress and Recovery* **8**: 215–229.

Kuch HM, Ballschmiter K. (2001). "Determination of endocrine-disrupting phenolic compounds and estrogens in surface and drinking water by HRGC-(NCI)-MS in the picogram per liter range." *Environ Sci Technol* **35** (15): 3201–3206.

Larsson D, Adolfsson-Erici M, Parkkonen J, Pettersson M, Berg AH, Olsson P-E, Forlin L. (1999). *Aquat Toxicol* (45): 91–97.

Liney KE, Hagger JA, et al. (2006). "Health effects in fish of long-term exposure to effluents from wastewater treatment works." *Environ Health Perspect* **114** (Suppl 1): 81–89.

Magliulo L, Schreibman M, Cepriano J, Ling J. (2002). "Endocrine disruption caused by two common pollutants at "acceptable" concentrations." *Neurotoxicol Teratol* **1** (24): 71–79.

Meyer VF. (2001). "The Medicalization of Menopause: Critique and Consequences." *Int J Health Sci* **4** (31): 769–792.

Norris DO. (2007). *Xenoestrogen Actions on Reproduction: Implications for Health of Wildlife And Humans.* American Water Resources Association 2007 Summer Specialty Conference, Emerging Contaminants of Concern in the Environment: Issues, Investigations and Solutions.

Nutter LM, Ngo EO, Abulhajj YJ. (1991). "Characterization of DNA damage induced by 3,4-estrone-ortho-quinone in human-cells." *J Biol Chem* **25** (266): 16380–16386.

Nutter LM, Wu YY, Ngo EO, Sierra EE, Gutierrez PL, Abulhajj YJ. (1994). "An O-quinone form of estrogen produces free radicals in human breast-cancer cells-correlation with DNA-damage." *Chem Res Toxicol* **1** (7): 23–28.

Petrovic M, Barcelo D. (2000). "Determination of Anionic and Nonionic Surfactants, Their Degradation Products, and Endocrine-Disrupting Compounds in Sewage Sludge by Liquid Chromatography/Mass Spectrometry." *Anal Chem* **72** (19): 4560–4567.

Purdom CE, Hardiman, PA, Bye VJ, Eno NC, Tyler CR, Sumpter JP. (1994). "Estrogenic effects of effluents from sewage treatment works." *Chem Ecol* **8**: 275–285.

Rajapakse N, Silva E. et al. (2002). "Combining xenoestrogens at levels below individual no-observed-effect concentrations dramatically enhances steroid hormone action." *Environ Health Perspect* **110** (9): 917–921.

Routledge EJ, Sheahan D, et al. (1998). "Identification of Estrogenic Chemicals in STW Effluent. 2. In Vivo Responses in Trout and Roach." *Environ Sci Technol* **32** (11): 1559–1565.

Safe SH. (2000). "Endocrine Disruptors and Human Health-Is There a Problem? An Update." *Environ Health Perspect* **108** (6): 487–493.

Sharpe RM, Fisher JS, et al. (1995). "Gestational and lactational exposure of rats to xenoestrogens results in reduced testicular size and sperm production." *Environ Health Perspect* **103** (12): 1136–1143.

Silva E, Rajapakse N, Kortenkamp A. (2002). "Something from "nothing"-eight weak estrogenic chemicals combined at concentrations below NOECs produce significant mixture effects." *Environ Sci Technol* **8** (36): 1751–1756.

Sohoni P, Sumpter JP. (1998). "Several environmental oestrogens are also anti-androgens." *J Endocrinol* **158** (3): 327–339.

Stachel B, Ehrhorn U. et al. (2003). "Xenoestrogens in the River Elbe and its tributaries." *Environ Pollut* **124** (3): 497–507.

Stevens JL, Northcott GL, Stern GA, Tomy GT, Jones KC. (2003). "PAHs, PCBs, PCNs, organochlorine pesticides, synthetic musks, and polychlorinated n-alkanes in UK sewage sludge: survey results and implications." *Environ Sci Technol* **3** (37): 462–467.

Sumpter J. (2003). "Endocrine disruption in wildlife: the future?" *Pure Appl Chem* **11–12** (75): 2355–2360.

Sumpter JP, Jobling S. (1995). "Vitellogenesis as a biomarker for estrogenic contamination of the aquatic environment." *Environ Health Perspect* **103** (Suppl 7): 173–178.

Ternes TA, Bonerz M, et al. (2007). "Irrigation of treated wastewater in Braunschweig, Germany: an option to remove pharmaceuticals and musk fragrances." *Chemosphere* **66** (5): 894–904.

Ternes TA, Kreckel P, et al. (1999). "Behaviour and occurrence of estrogens in municipal sewage treatment plants – II. Aerobic batch experiments with activated sludge." *Sci Total Environ* **225** (1–2): 91–99.

Thorpe KL, Cummings RI, Hutchinson TH, Scholze M, Brighty G, Sumpter JP, et al. (2003). "Relative potencies and combination effects of steroidal estrogens in fish." *Environ Sci Technol* **6** (37): 1142–1149.

Thorpe KL, Hutchinson TH, Hetheridge MJ, Scholze M, Sumpter JP, Tyler CR. (2001). "Assessing the biological potency of binary mixtures of environmental estrogens using vitellogenin induction in juvenile rainbow trout (Oncorhynchus mykiss)." *Environ Sci Technol* **12** (35): 2476–2481.

Toppari J, Larsen JC, Christiansen P, Giwercman A, Grandjean P, Guillette LJ, Jr, Jégou B, Jensen TK, Jounnet P, Keiding N, et al. (1996). "Male reproductive health and environmental xenoestrogens." *Environ Health Perspect Suppl* **104**: 741.

University of Michigan, C. f. S. S. (2005). U.S. Water Supply and Distribution.

USEPA. (2001). Removal of Endocrine Disrupting Chemicals in Drinking Water. D. Office of Research and Development Washington, Technology Transfer and Support Division, National Risk Management Research Laboratory, Office of Research and Development, U.S. Environmental Protection Agency, Cincinnati, Ohio 45268.

USEPA. (2004). Primer for Municipal Wastewater Treatment Systems. O. o. W. O. o. W. Management, Washington, DC.

Willingham E, Crews D. (1999). "Sex reversal effects of environmentally relevant xenobiotic concentrations on the redeared slider turtle, a species with temperature-dependent sex determination." *Gen Comp Endocrinol* 113: 429–435.

Zhou J, Liu R, Wilding A, Hibberd A. (2007). "Sorption of Selected Endocrine Disrupting Chemicals to Different Aquatic Colloids." *Environ Sci Technol* **41**: 206–213.

Challenges in the Devolution of Environmental Health Service Delivery in South Africa

Renay Van Wyk

Abstract Rapid urbanization is a critical challenge for those charged with service provision to urban areas in developing countries. Due to so many South Africans living under poor environmental conditions, our country is confronted by an enormous burden of disease. Providing adequate environmental health services continue to present many problems in South Africa. Despite the implicit knowledge expressed in policy documents, mission statements and broad aims of organisations striving to address environmental health service delivery, this knowledge is seldom found to be evident in current practice. There is an evident "separation of practice from theory". Investigating current practices and incorporate the development of the practitioners understanding of their current practice and transforming the practitioners' implicit knowledge into explicit knowledge.

Introduction

Internationally, there is growing recognition that Governments should devise a means of extending and improving access to services for the urban poor, regardless of whether the settlements in which they reside are unplanned and illegal (World Bank, 2004).

Providing adequate services to people living in urban poverty continues to present many problems in South Africa, where the speedy delivery of conventional formal housing with costly water-borne sanitation systems is constrained by a lack of resources (Lagardien and Cousins, 2003). In reality, such provision is a daunting development challenge, exacerbated by growing unemployment and the spread of unplanned informal settlements. Socio-economic realities facing both local authorities and communities in need make for difficult working conditions and poor cost recovery. This has inhibited effective action by either.

R. Van Wyk (✉)
Masters Environmental Health, University of Johannesburg, South Africa

G.A. Uzochukwu et al. (eds), *Proceedings of the 2007 National Conference on Environmental Science and Technology*, DOI 10.1007/978-0-387-88483-7_16,

Poverty in urban environments exists for that section of the population who are unable to pay for basic services or access development opportunities (Van Ryneveldt et al., 2003). The cost of providing services in such " abnormal circumstances must find alternative returns to the ability to pay for services. This is a much wider economic issue than public sector capacity building in "normal circumstances" (Rossouw et al., 2001).

Objective: The research process served to contribute to transforming the practitioners' implicit knowledge into explicit knowledge, aiming to make this available to their organisations through investigating current practices.

Methodology: The study adopted an action research approach to three case study sites of current sanitation provision practices, in order to respond constructively to the evident "separation of practice from theory" (Friedlander, 2001). Friedlander discusses the "false separation between action and research on the one hand and between practice and theory on the other", pointing out that the artificial separation between practice and theory is mirrored by separating research from action (Friedlander, 2001).

In accordance with this view, the action research approach that was applied engaged the practitioners involved in delivery (community beneficiaries, volunteers, contractors, and municipal officials) as research participants. Practitioners were actively engaged in the "quest for information and ideas to guide their future actions" (Whyte et al., 1991). The research process thus served to contribute to transforming the practitioners' implicit knowledge into explicit knowledge, aiming to make this available to their organisations.

Investigating current practices (action) in the case study sites thus incorporated the development of the practitioners understanding of their current practice (research). In the context of a theoretical framework for an approach that aims at achieving the ideal service delivery programme. In this context, principles and the processes for building their capacity to plan and act, based on their knowledge, were generated. Such processes were adjusted and refined by testing, adapting and validating with the practitioners themselves to cross-check relevance and usefulness in practice.

During the data gathering activities at case study sites, it became apparent that where information and knowledge management practices were inadequate, the sharing and application of knowledge and hence capacity to deliver was limited. The action research methodology consisted of field visits, questionnaires, interviews, focus group discussions and workshops in partnership with city officials and beneficiaries.

Case study data was collected, analysed and validated at appropriate stages of the study, with the assistance of research reference groups from the City of Cape Town, eThwekwini (Durban) and Joburg Water for feedback and validation. Internal project reports provided by participating environmental health practitioners were augmented by telephonic and email communication and correspondence between researchers and municipal practitioners.

Specific research activities included:

- Investigating capacity requirements for integrated service delivery
- Assessing capacity for integrated sanitation service delivery

- Identifying and planning strategic improvements to sanitation service delivery programme.

Investigating the principles of integrated service delivery involved an extensive literature review that included International service sector approaches and lessons; national social development and poverty alleviation imperatives for infrastructure delivery; case studies and reports of local public works programmes and directed procurement for poverty alleviation; documents and reports relating to policy and approaches to delivery of basic sanitation services including at national and local level.

Assessing the capacity for integrated service delivery was done through a comparative analysis of the case studies through reviews of audits on service delivery backlogs and related policies, delivery plans and procurement strategies; interviews with appointed consultants and contractors; field trips to case study sites; interviews with officials in relevant departments; workshop to validate the Review Table as a rapid capacity assessment tool with practitioners.

Identifying and planning improvements to service delivery programme involved (a) reviewing case study data to develop an assessment tool as a common structure for the collation and assessment of strategic information of a sanitation service delivery programme in any context; (b) application of the assessment tool by practitioners as a structure for integrated knowledge management at case study sites; (c) an assessment tool workshop with participating practitioners from the case studies to confirm the process requirements for identifying gaps and guiding improvements in service delivery capacity and (d) developing a guideline for planning improvements.

The workshops provided the opportunity to confirm the principles underlying integrated programme design and capacity development. In addition, the preparations for the workshops and subsequent facilitation of the workshops also tested the process by which strategic improvements to service delivery can be initiated, specified and planned in an integrated manner. This provided the basis for the guideline developed as part of the study.

Results and Discussion

In response, two tools were developed. A review table incorporating the integration and capacity principles was developed as a rapid capacity assessment tool. Subsequently a detailed programme assessment tool based on the review table was developed to specify the knowledge requirements, to record the organizational information especially the tacit knowledge and therefore provide structure for the planning of improvements. Based on the application of the review table and the assessment tool the following process is proposed for planning improvements.

If a strategy is a systematic way of tackling a problem or working towards an objective, a strategic approach in this context should respond to the following.

What are the principles that underlie an integrated program and capacity for successful implementation? It is necessary to establish the principles that are fundamental to improving sanitation services and if efforts are to achieve overall objectives and thus be truly strategic in nature.

What are the processes to be followed to plan improvements in service delivery? Strategic plans need to be flexible and adaptable, with interventions influenced by an understanding of the principles of good practice and knowledge of current practice. This suggests the need for a stepwise approach to setting, refining and working towards objectives.

Development Sector Lessons – A Learning Approach

> *Neither the researcher, administrator, nor beneficiary or any other stakeholder is likely to achieve his or her potential for contribution to development until they join as partners in a mutual learning process, committed not to the search for magical blueprints, but to the building of new capacities for action* (Korten 1980).

This statement was based on a 1980 Ford Foundation study of development assistance programs in India, Bangladesh, Thailand and the Phillipines; many of which had failed. Korten argued that success in development depended on achieving a fit between beneficiary needs, program outputs and the competence of the assisting organization. He contended that "fit" is achieved when the program model is responsive to the beneficiary needs at a particular time and place and a strong organization capable of making the program work is in place. A triangle (Fig. 1) illustrating the concept consists of beneficiary and supporting organization at the base and the program at the apex, providing a visualization of the relationships required for effective program design and how fit is achieved.

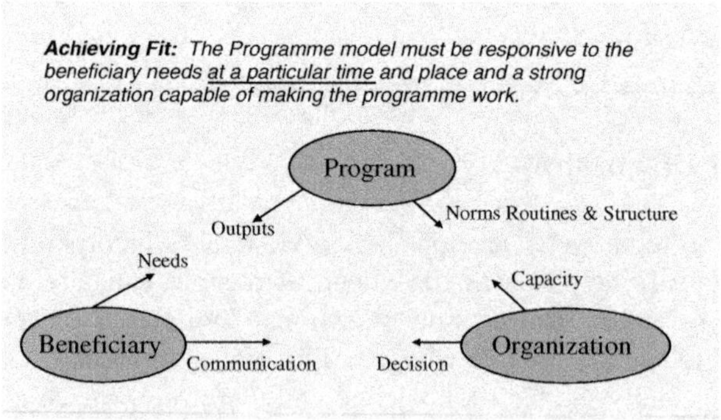

Fig. 1 Program Model (Korten, 1980)

Fit is achieved in three ways:

- Fit between beneficiary needs and particular resources made available by the program.
- Fit between the needs that the beneficiaries are able to define and communicate and the process by which the organization makes decisions.
- Fit between program task requirements and the distinctive competence of the organization.

The distinctive competence refers to the structures, routines and norms that govern functioning and the technical and social capabilities of the organization. He further noted that the fit is achieved in a learning approach that proceeds through three stages. Each stage emphasizes a different learning focus.

Stage 1: **Learning to be effective** focuses on the fit between beneficiaries and the working program model. In this stage the capabilities of the action research team, which includes all the stakeholders, in understanding the community dynamics, asking relevant questions and embedding the learning approach, is important.

Stage 2: **Learning to be efficient** is focused on reducing input requirements per unit output, eliminating extraneous non-essential activities and reutilizing important activities. Achieving a fit between program requirements and realistically attainable organizational capacities is a requirement.

Stage 3: **Learning to expand** stage focuses on organizational capacity for replication and logistics for larger scale operation while maintaining fit.

In this "Learning Approach", embracing error, learning with people and building new knowledge and institutional capacity through action is encouraged. The program and organization are developed concurrently.

Conclusion

With regard to the process for planning improvements in service delivery, the workshops conducted with environmental health practitioners highlighted the following as key steps in the development of practical guidelines.

- Developing consensus on the programme and capacity requirements.
- The identification and analysis of the organization's available and required knowledge assets and related processes.
- The planning of actions to fulfill and to improve service delivery.

The Assessment Tool facilitated the process by specifying the knowledge requirements through questions, by recording current practice (especially the tacit knowledge) and by providing a structure for the sharing of knowledge and planning the improvements.

References

Ashley, C., and Carney, D. 1999. *Sustainable Livelihoods: Lessons from Early Experience*. Department for International Development (DFID). United Kingdom, London.

City of Cape Town Health Working Group. 2003. *Draft City Health Directorate Strategic Plan for Servicing of Informal Areas*. Cape Town, South Africa.

Friedlander, F. 2001. *Participatory Action Research as a Means of Integrating Theory and Practice*. Fielding Graduate Institute, Action Research Symposium Proceedings, Alexandria, VA

Korten, D. September 1980. *Community Development and Rural Development: A Learning Process Approach*. Public Administration Review, 40(5): 480–511

Lagardien, A., and Cousins, D. 2003.*Sanitation Demand and Delivery in Informal Settlements-Planning and Implementation Support*. Water Research Commission, Pretoria, South Africa.

Rossouw, W, A.M.M., and Crous, P.C. 2001. *Rapid Capacity Building for Water and Waste Management at Local Authority and District Council Level*. Water Research Commission, Pretoria, South Africa.

Van Ryneveldt, P. Parnell, S. and Muller, D. 2003. *Indigent Policy. A Pro-Poor Income Strategy*. City of Cape Town, Cape Town, South Africa.

Whyte, W.F., Greenwood, D.J., and Lazes, P. 1991. *Participatory action research*. Newbury Park, CA: Sage.

Wright, A. M., 1997. *Towards a Strategic Sanitation Approach, Improving Sustainability of Urban Sanitation in Developing Countries*. World Bank, Washington, DC.

World Bank, 2004. *Making Services for the Poor People*. Development Report. http://econ.worldbank.org/wdr/wdr2004/

A Review of Health and Hygiene Promotion as Part of Sanitation Delivery Programmes to Informal Settlements in the City of Cape Town (South Africa)

Renay Van Wyk

Abstract Good sanitation includes appropriate health and hygiene promotion. This implies that proper health and hygiene promotion would have the desired effect as part of sanitation service delivery. However, lessons learnt worldwide show that in the promotion of health and hygiene, it is not enough simply to provide facilities, because if people do not use the available facilities properly, conditions do not improve or the system breaks down. The Ottawa charter (WHO, 1986) suggests that effective health and hygiene promotion require the empowerment of local communities, collaborative partnerships and a supportive national policy environment. Against this background, the focus of this study is the extent to which health and hygiene promotion forms part of sanitation delivery programmes to informal settlements in the City of Cape Town. The investigation is confined to a comparative review of approaches to health promotion in 4 case study sites (Khayelitsha, Joe Slovo, Kayamandi and Imizamu Yetho) in respect of community and household capacity to take responsibility for community based programmes, role-players forming collaborative partnerships across departments and implementation of health and hygiene aligned with national policy. Analysis of the case studies highlights the ineffectiveness of once-off awareness campaigns and the need for a more comprehensive approach to health and hygiene promotion in line with the Ottawa Charter.

Introduction

Experience suggests that sanitation facilities alone do not guarantee the improvement of health, thus emphasizing the need for a broader approach to sanitation implementation where health and hygiene promotion is included as an integral part of sanitation delivery programmes. The development of community capacity,

---

R. Van Wyk (✉)
Masters Environmental Health, University of Johannesburg, South Africa

G.A. Uzochukwu et al. (eds), *Proceedings of the 2007 National Conference on Environmental Science and Technology*, DOI 10.1007/978-0-387-88483-7_17,

institutions and policies to manage and sustain hygiene, water and sanitation programmes is essential for long-term success.

Many well-constructed water and sanitation facilities do not have the desired health impact. Why? The answer is simple: they do not guarantee changes in hygiene practices. Is it possible to guarantee such changes? Maybe not, but we can certainly motivate users to adopt elements to enhance the health impact water and sanitation facilities by developing and implementing approaches that motivate people to change practices.

Sustainable improvements in hygiene behaviour patterns require more than promotion activities. Water, sanitation and hygiene facilities, while not sufficient on their own to improve health, are critical components in a comprehensive health and hygiene promotion programme.

The White Paper on Basic Household Sanitation, 2001 suggests that good sanitation include both appropriate facilities and appropriate health and hygiene promotion. The provision of facilities is not enough, households share toilet facilities and therefore systems are not acceptable to users. Failure of sanitation systems occur because people do not use facilities properly. Poor sanitation and unhygienic practices is not appropriate and have an adverse effect on household, community health and the environment.

The problem statement is supported by the Ottawa charter (WHO, 1986) which suggests that effective health and hygiene programmes require supportive environments through empowered communities and role-players across departments in taking responsibility for promoting sanitation and environmental health.

A more a comprehensive approach to health and hygiene promotion is needed to strengthen local communities, to assist in reorienting health services and to suggest enabling and supportive policy environments. This study is important to role-players such as municipalities, communities and households involved in the implementation of appropriate health and hygiene promotion to assist local authorities in improving sanitation service delivery.

The investigation is confined to a comparative review between local authority approaches to sanitation and health promotion in 4 case study sites (Khayelitsha, Joe Slovo, Kayamandi and Imizamu Yetho). The research did not investigate the provision of acceptable, affordable and sustainable 'hardware' such as toilets, sewers, water pipes and handwashing facilities.

Objectives: The purpose of the investigation of approaches to health and hygiene promotion in four case study sites in the City of Cape Town is:

- to identify targets and find out if they have the capacity to execute community based approaches in health and hygiene promotion,
- to clarify involvement and relationships between role-players across departments in health and hygiene promotion as part of sanitation service delivery to promote collaborative partnerships,
- to determine the extent to which implementation of appropriate health and hygiene promotion is aligned with national policy to improve health and hygiene promotion programmes.

Methodology: The action research methodology engaged the stakeholders in reflecting on the outputs of research in order to encourage their own learning and inform their future practice. The research activities included: collection, collation, synthesis and validation of data.

The study followed a qualitative research design and the research was conducted in a natural setting of stakeholders. Four case study sites were identified by stakeholders and a comparative review of the case studies was done through intensive investigation of each approach as part of sanitation delivery.

The qualitative approach consisted of the following two steps:

Step 1: A comprehensive review of relevant literature provides the background of the broader context and current theories on effective health and hygiene promotion.

Step 2: An investigation of approaches to health and hygiene promotion as part of sanitation delivery in four case study sites in Cape Town by using multiple sources of data.

Data Collection: Data on sanitation delivery programmes was collected. Each of the four case study sites had a different approach to health and hygiene promotion as part of the sanitation delivery programme. The data included the conceptualized health and hygiene promotion programme, actual implementation of such a programme in relation to a particular technology option.

A group of research assistants were involved in research activities to collect, collate, synthesize and validate data. Methods for collecting case study data included interviews (telephonic and onsite) with respondents to collect and validate data on targets and content of programmes, site visits and observations to collect data on implementation of sanitation delivery programmes, review of policy documents, health and hygiene materials and pamphlets. The reference group played a pivotal role in the collection and validation of data.

Results and Discussion

Approaches to health and hygiene is compared and reviewed in respect of targeted communities, role-players and implementation of programmes (see Table 1). Themes for health and hygiene promotion were derived from the background theory in order to:

- determine the capacity of communities and households and encourage a community based approach
- clarify relationships and involvement of role-players in health and hygiene promotion as part of sanitation service delivery
- review the extent to which implementation of health and hygiene promotion and the provision of facilities is aligned with national policy

Capacity-building activities are not being extended beyond education and training that is limited to hygiene messages, technical guidance and operation skills, to developing community-level roles and responsibilities.

Table 1 Findings based on comparative review of 4 case study sites (Khayelitsha, Joe Slovo, Kayamandi and Imizamu Yetho)

Research questions	Findings
Who are the targets?	The targets are children, youth, and households living adjacent to facilities.
Is their capacity being built through health and hygiene promotion programmes?	Capacity was build through participatory health and hygiene training workshops but not through door to door pamphlet.
Have role-players from the different departments been involved in programmes?	Role-players from different departments were involved but in most cases there was no collaboration between the role-players.
How are these relationships manifested?	Relationships between role-players were not clear and definite. Ad-hoc relationships between role-players were established.
To what extent is the implementation of health and hygiene promotion programmes aligned is with national policy?	Once off campaigns and education drives were not aligned with national policy, it merely created short term sensitization.
What do these programmes entail?	Pamphlets were distributed door to door.

There is an assumption within each approach that users of facilities will take responsibility for maintaining the facilities they use as an outcome of education and 'awareness' drives. Accepting that local perceptions of services will play a major part in their efficacy, if the level of service does not meet the needs of the users, the technical objectives will not be met.

In all but one site (Kayamandi); the extent to which community responsibility for sanitation is translated into active roles and responsibilities during planning and decision-making is limited to a representative committee with a facilitation function in the project. Community mobilization, awareness and 'education drives' appear to be 'synonymous with capacity building' (Pybus et al., 2001), while clearly defined functions, specific roles and responsibilities located within the community that contribute to sustainability of the systems, is lacking.

Lessons learned through the comparative review of case studies

- Despite intervening through engineering as well as short hygiene education intervention, sanitation health and hygiene still did not improve
- Long-term ongoing health and hygiene promotion should form part of sanitation delivery programmes
- health and hygiene promotion interventions needs to be strengthened by stronger community involvement

Awareness programmes are short term, often once off activities, that will, at best, create short-term sensitization of sanitation health and hygiene issues whereas education imparts knowledge and skill, with long-term impacts relating to attitude and behaviour (DOH, 2004). There is a clear differentiation between awareness and education. Health and hygiene promotion includes both awareness and education thus education cannot be replaced by awareness or promotion. Education imparts knowledge and /or skill whereas awareness imparts elementary understanding or sensitization.

Participatory approaches (WHO, 2003) to development interventions deliberately seek out relevant user groups such as women, youth or children, to engage the inputs and perspectives of target groupings within a community for planning sanitation improvements that are appropriate to the specific, local context (WHO, Soul City and PHAST). As the provision of a particular technology option becomes part of the local context, particular targets for health and hygiene promotion programmes are suggested. These targets relate to essential user components of functioning sanitation systems. (*In all but one of the case study sites the targets for health and hygiene programmes were properly identified but their capacity was not built due to inappropriate programmes.)*

An environment that enables and supports the improvement of hygiene must be continuously promoted at all levels – from the household at local level, to the role-players in delivering sanitation services. (*Therefore there is a need for clear roles and collaborative relationships between departments and their involvement in health and hygiene promotion.*)

Long-term ongoing health and hygiene promotion should form part of sanitation delivery programmes. Health and hygiene promotion interventions like these need to be strengthened by stronger community involvement. The impact of promotion programmes will ultimately manifest in the effective functioning of the sanitation system.

Conclusion

Sustainable improvements in health and hygiene behaviour patterns require more than promotional activities. Water, sanitation and hygiene facilities, while not sufficient on their own to improve health, are critical components in a comprehensive health and hygiene promotion programme. The development of community capacity, institutions and policies to manage and sustain hygiene, water and sanitation programmes is essential for long-term success.

Analysis of the case studies highlights the ineffectiveness of once-off awareness campaigns and the need for a more comprehensive approach to health and hygiene promotion in line with the Ottawa Charter. Therefore appropriate health and hygiene promotion as part of sanitation delivery requires:

- Communities to participate in an informed and empowered manner, take responsibility for promoting sanitation and health.
- Collaborative partnerships between role-players responsible for health and hygiene promotion.
- Health and hygiene promotion programmes to be aligned to national policy.

References

Department of Health. 2004. *National Health and Hygiene Education Strategy*. Department of Health, Pretoria, South Africa.

Department Water Affairs and Forestry. September 2001. *White Paper on Basic Household Sanitation*. Pretoria, South Africa.

Pybus, P., Schoeman, G., Hart, T. May 2001. *The Level of Communication between Communities and Engineers in the Provision of Engineering Services*. Report No. TT 133/00, Water Research Commission, Pretoria, South Africa.

Simpson, M., Wood, S. 1998. *Sanitation Promotion: WSCC Working Group on Promotion of Sanitation*. WHO, Geneva, Switzerland.

Water Supply and Sanitation Collaborative Council (WSSCC). 2001. *Iguacu Action Plan: Water, Sanitation and Hygiene for All*. WHO, Geneva, Switzerland.

World Health Organization, 1986. *Ottawa Charter for health promotion: an International Conference on Health Promotion, the move towards a new public health*. November 17–21, 1986, Ontario, Canada.

World Health Organization, UNICEF, WSSCC. 2000. *Global Water Supply and Sanitation Assessment 2000 Report*. WHO, Geneva, Switzerland.

World Health Organization. 2003. *Participatory Hygiene and Sanitation Transformation: A new approach to working with communities*, WHO/EOS/96.11 (http://www.who.int/docstore/water_sanitation_health/Environmental_sanit/PHAST/phast96-11/Phast1.htm).

Pollution, Environmental Justice, and the North Carolina Pork Industry

Osei Yeboah, Terrence Thomas, Timothy Foster, and Edward Fosu

Abstract North Carolina ranks as the 2nd largest hog producer in the United States. Currently, the state houses over 10 million hogs. It is estimated that about 19 million tons of hog waste was produced in 2004 by the 10 million hogs in the state (scorecard.com). Despite the increase in waste output due to increased production, concerns are that majority of these producers have not improved their waste management practices. Most importantly, all the big factories are located in counties in the Black Belt region of the state. These are counties where the percent Black population is greater than the state and national averages of 25.4% and 12% respectively. The existence of pollution, poverty, and minority populations in a given area, more often than not, raise questions about environmental justice. Statistical and geographical information systems (GIS) analyses are performed at the zip code level to whether relatively poor, nonwhite populations are disproportionately exposed to hog waste directly or indirectly. The results indicate that these populations are not directly targeted but their association with poverty and also being rural dwellers help explain their disproportionate exposure.

Introduction

North Carolina ranks as the 2nd largest hog producer in the United States. The sector has experienced exponential growth in the 1990s. Within a decade, the hog population in North Carolina increased from 2.6 million hogs in 1988 to over 8 million by 1997. Currently, the state houses over 10 million hogs. However, the number of hog farms in the state has decreased from 15,000 in 1986 to 3,600 in the year 2000 (NC and the Global Economy). Big hog factories have taken over the hog production business. Over 92% of the 10 million hogs are raised on factory operations of at least 2,000 hogs. An average factory in the state has about 3,700

O. Yeboah (✉)
North Carolina A&T State University, Greensboro, North Carolina

G.A. Uzochukwu et al. (eds), *Proceedings of the 2007 National Conference on Environmental Science and Technology*, DOI 10.1007/978-0-387-88483-7_18,

hogs and produces 38,480 pounds of feces and urine every day. It is estimated that about 19 million tons of hog waste was produced in 2004 by the 10 million hogs in the state (scorecard.com). Despite the increase in waste output due to increased production, concerns are that majority of these producers have not improved their waste management practices. As a result of the environmental concerns, North Carolina state government passed a bill in 1997 that placed a moratorium on construction of any new farms with over 250 hogs (S.L. 1997-458/H515). The bill which was slated to be lifted in September 2007 was adopted permanently in July 2007 (S.L. 2007-523/S1465). The 2012 expiration date for the "Study Waste-to-Energy" was intended to buy time in order to complete a study of incentives designed to encourage potential purchasers and producers of renewable energy to produce and use renewable energy or other marketable by-products derived from animal manure; while the 2019 Funds would ask for $10 million in state money to begin a cost-share program (NC Pork Council, 2006)

Most importantly, all the big factories are located in counties in the Black Belt region of the state. These are counties where the percent Black population is equal to or greater than the state and national averages of 25.4% and 12%. The statistics from the N.C. Department of Commerce ranked these counties among the lowest in terms of key economic and social status indicators. They are located in proximity to the I-95 Interstate system, with market potential to the southeast and northeast U.S. and bode very well for future commerce and industrialization strategies. Although most of the counties have local development corporations and chambers of commerce, there has been little development along this corridor; partly because of historical patterns of behavior. For example, the communities lack comprehensive economic development plans, there are few new manufacturing initiatives to boast of, and jobs command very low wages on the average. Pork and pork products production are very labor-intensive and are likely to be located in low wage communities with high unemployment, where firms can take advantage of cheap labor supply.

The existence of pollution, poverty, and minority populations in a given area, more often than not, raises questions about environmental justice. Environmental justice is a concept which emerged in the late 1980s in response to mounting evidence that minorities were burdened with a disproportionate share of the nation's hazardous waste treatment, storage, and disposal facilities. A number of studies have sought to challenge the role of race in the distribution of such facilities, but the general consensus seems to be that the issue of environmental justice is a topic that deserves serious consideration.

Environmental justice attempts to address concerns that environmental and public health costs are borne by those least able to defend themselves against such costs (Taylor, 2000). The environmental justice movement began in 1982 during protests over siting of a landfill for polychlorinated biphenols (PBCs) near an African-American community in Warren County, NC. This drew public attention to the pattern of hazardous waste facilities being sited in African-American communities in the South (U.S. GAO, 1983) and the nation (United Church of Christ, 1987). Some authors (Bullard, 1983) argued that the root causes for this pattern were racism and the political marginality of African-American communities in the United States.

There are two main theoretical formulations that explain how institutionalized racism results in the targeting of minority communities as sites for environmentally hazardous industries and facilities. The first perspective emphasizes the functional link between racism, poverty, and powerlessness. Minority communities are targeted for siting environmentally hazardous industries because they are poorer, less informed, less organized, and less politically influential (Bullard, 1983). To save time and money, companies seek to locate environmentally hazardous industries in communities which will put up the least resistance – those who are less informed and less powerful politically. There are two central questions in the debate over the causes of environmental inequality. First, is whether the differential levels of environmental quality are the result of class factors or racial dynamics? Second, is whether the bias of distribution of environmental hazards is a function of poverty rather than race (Mohai and Bryant, 1992)? A follow-up question to the previous questions is this one: Aren't minorities disproportionately targeted and impacted because they are disproportionately poor?

The association between environmental quality and a locality's racial and income characteristics has been recognized in the environmental economics literature that examines the distribution of air quality and benefits. In an early study, Zupan (1973) found a negative correlation coefficient between pollution levels and income. A correlation analysis by Asch and Seneca (1978) determined that relatively poorer, less educated or nonwhite households dominates areas with higher pollution. Contrary to the previously sited findings, Walker and Calzonetti, (1989); and Wheeler and Muller (1986) concluded that industrial location factors are more important in determining the location of industries. New establishments often choose manufacturing locations based on site characteristics that affect their profitability.

Differential exposure to environmental risks may occur for reasons that have no relation with discrimination against minority or low-income populations. Companies that emit toxics determine their location based, in large part, on positive factors that make a community a good place to do business, and proximity to an interstate highway is one of the major factors considered. There are a few studies, Hamilton (1995); Stretesky and Hogan (1998); Hird (1994); that consider land values, political power, and industrial location factors such, as employment, population density, percent of urban population, and the proximity of interstate highway as other important explanatory variables for hazardous waste siting.

Objective: The objective of the study is to apply statistical and GIS analyses to determine whether relatively poor, nonwhite populations are disproportionately exposed to hog waste directly or indirectly.

Data and Methods

The Unit of Analysis: Geographical scale is one of the major reasons for the conflicting findings in previous environmental justice studies. Past research has examined environmental justice at the county level, (Hird, 1993); city level, (Mohai and Bryant, 1992); ZIP Code level, (Hamilton, 1995); census tracts level, (Pastor et al.,

2001). County, city, and ZIP Code levels are less appropriate scales for investigating environmental justice issues, because nonwhite and low-income populations are typically concentrated in small geographic areas (U.S. EPA, 1994). However, the smallest units of analysis, such as census blocks or tracts, do not have income reported, and that makes it difficult to include low-income populations in environmental justice studies.

Statewide county level data of currently permitted swine facilities with latitude and longitude coordinates were obtained from the Aquifer Protection Section, Division of Water Quality, North Carolina Department of Environment & Natural Resources. Using ESRI ArcGIS 9.2 Software, the data was translated into X-Y coordinates. The "Add XY Data" Tool was employed to convert the coordinates into spatial point locations. An existing coverage of all ZIP Codes within the United States was clipped to only include ZIP Codes within the state of North Carolina. The XY Spatial data was then overlaid onto the ZIP Codes shape file within ArcMap, and a spatial join was executed. The spatial join combines data from all points that fall within a certain geographic polygon (in this case, ZIP Code). All the numerical codes in the data set were summed (in this case sum of animals). The resulting data were exported to produce a new shape file with swine data associated by ZIP Code. The swine data are then displayed to show concentrations of swine activity and the number of swine in a spatial format.

From scorecard.com, it is estimated that about 19 million tons of hog waste was produced in 2004 by the 10 million hogs in the state. Thus an average hog produces 1.9 tons of waste per year. This factor is used to multiply by the number of hogs to generate the waste dependent variable in tons.

Estimated Model: The ordinary least squares (O.L.S.) procedure was used to establish the relationship between the location of hog waste and the explanatory variables as follows:

$$\ln(W_{it}) = \alpha + \ln(BP_{it}) + \ln(RU_{it}) + \ln(NW_{it}) + \ln(PI_{it}) + \ln(PD_{it}) + \ln(UR_{it}) + e_{it}$$

where, *ln* is natural log; W_{it} is hog waste in tons in a ZIP Code ***i***; α is a constant intercept term; and e_{it}, an error term for a ZIP Code in a particular year. Based on past EJ research, several explanatory variables are chosen that are believed to be good indicators in explaining the level of hog waste in a ZIP Code. In our analysis, ***NW*** is proportion of non Whites in a ZIP Code. Three economic discrimination indicators were also selected: percent of household living below the federal poverty line ***(BP)***; per capita income ***(PI)***; and the unemployment rate (*UR*). In addition, two industrial location variables were included: number of persons per square mile or population density ***(PD)*** and the percent rural population ***(RU)***. Data on these explanatory variables were drawn from the U.S. Census Bureau, 1990 and 2000 Decennial Census through the www.census.gov website.

Analysis and Results: Tables 1 and 2 display the comparative descriptive statistics of the demographic and socioeconomic variables used in the 1990 and 2000 models, respectively. The percent average non-white population in 1990 was 78.4 (ranging from a minimum of 59, to a maximum of 84.1. This is considerably higher

Table 1 Distribution of demographic and socioeconomic data compared to state: 1990

Variable	N	Minimum	Maximum	Mean	State mean
BP	250	0.035	0.369	0.153	0.13
NW	250	0.59	0.841	0.784	0.244
PI	250	3083	16889	10444	17311

Table 2 Distribution of demographic and socioeconomic data compared to state: 2000

Variable	N	Minimum	Maximum	Mean	State mean
BP	299	0.041	0.386	0.158	0.123
NW	299	0.614	0.959	0.841	0.279
PI	299	9096	26791	16367	20307

than the national and the state's average of 12.5% and 25.4%, respectively (Bureau of the Census, 1990). The mean per capita income in 1992 U.S. dollars across ZIP Codes in the region is $10,444 (ranging from a minimum of $3,083.30, to a maximum of $44,875, with a standard deviation of $3,286) compared to the state average of $17,311. Similarly, there are more people living below the national poverty line in these areas than the rest of the state. As high as 15.3% of the people in the region lives below the federal poverty line compared to a state average of 13% in 1990.

The mean proportion of non-whites in 2000 is 84.1% (ranging from a minimum of 61.4% to a maximum of 95.9%). The mean per capita income in 1992 U.S. dollars is $16,367, and 15.9% of the population lives below the federal poverty line, while the state averages are $20,307 and 12.3%, respectively.

Bivariate O.L.S. Analysis: Three ordinary least squares regression analyses were performed: the first is a bivariate model, followed by the analysis of two multivariate models of both data sets. Many studies have failed to examine the independent effects of race, ethnicity, poverty, and income while controlling for other variables (Clark et al., 1995). In each analysis, we control for income, poverty, and the unemployment rate to see if race has an independent effect. We then control for race, and determine the effect of income separately and then jointly with race.

Multivariate O.L.S. Analysis: These same data sets were then examined with the models including income and race separately, and then with all the non-discriminating independent variables- population density, percent rural population, and unemployment rate.

Longitudinal Analysis: It may happen that the siting of hazardous waste detracts from neighborhood livability and diminishes land values, resulting in the exodus of middle-class (often Anglo) homeowners and influx of lower-class (often minority) residents as documented by Hamilton (1995). Statistical differences between means of the discriminating variables in the data sets were tested using the student t-statistic.

Results and Discussions

Figure 1 shows the distribution of waste in the region, while Figs. 2 and 3 present the distribution of non White population in 1990 and 2000. The results for the bivariate regression analyses are presented in Tables 3 and 4. The coefficients of the non White variable are significant at ($p = 0.000$) in both years. While the 1990 model has an expected positive sign, the coefficient in the 2000 model has negative sign. We note that this latter finding is consistent with (Anderton et al., 1994a,b; Anderton, 1997); (Bowen et al., 1995); and (Yandle and Burton, 1996) that suggest that the spatial association between race and hazardous waste is weak at best. The income coefficients are also significant at ($p = 0.000$) with expected negative signs. Tables 5 and 6 present the results of the multivariate analyses. Table 5 shows the results

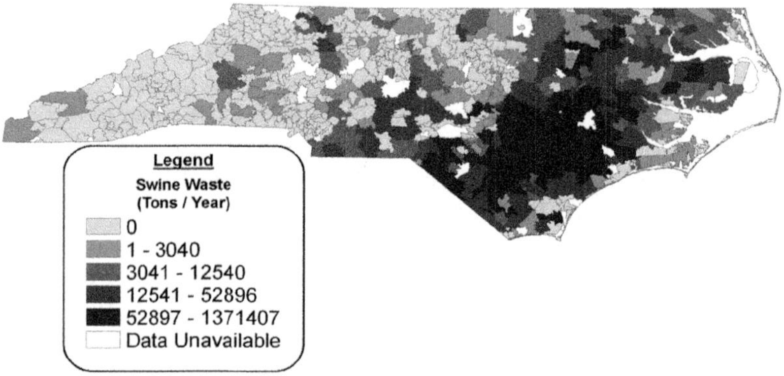

Fig. 1 Distribution of hog waste by ZIP Code (Tons Per Year) North Carolina, 2000

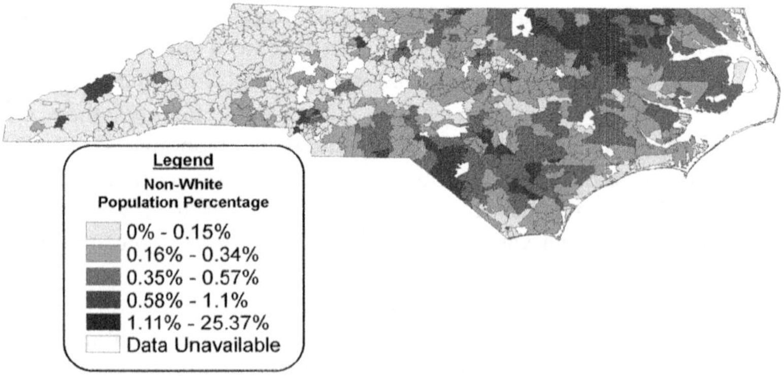

Fig. 2 Percentage of non-white residents by ZIP Code North Carolina, 1990

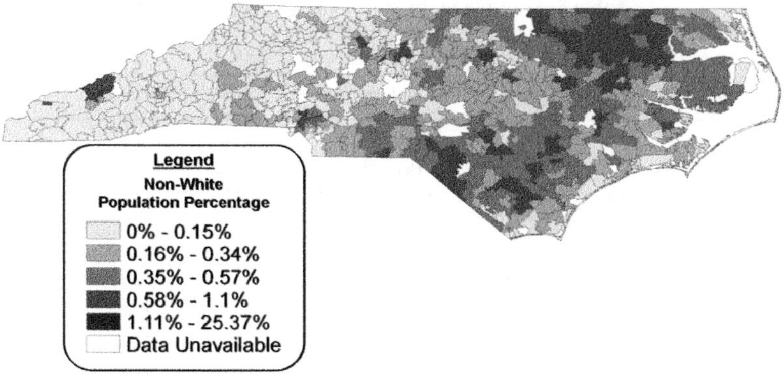

Fig. 3 Percentage of non-white residents by ZIP Code North Carolina, 2000

Table 3 Results of the bivariate regression with race

	1990		2000	
Variable	Estimated coefficient	P-value	Estimated coefficient	P-value
Constant	10.485	0.000*	7.574	0.000*
NW	1.009	0.000*	–10.035	0.000*
R-Square	0.13		0.11	

* Refers to significance at 1% level.

Table 4 Results of the bivariate regression with income

	1990		2000	
Variable	Estimated coefficient	P-value	Estimated coefficient	P-value
Constant	27.463	0.006*	48.357	0.000*
PI	–2.002	0.000*	–4.029	0.000*
R-Square	0.075		0.11	

* Refers to significance at 1% level.

of the multiple regression analysis with the effects of race and income variables determined simultaneously. The race coefficient is significant in both years but the sign is negative in the 2000 model as in the bivariate analysis. Income on the other hand has an expected negative sign which is insignificant in the 2000 model. Table 6 presents the results of the overall model. In these analyses, the race coefficients have expected positive signs, but while the p-value is (0.039) in the 2000 model, it is only significant at ($p = 0.133$) in the 1990. The elasticity of 11.97 implies a 1% increase in the non non-white population in 2000 will cause the residents to experience a 11.97% increase in the level of hog waste.

Table 5 Multiple regression using race and income

	1990		2000	
Variable	Estimated coefficient	P-value	Estimated Coefficient	P-value
Constant	23.583	0.000***	30.85	0.08*
NW	0.869	0.002***	–10.164	0.016***
PI	–1.444	0.042**	–2.416	0.167
R-Square	0.162		0.21	

* Refers to significance at 10% level.
*** Refers to significance at 1% level.

Table 6 Regression results of the overall model

	1990		2000	
Variable	Estimated coefficient	P-value	Estimated coefficient	P-value
Constant	17.681	0.004**	20.639	0.033*
BP	2.474	0.001**	4.207	0.000**
PD	–0.335	0.042*	0.758	0.625
RU	0.617	0.042*	0.925	0.002**
NM	0.536	0.133	11.977	0.039*
UE	–0.917	0.224	0.659	0.663
PI	–0.368	0.595	–0.120	0.906
R-Square	0.35		0.21	
Durbin-Watson	1.610		1.731	

* Refers to significance at 5% level.
** Refers to significance at 1% level.

Figures 4 and 5 present the distribution of population living below the poverty level in 1990 and 2000. The percentage of the population living below the federal poverty level is significant in both models at ($p = 0.0001$) and positive. The elasticities for proportion of the population living below the federal poverty line in 2000 and 1990 are 2.47 and 4.21, respectively. Thus, 1% increase in the people living under poverty will see an increase of hog waste by 2.47% and 4.21% for those years. These findings are consistent with Glickman (1994); Pastor et al. (2001). Glickman (1994) for example, applied a geographical information system to analyze environmental risk in Allegheny County, Pennsylvania. The results indicate that the percentage of non-whites and the percentage of the Poor living within one mile of hazardous facilities are greater than the percentage of those living farther away. Thus, poverty is always associated with minority populations.

The industrial location factors- population per square mile and percent rural population are each significant at ($p = 0.042$) in the 1990 model. Rural population has an expected positive sign as all the hog facilities or farms are in rural areas. The elasticities for rural population are 0.62 and 0.92 for 1999 and 2000 which translates

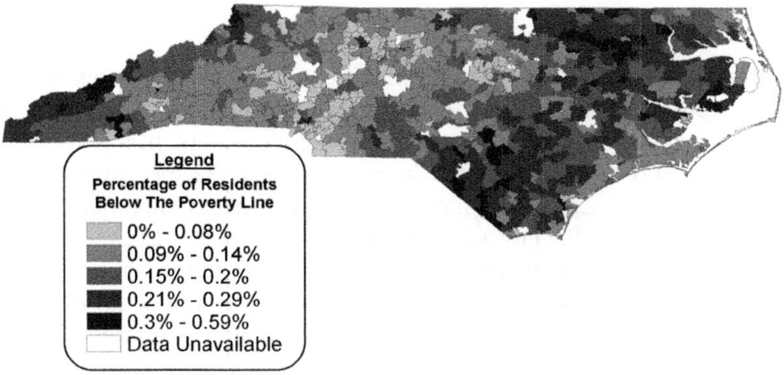

Fig. 4 Percentage of residents living below the poverty line North Carolina, 1990

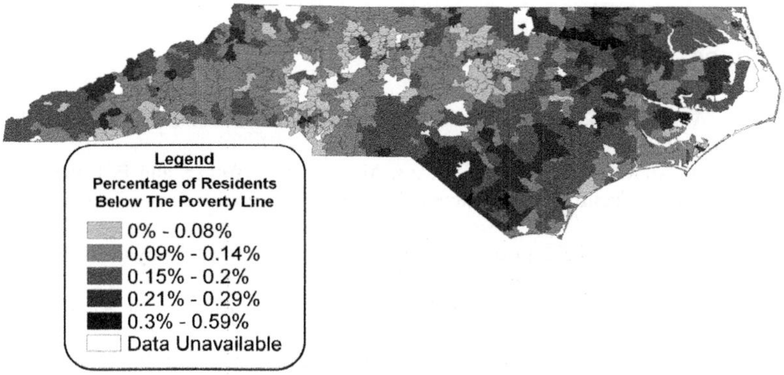

Fig. 5 Percentage of residents living below the poverty line North Carolina, 2000

into 0.62% and 0.92% increase in hog waste for 1% increase in the rural population. Population density also has an expected negative sign and this is consistent with (Stretesky and Hogan, 1998). Higher population density is expected to be associated with lower toxic emissions, since polluting firms should prefer a location where their activity is noticeable to fewer people. Rural population is also significant and positive at ($p = .002$) in the 2000 model, but population density is not.

Income and the unemployment rate are not significant in any of the two overall regression models. Similar results were established by (Mohai and Bryant, 1992). Mohai and Bryant show that race and ethnicity are the most salient factors in predicting the location of hazardous waste sites.

The insignificance of the differences between means of the discriminating variables in the longitudinal analysis rejects the theory of minority influx into the neighborhood due to diminished land values caused by pollution.

Conclusion

The regression analyses suggest that minorities might not have been directly targeted for exposure to hog wastes. Their exposure to hog waste may be explained by their association with poverty as also being rural dwellers. In this study, poverty and proportion of rural population are the most important factors explaining the levels of hog waste in a ZIP Code. Income is not seen as an important determining factor as it is only significant independently or with race. Also, race though significant in all analyses has inconsistent signs, especially in the bivariate analyses. Most importantly, the fact that most of the factories were built in the 1990s, and the poverty, rural population, and population density variables performed better in the 1990 models in terms of significance and consistent signs may lead one to conclude that the siting was mainly due to space and also where the investors would likely face the least resistance.

References

Anderton, D., A. Anderson, J. Oakes, and M. Fraser. 1994a. "Environmental Equity: The Demographics of Dumping." *Demography* 31: 229–248.

Anderton, D., A. Anderson, P. Rossi, J. Oakes, M. Fraser, E. Weber, and E. Calabrase. 1994b. "Hazardous Waste Facilities: Environmental Issues in Metropolitan areas." *Evaluation Review* 18: 123–140.

Anderton, D. 1997. "Methodological Issues in the Spatiotemporal Analysis of Environmental Equity." *Social Science Quarterly* 77: 508–515.

Asch, P., and J. J. Seneca. 1978. "Some Evidence on the Distribution of Air Quality." *Land Economics* 54(3): 278–297.

Bowen, W., M. Sailing, K. Haynes, and E. Cyran. 1995. "Toward Environmental Justice: Spatial Equity in Ohio and Cleveland." *Annals of the Association of American Geographers*85: 641–663.

Bullard, R. 1983. "Solid Waste Sites and the Black Houston community." *Sociological Inquiry* 53: 273–288.

Clark, R., S. Lab, and L. Stoddard. 1995. "Environmental equity: A Critique of The Literature." *Social Pathology* 1: 253–269.

Environmental Defense Fund/Green Media Toolshed 2005. (http://www.scorecard.org).

Glickman, T. S. 1994. "Measuring Environmental Equity with Geographical Information Systems." *Resources*116: 2–6.

Hamilton, J. T. 1995. "Testing for Environmental Racism: Prejudice, Profits, Political Power?" *Journal of Policy Analysis and Management* 14: 107–132.

Hird, J. A. 1993. "Environmental Policy and Equity: The Case of Superfund." *Journal of Policy Analysis and Management* 12: 323–343.

Hird, J. 1994. "Superfund: The Political Economy of Environmental risk." Baltimore, Maryland.

Mohai, P., and B. Bryant, 1992. "Environmental Racism: Reviewing the Evidence." In Race and the Incidence of Environmental Hazards, ed. Bunyan Bryant and Paul Mohai, Boulder, Colorado: Westview Press.

North Carolina Pork Council. 2004. "North Carolina Pork and the Global Economy." North Carolina Pork Council, 2006, "Weekly Briefing". *The Porkline* 8: 30.

Pastor, M., J. Sadd, and J. Hipp. 2001. "Which Came First? Toxic Facilities, Minority Move-in and Environmental Justice." *Journal of Urban Affairs* 23(1): 1–21.

Stretesky, P., and M. J. Hogan 1998 "Environmental Justice: Analysis of Superfund sites in Florida." *Social Problems* 45(2): 268(20).
Taylor, D. 2000, "Introduction, Advances in Environmental Justice: Research, Theory, and Methodology." Special Issues of *American Behavioral Scientists* 43(4): 504–507.
U.S. Environmental Protection Agency. 1994. "Environmental Justice Initiatives, 1993." Washington DC: U.S. Government Printing Office. EPA 200-R93-001.
U.S. Bureau of Census. 1990. (http://www.census.gov).
U.S. Bureau of Census. 2000. (http://www.census.gov).
U.S. General Accounting Office. 1983. "Siting of Hazardous Waste Landfills and their Correlation with Racial and Economic Status of Surrounding Communities." Washington, DC: U.S. GAO.
United Church of Christ. 1987. "Toxic Waste and Race in the United States." New York: United Church of Christ.
Walker, R., and F. Calzonetti. 1989. "Searching for new Manufacturing Plant Locations: A study of Location Decisions in Central Appalachia." *Regional Studies*25: 15–30.
Wheeler, J. O., and P. O. Muller. 1986. "Economic Geography" 2nd. Ed. New York: John Wiley and Sons.
Yandle, T., and D. Burton. 1996. "Reexamining Environmental Justice: A Statistical Analysis of Historical Hazardous Waste Landfill Siting patterns in Metropolitan Texas." *Social Science Quarterly* 77: 479–492.
Zupan, J. M. 1973 "The Distribution of Air Quality in the New York Region." Baltimore: John Hopkins.

Reducing Community Exposure to Toxics: A Critical Component in Building Healthy and Sustainable Communities

Robert C. Wingfield

Abstract Reducing the community's exposure to environmental toxics is very critical to achieving success in building healthy and sustainable communities. Reduced exposure to environmental toxics is quite a challenge because of the complexity of life in our communities. The reality is that certain populations have historically suffered and continue to suffer disproportionate exposure to environmental toxics. Often these same populations suffer disproportionately from poor health outcomes from diseases in which there are disparities in the incidence of their occurrence. The problem is further complicated by the many social, economic and political factors operative in these impacted communities. It is most important that the voice of the community be heard in addressing these issues.

The work of many researchers has shown the presence of toxic materials in the bodies of community residents worldwide. The toxicology of many of these substances has been studied, whereas the health impact of the various mixtures of toxic substances remains under investigation. Children represent a very vulnerable population. Despite programs and efforts to reduce exposure to toxic substances, toxic substances still are still released into the environment in millions of pounds a year. These toxic substances move throughout the environment and undergo chemical transformations sometimes to less desirable chemical substances. New environmental toxics are being identified each year as new materials are being introduced into use by our society. Changing lifestyles such as the move from rural to urban communities as well as urban sprawl is affecting the exposure of populations worldwide. Access to clean and sustainable water now ranks as high as the need for sustainable energy in importance to the development of communities and countries.

R.C. Wingfield (✉)
Department of Chemistry, Fisk University, Nashville, TN 37208, USA

G.A. Uzochukwu et al. (eds), *Proceedings of the 2007 National Conference on Environmental Science and Technology*, DOI 10.1007/978-0-387-88483-7_19,

Building Sustainable Communities

If we are truly going to build healthy and sustainable communities we must reduce the exposure of the residents and the ecosystem to environmental toxics. Sustainable communities are characterized by a healthy environment, a strong economy and social justice. These communities are built on the premise that while we are addressing our needs that we are making sure that the needs of future generation are not negatively impacted. We not only want strong healthy communities today but for generations to come. (World Commission on Environment and Development 1987, Grassian 2007).

In order to obtain this vision, mankind must make a paradigm shift. We must give up our present way of life which is characterized by extensive depletion of natural resources, an insatiable desire to consume, poisoning of the ecosystem, and generation of massive amounts of waste. We must reconnect with nature and realize that as we destroy our environment we are putting our own survival in peril (Anderson 2003, Cortese 2003, and Chemistry for a Sustainable Future 2006).

"Environmental health is the field of science that studies how the environment influences human health and disease. Environment in this context means things in the natural environment like air, water and soil, and also all the physical, chemical, biological and social features of our surroundings. The interaction of the environment with a person's genetic make-up can shift the balance between health and disease" (NIEHS 2005).

Social Factors, Disproportionate Exposures and Disproportionate Health Outcomes

Minorities and low income populations have been disproportionately impacted by mankind's extensive generation of toxic substances and toxic wastes. Too often we have seen landfills for disposal of toxic waste and facilities for handling and treatment of toxic waste located in minority and low income neighborhoods. (Bullard et al. 2007, IOM 1999). The Environmental Justice movement has fought vigorously over the last three decades to address these issues. They have been fighting to reduce the exposure of the community residents to these toxic substances through all means possible including clean-up of the toxic sites, reduction of toxic emissions in the community or, in some cases, relocation of the community (TDEC 2000).

These same community residents who have suffered these disproportionate environmental exposures have in many cases suffered disproportionately high incidents of chronic diseases such as cardiovascular diseases, diabetes, and cancer. Social science research has shown that there are many social determinates of the health of these communities. They include: lack of wealth, income, education, housing, and transportation (O'Hara 2006, Williams and Collins 2001). The impact of urban sprawl on the health of the urban community has been documented (Frumkin et al. 2004). Reports from the Institute of Medicine (IOM 1999, IOM 2002) have pointed to the importance of social and cultural factors for improving health and a better understanding of mechanisms linking social and cultural environment to a specific health outcome.

Social and cultural factors influence health by affecting exposure and vulnerability to disease, risk-taking behaviors, the effectiveness of health promotion efforts, and access to, availability of, and quality of health care.

These social factors will indeed influence the mechanisms by which the residents of these communities are exposed to environmental toxics. The primary routes of exposure are inhalation, skin/eye contact, injection and ingestion. The source of food and its preparation could influence exposure through ingestion. It is well documented that many foods from natural sources such as fish from contaminated surface water are also contaminated with many toxics. Food grown in contaminated soil is a source of toxics. Young children putting toys in their mouth have been exposed to lead. Lead containing paint has been found on the toys that were imported from foreign countries (U.S. Consumer Product Safety Commission 2007). Recently there has been a massive recall of such toys. Children may also be exposed to lead by putting lead paint chips in their mouth or exposure to lead in dust tracked into the home from outside or brought into the house on the work clothing of family members. Exposure through inhalation could occur during workplace exposure, through extensive exposure to motor vehicle exhaust through living in areas with high truck or bus traffic (especially idling vehicles) (Sheppard et al. 2002) or living in communities near industrial facilities with significant emissions of toxic materials.

Recently children's exposure to lead dust doing the renovation of homes and other facilities has become a major area of concern. The U.S. EPA is exploring supplemental rulemaking requiring lead safe work practices and worker training and certification for contractors and construction professionals conducting renovation activities on child occupied facilities (U.S. EPA 2007b). Too often minorities and low income populations live or work under conditions where these exposures occur.

Exposures to Toxic Substances (TRI)

The 2005 Toxic Release Inventory, prepared by the U.S. EPA, continued to show the release of millions of pounds of toxic materials into the environment in each state of the nation (U.S. EPA 2007a)Persistent bioaccumulative toxic (PBT) chemicals accounted for 477.4 million pounds or 11% of reported on- and off-site disposal or other releases in 2005. Of that total, lead and lead compounds accounted for 98% or 469.3 million pounds. Total disposal or other releases for mercury and mercury compounds were 4.4 million pounds and, for dioxin and dioxin-like compounds, they were 85,501 grams (189 pounds).

There were 179 known or suspected carcinogens on the TRI list for 2005. They accounted for 923.1 million pounds or 21% of reported on- and off-site disposal or other releases in 2005. Of that total, lead and lead compounds accounted for 51% and arsenic and arsenic compounds for 20%. Almost three-quarters (680.3 million pounds or 74%) were disposed of or otherwise released to some form of on-site land disposal. Styrene air emissions were 45% of the total 112.7 million pounds of air emissions of carcinogens.

The states of Tennessee and North Carolina respectively had 9.97 million and 7.03 million pounds of toxic fugitive air emissions and 71.5 million and 97.55 million pounds of point source air emissions. There were 3.3 million and 8.2 million pounds of toxic releases to surface water in TN and NC respectively. There also continued to be tons of production – related waste that were managed; Tennessee had 642 million pounds and North Carolina had 461 million pounds. Both states showed thousands of pounds of suspected neurotoxins released into the air and a Scorecard (Green Media Toolshed 2007) high cancer risk score for many counties in each states. Shelby County, TN which ranked number two on the list of counties in TN releasing suspected neurotoxicants into the air, showed releases of over eight million(8 M) pounds of toxic compounds into the air.

Haywood County, NC which ranked number three of counties on the Scorecard list of counties in NC releasing suspected neurotoxicants into the air showed releases of over two million (2 M) pounds of toxic compounds into the air. A matching of those counties with high TRI emissions with census data supports the hypothesis that those people who live in low income or minority communities are exposed to the highest level of toxic pollutants (U.S. Census Bureau 2007). An excellent case in point is Shelby County, TN which has a minority(non-white) population of approximately 54% and shows the high level of TRI releases described above. This compares with a TN state wide minority population of approximately 19%.

Persistant Bioacculumative and Toxic Chemicals Exposure

Persistent Bioacculumative and Toxic Chemicals being released into the environment are of special concern not just within the United States but globally. These toxic chemicals remain in the environment for long periods of time and they bioaccumulate by moving upward through the food chain. These include many mercury, lead, polychlorinated biphenyl (PCBs), dioxin, polyaromatic compounds and several organochlorine pesticides. The use of most of these pesticides has been banned for two or more decades yet they are still present in our environment. In the 2005 TRI, 94 thousand pounds of PBTs in Tennessee and 66 thousand pounds in North Carolina were released into the air or surface water. The presence of PBTs in surface water and their sediments is a major reasons for the fishing advisories posted many states like Tennessee, Michigan, etc.

The presence of mercury in the emissions of coal burning power plants continues to be an atmospheric pollutant of concern (Edgerton et al. 2006). Both elemental mercury and oxidized forms of mercury are emitted. Elemental mercury has a long atmospheric lifetime (months) and is transported over long distances from the point of emission. Oxidized forms of mercury are removed from the atmosphere relatively quickly. Possible in plume reduction of the oxidized forms of mercury to elemental mercury has to been postulated to explain the high levels of elemental mercury observed in the analysis of the emissions.

Emerging PBTs of Concern

Emerging toxic environmental pollutants continue to be of concern. Polybrominated diphenyl ethers (PBDE), which are used as flame retardants, are now being found throughout our environment. They are cropping up in people, animals and ecosystems. Polybrominated diphenyl ethers are currently used as additives to retard or prevent combustion in many commercial products including plastics, foam products, textiles, and printed circuit boards. Although production of some of the brominated diphenyl ethers has largely been phased out in North America and Europe, PBDEs are still present in consumer products (Streets et al. 2006)

PBDEs can enter the environment through a number of pathways including atmospheric emissions during manufacture, recycling wastes, containing PBDEs, volatilization from consumer products and leaching from disposal sites. Once in the environment, their physical-chemical properties predict that they would undergo long-range atmospheric transport to even remote regions of the world. PBDEs may exhibit toxicity to wildlife and humans through various mechanisms including neurobehavioral development, thyroid hormone levels, fetal toxicity/teratogenicity, and liver and kidney morphology.

The dominant PBDE exposure pathways to humans are expected to be through the inhalation of house dust and through the consumption of contaminated fish (Carrizo et al., 2007). The People in North America contain ten times the levels of PBDE of people in the rest of the world. Studies of the levels of PBDE in lake trout in Lake Superior showed that possible debromination of these compounds within the trout accounted for the lower percentage of the more highly brominated cogeners. Also it appeared that PBDE distribution throughout the food chain in Lake Superior differed significantly from that found in other ecosytems because of the lower temperature of the lake.

Despite a short half life in humans the more highly brominated BDE, BDE 209, is detected in the blood of workers with no occupational exposure, indicating continuous exposure. A recent survey of U.S. food concluded that meat was the most important source of dietary intake of the more highly brominated PBDEs by adults; whereas fish was the main contributor to the dietary intake of the lower brominated cogners in human in Europe.

Other emerging toxics of concern include a new generation of disinfectant byproduct (Krasner et al. 2006) and perfluorinated surfactants (Calafat et al. 2007). Polyfluoroalkyl compounds (PFCs) have been used extensively in commercial applications including surfactants, lubricants, paper and textile coatings, polishes, food packaging and fire retardant foams. Results of animal studies suggest potential adverse health effects including genotoxicity, reproductive and developmental toxicity, and carcinogenicity associated with exposures to perfluorooctane sulfonate

(PFOS) and prefluorooctanoic acid (PFOA). These key compounds and metabolites are persistent and have a tendency to bioaccumulate.

Evidence for Human Exposure to Toxic Substances

There is no more solid evidence of exposure than the presence of toxic materials, their metabolites or environmental transformation products in the human body as revealed by the 3rd National Report on Human Exposure to Environmental Chemicals released by the Centers for Disease Control and Prevention. This third report presents exposure data for the U.S. population for 148 environmental chemicals over the period from 2001 to 2002. The level of chemicals and their metabolites in blood and urine samples were measured from a random sample of the participants from the National Health and Nutrition Examination Survey (NHANES) (Centers for Disease Control and Prevention 2005)

This report showed reduced blood lead levels in children aged 1–5 years when compared to levels observed in previous studies. For the period 1999–2002, 1.6% of these children had elevated blood lead levels. This percentage has decreased from 4.4% in the early 1990s. The levels of continine (a metabolite of nicotine) show a decrease of 68% for children compared to the 1988–1991 data. A 68% reduction was observed for adolescents and about 75% reduction for adults. Non-Hispanic blacks have levels more than twice those of Mexican-Americans and Non-Hispanic whites. Children's levels are more than twice that of whites.

The work of other researchers show the importance of these CDC findings. Kahn and co-workers found exposure to prenatal tobacco and environmental lead are risk factors for attention deficit hyperactivity disorder (ADHD) in U.S. Children (Braun et al., 2006). Other studies have shown that neurotoxic effects of lead in children at blood levels lowers than 10 μg/dl the generally acknowledged action level.

The findings of this CDC report suggest wide spread exposure to pyrethroid insecticides since 3-phenoxybenzoic acid was found in much of the U.S. population. Synthetic pyrethoid are now widely used throughout the United States (Leng et al. 2006).

Lu and colleagues conducted a longitudinal study to assess the exposure of 23 elementary school-age children to pyrethroid pesticides using urinary pyrethroids as biomarkers (Lu et al. 2006). They found an association between the parent's self reported pyrethroid use in the residential environment and elevated pyrethroid metabolites in the children's urine. Children's ages appeared to be significantly associated with pyrethroid exposure, which is associated with the use of pyrethroids around the facilities where the older children engaged in outdoor activities.

Note that subsequent to the CDC report additional biomonitoring studies conducted by Calafat et al. showed the presence of the metabolites of polyfluoroalkyl compound in the U.S. population. (Calafat et al. 2007). These studies were conducted from a representative U.S. population participating in the NHANES study.

Perfluoroctanoic acid (PFOA) and Perfluorohexane sulfonic acid (PFHxS) and perfluorooctane sulfonamide were detected in all serum samples. Other metabolites were found in 90% of the samples. Mexican Americans had lower concentrations than non-Hispanic blacks and non-Hispanic whites. Higher education was associated with higher concentrations of PFOS and PFOA.

The Path Forward

Reviewing the multiple toxics and the multiple sources of exposures to toxics, it becomes increasing clear that the most important step that we can take to reduce exposure to toxics is to remove toxics from the environment. One very important step in this direction is full scale implementation of the United Nations Environmental Program (UNEP) Cleaner Production Initiative. Here we seek to remove all toxic materials from the products of commerce and to remove all toxic materials from use in the processes used to manufacture them. Thereforth there will be no opportunity to be exposed to toxic materials during the manufacture, application (use) or disposal. Whenever possible we should revert to natural systems to learn valuable lessons in design and materials. The Clean Production Research and Training Program of the Lowell Center for Sustainable Production at the University of Massachusetts Lowell follow this leads (Lowell Center for Sustainable Production 2007). Also the Clean and Just Production Initiative of the Deep South Center for Environmental Justice is based on a marriage of the concepts of Environmental Justice and Clean Production.

Green Chemistry and Green Engineering (Anastas and Warner 1998, Chemistry for a Sustainable Future 2006) offer a detailed approach that is useful in the design and manufacture of nontoxic materials. The Global Environmental Management Initiative (GEMI 2007) which has been adopted by several multinational corporations to incorporate sustainability into their operations has excellent potential to make a real impact on reducing exposure to toxic materials both in the workplace and the broader community. These corporations are dedicated to fostering global environmental, health and safety excellence through collaborative efforts.

Efforts must continue to use science to help achieve sustainable relationships between water, people and the environment. There is a clear need for sustainable water management globally. (Schiffries and Brewster 2004).

In addition, we must move quickly to remediate all sites where the presence of toxic substances represents a hazard to health existing and future generation. Education is important to maintain healthy indoor environment. This is especially important since most of our time is spent indoors (Mitchell et al. 2007). The importance of addressing mold is being emphasized (Wu et al. 2007) because of the significant health implications of being exposed to mold and the multitude of possible mold exposures along the Gulf coast and New Orleans after flooding due to hurricane Katrina (DSCEJ 2007).

Conclusion

There exist overwhelming evidence of continuing human and ecosystem exposure to toxic substances and materials. Local, national and global systemic changes must be made in order to reduce the level of toxics in the environment and achieve healthy and sustainable communities.

Acknowledgment The project described was supported in part by Grant Number U45ES006184 from the National Institute of Environmental Health Sciences. The content is solely the responsibility of the author and does not necessarily represent the official views of the National Institute of Environmental Health Sciences or the National Institute of Health.

References

Anderson, R. 2003. *A Call for Systemic Change, Plenary Lecture at the 3rd National Conference on Science, Policy and the Environment: Education for a Sustainable and Secure Future*. Washington, DC, www.ncseonline.org

Anastas, P., and J. Warner. 1998. *Green Chemistry: Theory and Practice*. Oxford University Press, New York

Braun J., R. Kahn, T. Froehlich, P. Auinger, and B. Lanphear. 2006. "Exposures to environmental toxicants and attention deficit hyperactivity disorder in U.S. children." Environ Health Perspect 114(12): 1904–1909

Bullard, R., P. Mohai, R. Saha, and B. Wright. 2007. Toxic Waste and Race at Twenty 1987–2007, *Grassroots Struggles to Dismantle Environmental Racism in the United States*. Justice and Witness Ministries, United Church of Christ, Cleveland, OH

Calafat, A. M., Z. Kuklenyik, J. A. Reidy, S. P. Caudill, J. S. Tully, and L. L. Needham. 2007. "Serum Concentrations of 11 Polyfluoroalky Compounds in the U.S. Population: Data from the National Health and Nutrition Examination Survey (NHANES) 1999–2000." Environ Sci Technol 41(7): 2237–2242

Carrizo, D., J. O. Grimalt, N. Ribas-Fito, J. Sunyer, M. Torrent. 2007. "Influence of Breastfeeding in the Accumulation of Polybromodiphenyl Ethers during the First Years of Child Growth." Environ Sci Technol 41: 4907–4912.

Centers for Disease Control and Prevention. 2005. *Third National Report on Human Exposure to Environmental Chemicals*. National Center for Environmental Health. Centers for Disease Control and Prevention, Atlanta, GA. Available: http://www.cdc.gov/exposurereport

Chemistry for a Sustainable Future. 2006. Report of NSF Workshop on Sustainability and Chemistry, May 30- June 1, 2006, Arlington, VA, www.chem.uiowa.edu/research/sustainability/index.html

Cortese, A. D. 2003. *The Critical Role of Higher Education in Creating A Sustainable Future*. Planning for Higher Education. The Society for College and University Planning, Ann Arbor, MI

Deep South Center for Environmental Justice. 2007. http://dsccej.org/

Edgerton, E., B. Hartsell, and J. Jansen. 2006. "Mercury Speciation in Coal-fired Power Plant Plumes Observed at Three Surface Sites in the Southeastern U.S." Environ Sci Technol 40: 4563–4570

Frumkin, H., L. Frank, and R. Jackson. 2004. Urban Sprawl and Public Health. Island Press, Washington, DC

Global Environmental Management Initiative. 2007. http://www.gemi.org/

Grassian, V. H. 2007. "Chemistry for a Sustainable Future." Environ Sci Technol 41: 4840–4846

Green Media Toolshed. 2007. Scorecard: The Pollution Information Site. http://www.scorecard.org/

Institute of Medicine. 1999. *Toward Environmental Justice: Research, Education and Health Policy Needs*. The National Academy Press, Washington, D.C.

Institute of Medicine. 2002. *Unequal Treatment: Confronting Racial and Ethnic Disparities in Health Care*. The National Academy Press, Washington, DC

Krasner, S. W., H. S. Weinberg, S. D. Richardson, S. J. Pastor, R. Chinn, M. J. Sclimenti, G. D. Onstad, and A. D. Thruston Jr. 2006. "Occurrence of a New Generation of Disinfection Byproduct."Environ Sci Technol 40: 7175–7185

Leng, G., W. Gries, and S. Selim. 2006. "Biomarker of Pyrethrum Exposure." Toxicol Lett 162: 195–201

Lowell Center for Sustainable Production. 2007. http://www.sustainableproduction.org/

Lu, C. D. B. Barr, M. Pearson, S. Bartell, and R. Bravo. 2006. "A Longitudinal Approach to Assessing Urban and Suburban Children's Exposure to Pyrethroid Pesticides." Environ Health Perspect 114: 1419–1423

Mitchell, C. S., J. Zhang, T. Sigsgaard, M. Jantunen, P. Lioy, R. Samson, and M. H. Karol. 2007. "Current State of the Science: Health Effects and Indoor Environmental Quality." Environ Health Perspect 115: 958–964

NIEHS 2005. http://www.niehs.nih.gov

O'Hara, P. 2006. *Social Inclusion Health Indicators: A Framework for Addressing the Social Determinants of Health*. Edmonton Social Planning Council, Edmonton AB TSH, 4H8, October 2006, http://www.inclusivecities.ca

Schiffries, C. M., and A. Brewster, (Eds.). 2004. *Water for a Sustainable and Secure Future*. A Report of the Fourth National Conference on Science, Policy and the Environment. National Council for Science and the Environment. Washington, DC

Sheppard, P. M., M. E. Northridge, S. Prakash, and G. Stover. April, 2002. *Preface:"Advancing Environmental Justice through Community-Based Participatory Research.*" Environ Health Perspect 110(Supplement 2), 139–140.

Streets, S. S., S. A. Henderson, A. D. Stoner, D. L. Carlson, M. F. Simcik, and D. Swackhamer. 2006. "Partitioning and Bioaccumulation of PBDEs and PCBs in Lake Michigan." Environ Sci Technol 40: 7263–7269

Tennessee Department of Environment and Conservation. 2000. *Environmental Justice in the State of Tennessee: A Strategic Plan for The Department of Environment and Conservation*. Tennessee Department of Environment and Conservation, Nashville, TN.

U.S. Census Bureau. 2007. http://quickfacts.census.gov/qfd/state/

U.S. Consumer Product Safety Commission. 2007. Guidance for Lead (Pb) in Consumer Products. http://www.cpsc.gov/busifo/leadguid.html

U.S. EPA. 2007a. Toxics Release Inventory (TRI) Program, 2005 TRI Public Data Release, (accessed 7/12/07) http://www.epa.gov/tri/tridata/tri05/index.htm

U.S. EPA. 2007b. Lead In Paint, Dust and Soils. Supplemental Notice of Proposed Rulemaking (accessed 9/9/07) http://www.epa.gov/lead/new.htm

Williams, D. R., C. C. Collins. 2001. "Racial Residential Segregation: A Fundamental Cause of Racial Disparities in Health." Public Health Reports 116: 404–416

World Commission on Environment and Development. 1987.*Our Common Future*. Oxford University Press, New York.

Wu, F., T. Biksey, and M. H. Karol. September 2007. "Can Mold Contamination of Homes Be Regulated? Lessons Learned from Radon and Lead Policies." Environ Sci Technol 41: 4861–4867

Environmental Ethics of Health Mission To Enugu Nigeria

Godfrey A. Uzochukwu, Mary E. Uzochukwu, and Ethelbert Odo

Abstract The Enugu USA Medical Mission is a humanitarian year round mission to provide free health care to the people of Enugu State of Nigeria who are underserved and less privileged. The purpose of the Enugu USA Medical Mission is to contribute to solutions that solve health issues. Toxic chemicals contaminate drinking waters, build up in human organs and bones, and become a time bomb that may trigger birth defects, physical disabilities, mental retardation, and other illnesses years after being consumed. After a rain event, raw sewage flows directly into creeks and streams and causes water pollution. A chemical leak produces poisonous gas cloud that pollutes the air. Illegal hazardous and toxic waste dumps are common in Enugu Nigeria. Empty plastic bottles are breeding grounds for mosquitoes.

Introduction

An environmental and health mission to Enugu Nigeria was completed in December 2005 by physicians, nurses, pharmacists, laboratory technicians, environmental health experts and dieticians who are licensed in the United States and Nigeria. Fourteen towns were visited in five days. Eleven thousand underserved and less privileged men, women and children were diagnosed and treated for malaria, worm infestations, arthritis, diabetes, infections, ulcer, high blood pressure, wound, cataract, glaucoma, burns, and general aches and pains. Adulterated medications and too many deaths are common. Poor economic conditions contribute to unaffordable medical care by residents.

G.A. Uzochukwu (✉)
Orient Home Care Services Inc., Grand Prairie TX, USA

G.A. Uzochukwu et al. (eds), *Proceedings of the 2007 National Conference on Environmental Science and Technology*, DOI 10.1007/978-0-387-88483-7_20,

Goals of the Health Mission

The goals of the EnuguUSA medical mission are to promote a healthier population, examine and treat residents who otherwise would not be able to get medical treatment., reduce disease morbidity/mortality rate and reduce the number of wake keepings in USA for people from Enugu.

Health Mission Activities and Tasks

The EnuguUSA Medical Mission activities include but not limited to: assessing and screening for high blood pressure, diabetes, pain and fever, arthritis, gastro-intestinal problems, dental care, wound care, eye care/prescription glasses, and pollution prevention education; promoting of preventive health care: assembling doctors, nurses and other paramedical professionals (pharmacists, medical technologists, etc.) practicing in the USA to travel to Enugu towns and distributing disease and pollution prevention pamphlets. The duration of the mission is four to seven days depending on available volunteers. The expected outcome of the medical mission include reduced disease morbidity and mortality rate in the state. Reduction in the number of people in pain and reduced cost of health care.

Towns Visited

Fifteen towns were visited on December 19–23, 2005 *Health data was based on history provided by patients, physician's documentation and impression. Further laboratory tests may be needed.*

Figure 1 below shows the number of people assessed, diagnosed and treated for arthritis, general aches, diabetes, hypertension, flu like symptoms, malaria symptoms, worm infestations, cataract, blurred vision, etc at Isi Ozalla, Agbani, Obuofia, Mpu, Ugbonabo, Obeagu-Awkunanaw, Opi, Ovoko, Emeora Neke, Ugwogo-Nike, Ezi-Ukehe, Umuabi, Amafia-Afia, Iwollo and Ndeaboh.

Environmental Pollution and Impacts on Human Health

> ..."In Achi, over 1600 citizens were either assessed, treated, or given medications. Prominent among the diseases are malaria and worm infestations that cause aches and pains in that order. ...." Ethelbert Odo, 2005 Mission Coordinator

The high incidence of malaria symptoms and worm infestations (Fig. 1) are linked to improper waste management and pollution. Standing waters and open containers are common in the towns visited. The standing waters and open containers are perfect breeding grounds for mosquitoes and worms. Mosquitoes transmit a deadly disease called malaria (Essig, 2007). Worm infestations are common because

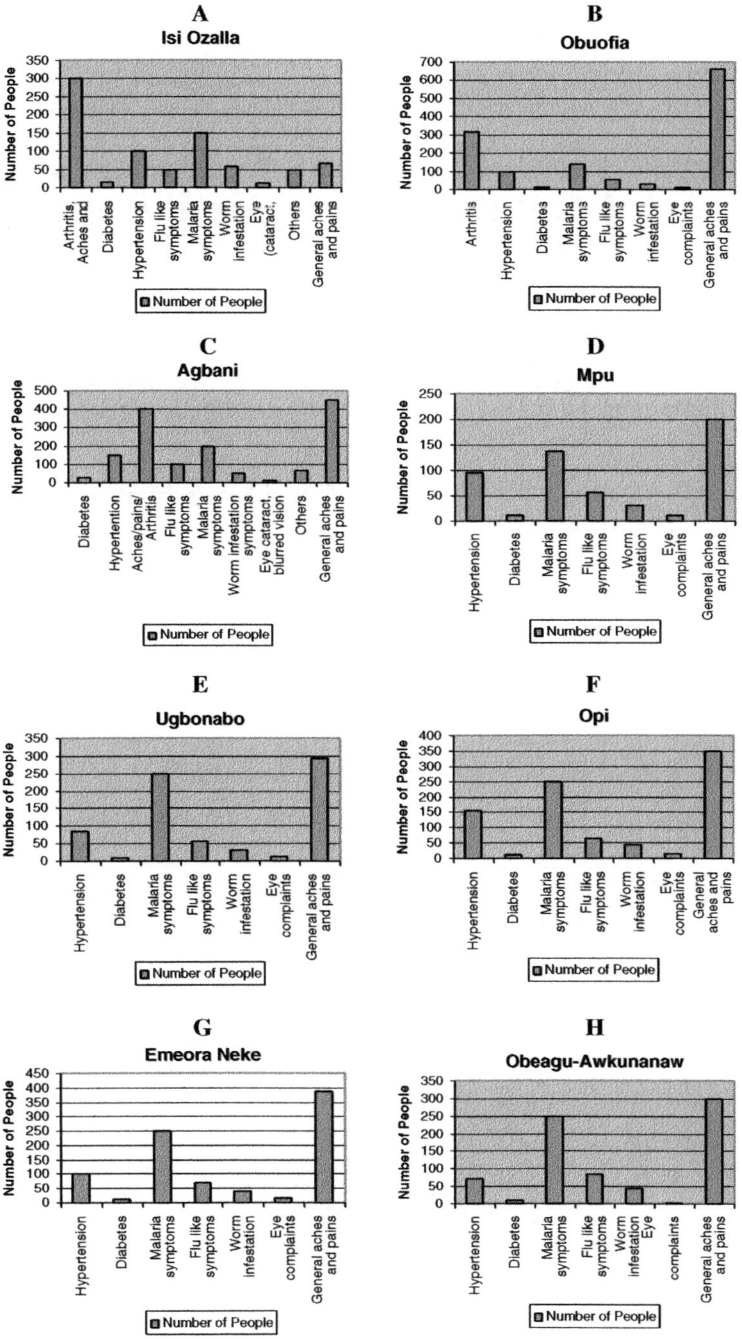

Fig. 1 (A–M) Number of people assessed, diagnosed and treated.

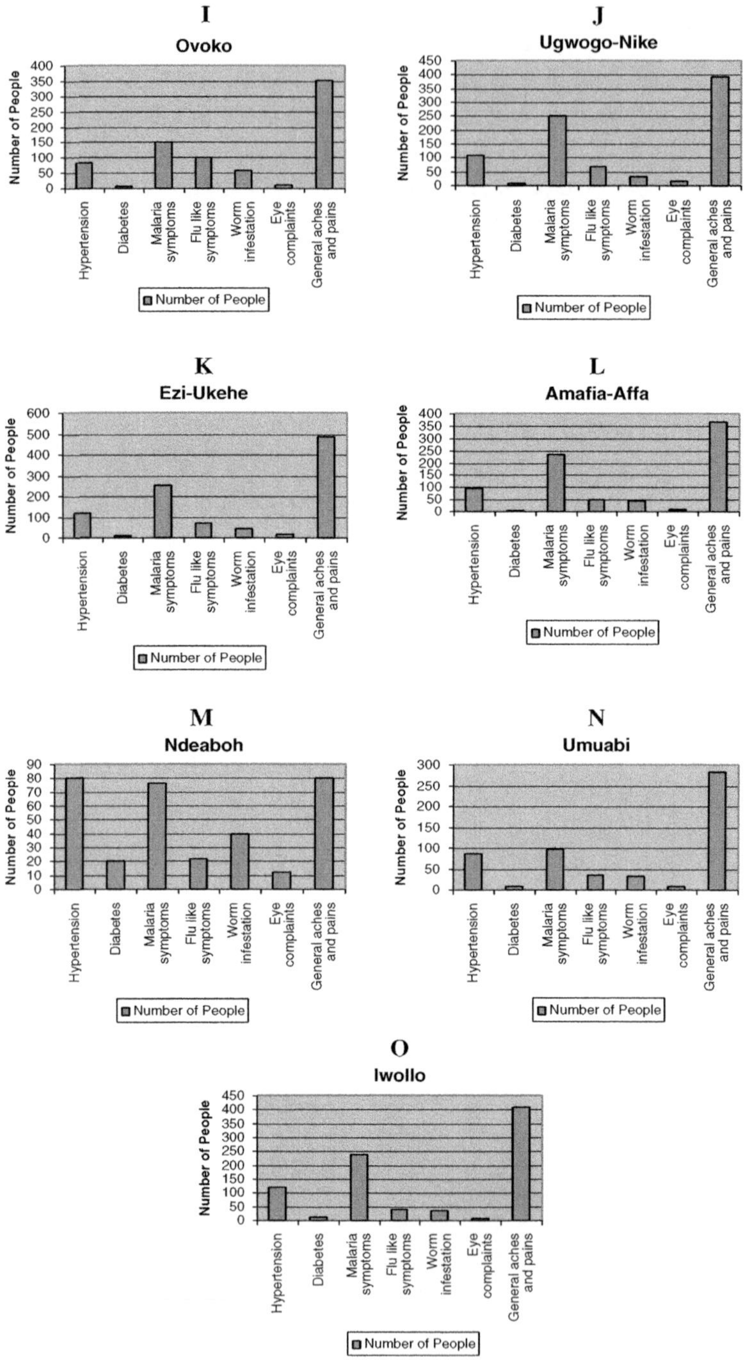

Fig. 1 (continued)

of contaminated drinking waters. Raw sewage flows directly into creeks and streams and causes water pollution after a rain event. The data presented suggest that malaria symptoms and worm infestations are linked to environmental pollution. Aches and pains are linked to malaria symptoms and worm infestations. A medical evaluation of disease symptoms is recommended. Water should not be allowed to accumulate in open containers and standing waters in concave areas and depressions should be drained. These actions will make it difficult for mosquitoes to breed. Other measures for malaria prevention are described by Croft, 2006. Methods of preventing worm infestations are described by Shanti-Village, 1998–2006 (www.shanti.com.au).

Environmental Ethics

Improper waste management is one of the main causes of environmental degradation and pollution in Enugu State. There are no regulations and proper waste disposal facilities for infectious and toxic wastes. Data presented suggest that residents of towns visited during the health mission are exposed to deadly worm infestations and mosquitoes that transmit malaria. The ethical challenge is how to balance the environmental needs versus walking away. Public involvement that emphasizes listening, learning and leading is needed to address the risks associated with pollution issues. Improving public knowledge of environmental and health risks associated with alternatives including the risk of doing nothing is a step in the right direction. Individuals, federal officials, state officials and local government officials have an ethical responsibility to do more to prevent pollution and improve the quality of life and protect the environment.

Conclusion

Environmental pollution has been linked to public health. Malaria symptoms and worm infestations are common in towns visited by the Enugu USA Medical Mission. A medical evaluation of residents to confirm high exposure to worm infestations and malaria symptoms is suggested. Public environmental education campaign to address pollution prevention and health issues is recommended.

References

Croft, A. Am, (2006), Malaria: Prevention in Travellers. Clinical Evidence. Volume 15: 1088–1106. Healthcare Assist Preferred Partner

Essig, M. G. (2007), Protecting Yourself from Mosquito Bites. Healthwise.

Odo, E. (2005), Personal Communication. Orient Home Care Services.

Shanti-Village, (1998–2006), www.shanti.com.au.

Part IV
Fate and Transport of Contaminants

Use of Kalman Filter and Particle Filter in a One Dimensional Leachate Transport Model

Shoou-Yuh Chang and Sikdar M.I. Latif

Abstract Modeling the behavior of pollutants during the flow of leachate through soil is important in predicting the fate of the pollutants. A one dimensional transport model with advection and dispersion is used as the deterministic model of benzene leachate transport from an industrial landfill. A Particle filtering with sequential importance re-sampling (SIR) filter and a classical Kalman filtering were used to predict the benzene plume transport. A traditional root mean square error (RMSE) is used to compare the effectiveness of the Kalman filtering and Particle filtering. The results show that Kalman filter performs better than Particle filter at the initial several time steps. Both Kalman filter and Particle filter can reduce up to 80% error in comparison to a conventional numerical approach.

Introduction

Benzene is common soil contaminants in leachate from industrial landfills. In the U.S. alone there are approximately 100,000 different sites which have benzene soil or groundwater contamination. Subsurface contaminant transport model plays an important role in explaining how a plume of contamination evolves, in evaluating the likely behavior of systems for contaminant remediation, in predicting how it will behave in the future, and in assessing the risk of contaminant properly.

For many water quality problems, filtering technique combined with traditional numerical approach can provide more cost-effective solutions where the construction of observation wells are very costly and some times impossible. Contaminant state estimation can be considered as an optimal filtering problem within a Bayesian framework. The most well-known Bayesian state-estimation algorithm is

---

S.-Y. Chang (✉)
North Carolina A&T State University, Greensboro, North Carolina 27411, USA

G.A. Uzochukwu et al. (eds), *Proceedings of the 2007 National Conference on Environmental Science and Technology*, DOI 10.1007/978-0-387-88483-7_21,

the Kalman filter which is good for linear systems with a Gaussian distribution. Particle filter is generally applied for non-linear systems with non Gaussian distribution. Since several unknown and random hydro geological factors and parameters are associated with the contaminant migration in the subsurface, the system behavior cannot be categorized in a specific linear or non-linear system. Therefore, the evaluation of both filtering techniques in association with a traditional deterministic numerical model is conducted in this study.

Methodology

The one-dimensional form of the advection-dispersion equation for benzene leaching in saturated, homogeneous, isotropic materials under steady-state uniform flow is described by the following partial differential equation:

$$\frac{\partial C}{\partial t} = \frac{D_y}{R}\frac{\partial^2 C}{\partial y^2} - \frac{V}{R}\frac{\partial C}{\partial y} \tag{1}$$

Where, C = Solute concentration, mgl^{-3},

t = Time, s,
y = Cartesian coordinate direction along the flow line, m,
D_y = Dispersion coefficients in y direction, m^2s^{-1},
V = Linear velocity of flow field in the y direction, ms^{-1},
R = Retardation factor, dimensionless.

The one-dimensional subsurface transport equation can be solved numerically by forward time centered space (FTCS) difference method. Assuming 10 discrete nodes, the state-space model becomes

$$x_{t+1} = A_t x_t \tag{2}$$

Where x_{t+1} and x_t are the state variables described as vectors of contaminant concentration at all nodes in the problem domain at time (t+1)☆t and t☆t, respectively; $A_t$is the state transition matrix which gives a finite difference scheme to improve next step concentrations.

One dimensional leachate contaminant plume can be simulated assuming the system as stochastic with uncertain sources of errors. Process or system equation can then be expressed as,

$$x_{t+1} = A_t x_t + p_t,\ t = 0, 1, 2, 3, 4, \ldots\ldots \tag{3}$$

Here, P_t is the model system error or process noise. The error vector P_t is assumed as white noise having a covariance matrix Q_t in this case. The model error P_t is taken from a Gaussian distribution which has zero mean and standard deviation 10 mg/L.

On the other hand, observation or measurement equation can be expressed as,

$$z_t = Hx_t^T + O_t, \; t = 0, 1, 2, 3, 4, \ldots\ldots\ldots \tag{4}$$

Observation Z_t is a vector having the observed values of state of all nodes at time step t. Superscript T over state vector x_t denotes the true value of the state. Observation Z_t can be simulated by data pattern matrix H and observation error O_t. If we have the same data at all the nodes we are dealing with, H will be a $N \times N$ identity matrix, where N is the total number of nodes in the model domain. The observation vector O_t is constructed by a vector having the elements from a Gaussian distribution having a mean of zero and a standard deviation of 5 mg/L. The error vector O_t is assumed to have a covariance matrix R_t.

Figure 1 illustrates the sequential state estimation by Kalman filtering. There are two major steps in Kalman filtering operations: prediction and correction or update. With the initial state the filter estimates the Kalman gain. Then it updates the optimal error covariance matrix before going to the prediction step. The cycle goes further for the estimation of the next state.

The basic idea of Particle filtering is to approximate the distribution, $p(x_t|z_{1:t})$ using a set of random samples called particles. Let $z_{1:t} = (z_i, \, i = 1, \ldots\ldots\ldots\ldots, t)$ be the measurement sequence and assuming the prior distribution $p(x_o)$ is known, the posterior probability can be obtained sequentially by prediction and update. Particle filters approximate the posterior probability distribution, $p(x_t|z_{1:t})$ by a set of supporting random samples x_t^i, $i = 1\ldots\ldots\ldots N$, with associated weights, $w^i{}_t$:

$$p(x_t|z_{1:t}) \approx \sum_{i=1}^{N} w_t^i \delta(x_t - x_t^i) \tag{5}$$

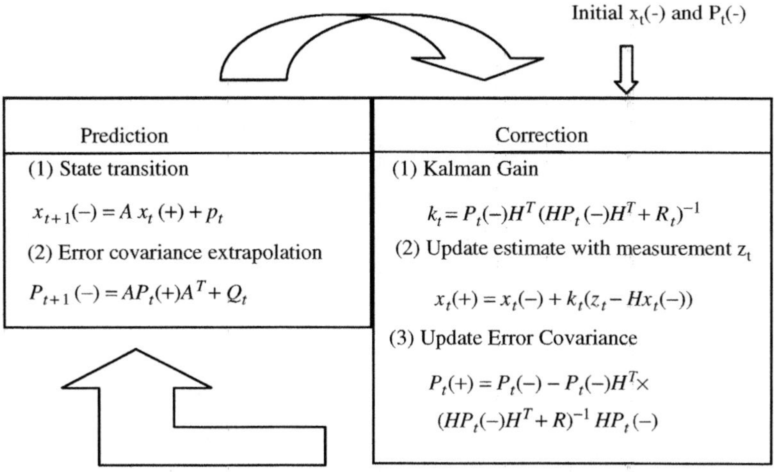

Fig. 1 Recursive operation of Kalman filtering

Where $\delta(x)$ is an indicator function known as Dirac's delta function. The weights sum to zero. The filtered state is taken as the mean of posterior density. The next issue is how to decide the weights of the particles. Since $p(x_t|z_{1:t})$ is not in the form of traditional probability density function, weights can not be assigned by direct sampling. The weights are decided using *importance sampling.*

With these particles and associated weights, the mean of the states can be approximated by $\overline{x_t} = \sum_{i=1}^{N} w_t^i x_t^i$. Re-sampling can be used to eliminate those particles with small weights thereby focusing on particles with large weights. Re-sampling generates set $\langle x_k^i, i = 1..........N \rangle$with Pr $(x_t^j = x_t^i) = w_t^i$. Here j is the particle index after resampling. The parent relationship is denoted, parent$(j) = i$. A particle filter with this importance density and re-sampling step is called a sequential importance re-sampling (SIR) filter.

In this study, an one dimensional leaching of benzene from an industrial landfill site is simulated by Kalman filtering and Particle filtering data assimilation scheme. Figure 2 shows that vertically 10 nodes with 0.02 m spacing are taken in model space domain. Each time interval is taken as 0.01 day.

In this model, the benzene contaminant mass at the source, M_o is 16.5 gm at node 1 (y = 0) and t = 0 in a column with a cross-sectional area of 0.01 m^2. The longitudinal dispersivity, α is 1meter. A water flux of 0.2 m^3/d is maintained continuously at y = 0. The porosity, η of the leaching soil medium is 0.25 which gives tortuosity of porous medium, $\Gamma = 0.63$. Flow velocity, V is calculated as 0.8 m/day vertically downward. Molecular diffusion D_m = 1.44E-06 m^2/day. It gives the dispersion of the contaminant, D_y = 0.008001 m^2/day. Fraction organic carbon = 0.0001, soil solid specific gravity = 2.65, organic carbon/water partition coefficient = 83 L/Kg,

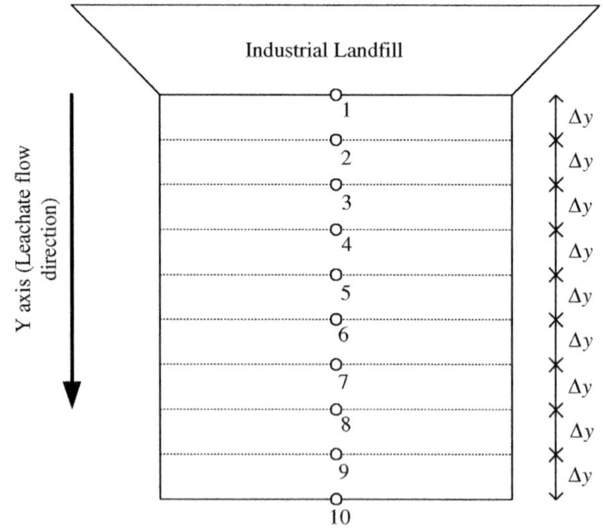

Fig. 2 Schematic diagram of benzene leachate flow showing 10 discrete nodes

bulk density = 1.99 g/cm^2, and retardation factor = 1.1 are used in the study. For industrial landfill leachate, typical concentration of benzene is about 2000 mg/L. Initial concentration, C_o is calculated as about 1985 mg/L for pulse or instantaneous source which is injected at the top node.

Results and Discussion

The contaminant state for a fixed position is represented in Fig. 3. The profile constructed by the FTCS numerical procedure gives a smooth curve representing the theoretical deterministic behavior of the solution process. On the other hand, concentration profiles by filtering procedure give irregular shapes. These shapes represent the true field behavior. Due to the randomness and uncertainties in the true field, the true field concentration profile generally gives oscillated curves rather than smooth ones. The peak concentration for FTCS prediction occurs one time step after the true value. The magnitude differs by 250 mg/L. Profiles with filtering fit well with the true value resulting negligible differences. By this way, filtering provides better quantitative results than FTCS predictions.

The experiment is designed and constructed in such a way that every time the program is run, different random numbers are generated to simulate the random behavior of the subsurface transport plume. The RMSE (root-mean-square error) prediction curve does not vary too much and the concentration profile is similar for different runs. The RMSE prediction is evaluated with one time run. At time step 5,

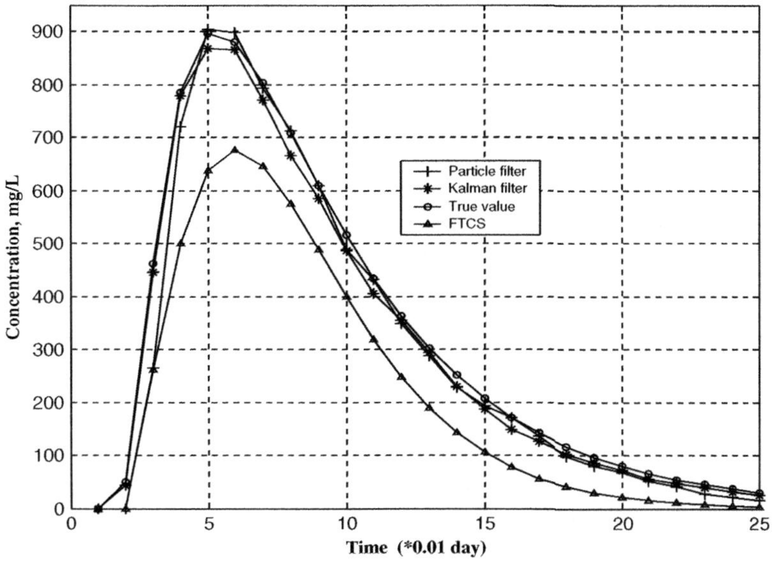

Fig. 3 Benzene concentration profile at node 3

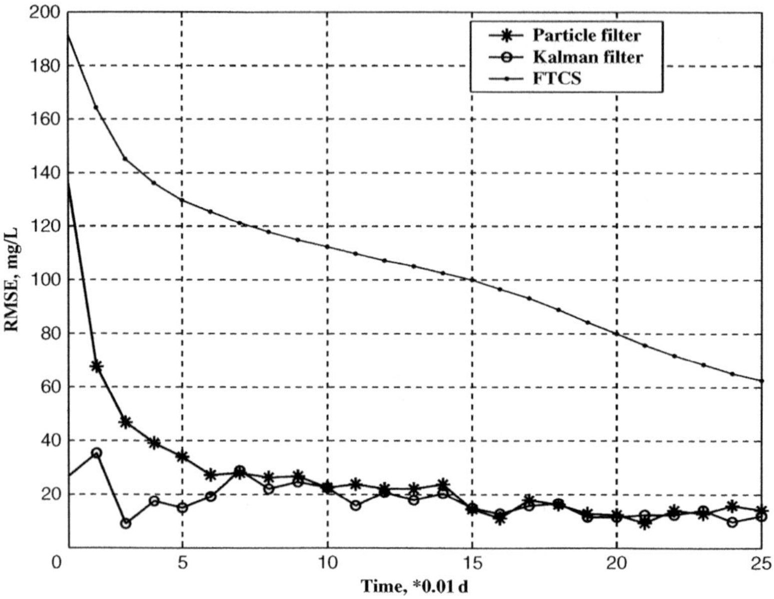

Fig. 4 Prediction error (root-mean-square error) for data assimilation with deterministic model (FTCS), Kalman filtering and Particle filtering

the estimated RMSE of numerical solution is 7.5 times greater than that of Kalman filtering prediction and 3.5 times greater than that of Particle filtering prediction. As shown in Fig. 4, RMSE value for Particle filter prediction gradually decreases with the increase of time steps. The RMSE vs. time step graph for Particle filter prediction takes a hyperbolic shape expressing the perfect merging with true value. Such shape signifies the successful data assimilation. Particle filter prediction reduces 30% to 80% error from time step 1 to time step 22, respectively. On the other hand, RMSE for Kalman filter prediction varies from 40 to 4 mg/L. Kalman filter reduces 70–90% error. That means the range of error reduction in case of Kalman filter varies within 20% whereas error reduction of Particle filter varies within 50%. Up to time step six, the prediction error of Kalman filter is 30–50% less than that of Particle filter.

Conclusion

True value of state takes place in a space for only one time due to randomness of the field behavior. It is impossible to relate all hydro-geological parameters mathematically in a deterministic model so that it can give the true solution. Kalman filtering and Particle filtering can provide more realistic solutions as is illustrated in this study. Both Kalman filter and Particle filter can reduce up to 80% error in comparison to conventional numerical approach. The approach can also be used in a two and three-dimensional groundwater model.

References

Arulampalam, S., Maskell, S., Gordon, N., and Clapp, T. 2002. "A tutorial on particle filters for on-line non-linear/non-gaussian Bayesian Tracking." *IEEE Transactions on Signal Processing*, 50, 174–188.

Chen, T., Morris, J., and Martin, E. 2005. "Particle Filters for state and parameter estimation in batch processes." *Journal of Process Control*, 15, 665–673.

Chang, S.Y., and Jin, A. 2005. "Kalman filtering with regional noise to improve accuracy of contaminant transport models." *Journal of Environmental Engineering*, 131, 971–982.

Organic Pollutant Transport in Groundwater Using Particle Filter

Shoou-Yuh Chang and Xiaopeng Li

Abstract In this study, the Particle Filter approach was introduced for the estimation of BOD concentration and its first-order decay constant. A two-dimensional subsurface transport model with decay factor was developed. Random Gaussian errors were added to the numerical method result to simulate the observation data. Then, the Particle Filter recursive process was applied for the data assimilation. The performance consistency of the Particle Filter approach was evaluated by the root mean squared error (RMSE) for 20 runs. Furthermore, two sensitivity analyses were done to show the effects of observation error and the true value of k on the estimation. The results show that the Particle Filter approach can give accurate estimations of the BOD concentration and the first-order decay constant and that its performance is consistent for 20 runs.

Introduction

Particle Filter, a type of Bayesian state space model (Doucet et al., 2000), was used for both state and parameter estimation processes in various engineering areas. However, there is limited application of the Particle Filter approach in environmental engineering and water quality modeling. During the last decade, the Particle Filter has been introduced to state estimation in nonlinear systems and successfully applied in a number of different areas such as electronic engineering, ecosystem modeling, structure modeling, and chemical process modeling. Ching et al. (2006) applied the Particle Filter approach for state and parameter estimation in nonlinear structure model. Chen et al. (2005) applied the Particle Filter to the highly nonlinear batch chemical process. Yu and Cheng (2005) developed the Particle Filter model for mobility tracking. Kivman (2003) and Svetlana et al. (2003) used the Particle Filter approach for data assimilation in high-dimensional nonlinear ocean models.

S.-Y. Chang (✉)
North Carolina A&T State University, Greensboro, North Carolina 27411, USA

G.A. Uzochukwu et al. (eds), *Proceedings of the 2007 National Conference on Environmental Science and Technology*, DOI 10.1007/978-0-387-88483-7_22,

Particle Filter is the extension to point-mass filters. The basic idea is that a large number of samples (particles) are adaptively concentrated in regions of high probability. Because of the limited application of the Particle Filter approach in water quality modeling, the purpose of this study is to investigate the performance of the Particle Filter approach when it is applied to a 2-dimensional subsurface organic contaminant transport model for both BOD concentration (state) and first-order decay rate (parameter) estimation.

Methodology

The transporting of biodegradable organic contaminant in a two-dimensional uniform saturated groundwater flow field can be described as the following equation:

$$\frac{\partial C}{\partial t} = \frac{D_x}{R}\frac{\partial^2 C}{\partial x^2} + \frac{D_y}{R}\frac{\partial^2 C}{\partial y^2} - \frac{V}{R}\frac{\partial C}{\partial x} - \frac{kC}{R} \tag{1}$$

where C = Solute concentration, mgl^{-3}; t = Time, s; x, y = Cartesian coordinates, m; D_x, D_y = Dispersion coefficients in x and y direction respectively, m^{2}s^{-1}; V = Linear velocity of flow field in the x direction, ms^{-1}; R = Retardation factor, dimensionless; and k = First-order degradation rate, d^{-1}.

Chang and Jin (2005) applied the Forward-Time and Central-Space (FTCS) differencing scheme to develop the state-space form of system equation for a two-dimensional transport problem. The same numerical method was used in this study for system equation development. Due to the biodegradable contaminant considered in this study, the decay factor k was included in the FTCS scheme. In this study, the Gaussian errors with $\sigma_{sys} = 5\%$ were chosen for indicating the uncertainty in system equation. After defining the errors associated with state variables, the system equation can be written as:

$$C(i, j, t+1) = M(k)C(i, j, t) + v_t \cdot I \tag{2}$$

where $C(i, j, t)$ is the state variable matrix, containing contaminant concentrations at all nodes at time t; $C(i, j, t+1)$ is the state variable matrix, containing contaminant concentrations at all nodes at time t + 1; $M(k)$ is the state transition matrix, containing the degradation decay rate k. Note because k is not a known parameter but a variable in this model, $M\,(k)\,C\,(i,\ j.t)$ becomes a nonlinear function of k and$C(t)$. v_t is the Gaussian error with $\sigma_{sys} = 5\%$; and I is the identity matrix.

When actual field data are used as observation in the Bayesian state space model, the observation error could be abstracted from the field data directly. In this study, we used the numerical random scheme to create the "observation data" as Chang and Jin (2005) suggested. The random errors, from Gaussian noise with $\sigma_{obs} = 5\%$, were imposed to the numerical result to get the observation equation.

$$z(i, j, t+1) = C(i, j, t) + \varepsilon_t \cdot I \tag{3}$$

Where $z(i, j, t+1)$ is the observation in time step t+1; $C(i, j, t)$ is the state variable in time step t; ε_t is the Gaussian error with $\sigma_{obs} = 2.5\%$; and I is the identity matrix.

Particle Filter is the extension to point-mass filters. The basic idea of particle filters is to approximate $p(c_t | z_{1:t})$ using a set of random samples (also called particles) $\{c_t^i, i = 1....N\}$ with associated weights $\{w_t^i, i = 1....N\}$ where $\sum_{i=1}^{N} w_t^i = 1$ (Doucet et al., 2000):

$$p(c_t | z_{1:t}) \approx \sum_{i=1}^{N} w_t^i \delta\left(c_t - c_t^i\right) \tag{4}$$

where $\delta(x)$ is an indicator function which is equal to unity if $x = 0$; otherwise it is zero. The location and weight of each particle reflect the value of the density in that region of the state space. With these particles and associated weights, the estimated state vector, $\hat{C}_t$, is the mean of $p(c_t | z_{1:t})$ and is calculated as

$$\hat{c}_t = \sum_{i=1}^{N} w_k^i c_i^i \tag{5}$$

Then the Particle Filter updates the particle locations and the corresponding weights recursively with each new observation. Resampling step was introduced to avoid the divergence problem (Gordon et al., 1993).

If both states and model parameters are to be estimated, Baye's rule gives the following joint posterior distribution (Chen et al., 2005):

$$p(C_t, k_t | z_{1:t}) \propto p(z_t | C_t, k_t)\, p(C_t | k_t, z_{1:t-1})\, p(k_t | z_{1:t-1}) \tag{6}$$

where k_t is the vector of the model parameter.

Joint state and parameter estimation is achieved through the augmentation of the state space with the parameter vector. A Gaussian random walk for the parameters can then be specified to enable their adaptation to new data:

$$k_t = k_{t-1} + \zeta_{t-1} \tag{7}$$

where $\zeta_t \sim G(0, W_t)$ satisfies a Gaussian distribution with zero mean and predefined covariance matrix W_t. Then, $p(k_t | z_{1:t-1})$ is approximated by

$$p(k_t | z_{1:t-1}) \approx \sum_{i=1}^{N} w_{t-1}^i G\left(k_t | k_{t-1}^i, W_t\right) \tag{8}$$

The effectiveness of the Particle Filter can be shown by comparing the estimated concentration with the simulated observation for each time step. The root mean squared error (RMSE) is used to show the difference between these two data sets.

Results and Discussion

The model grid Gm is defined on a two-dimensional plane domain (Fig. 2). The two-dimensional plane has 400 grid points with grid interval dx = 1.524 m and dy = 1.524 m. The set of parameters suggested by Zou an Parr (1995) were adopted in this study. The horizontal dispersion Dx is assumed as 1.554 m^2/day, Dy as 0.4662 m^2/day, porosity as 0.30, velocity as 1.054 m/day and aquifer thickness as 6.1 m. In the numerical scheme, the time interval per each time step is 0.20 day and the total time step is 50. The total number of grid points in the two-dimensional plane is 400. For this study, the first order decay of the contaminant was included into the modeling as an unknown parameter which will be estimated through the data assimilation. An instantaneous contaminant with an initial BOD concentration of 10000 mg/l was injected into the grid point with coordinates (5, 10). The Particle filter predictions and simulated observation data at time step 50 are shown in dotted lines in Fig. 1. The estimated concentration contours are similar to observation data contours indicating the effectiveness of the Particle filter. The estimation for the

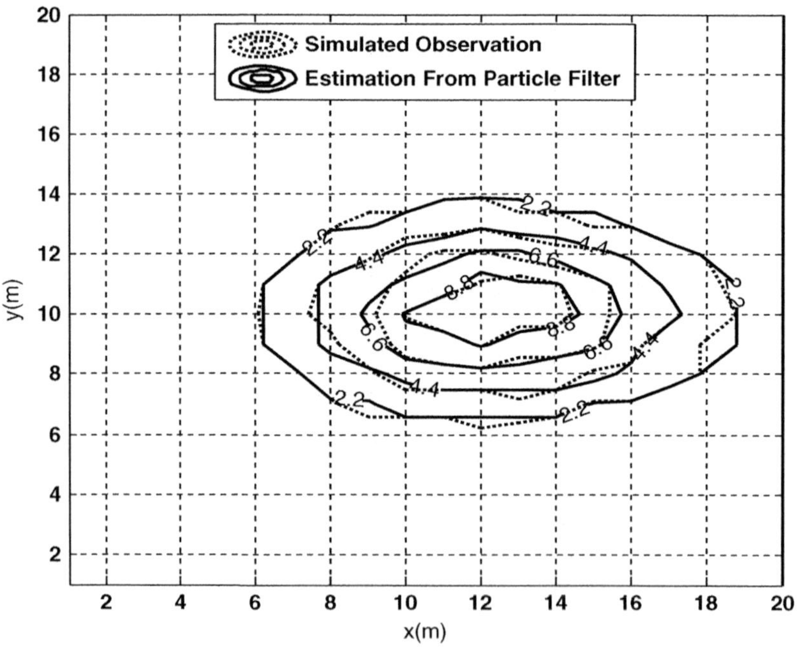

Fig. 1 Comparison of simulated observation and Particle Filter estimation at time step=50, mg/l

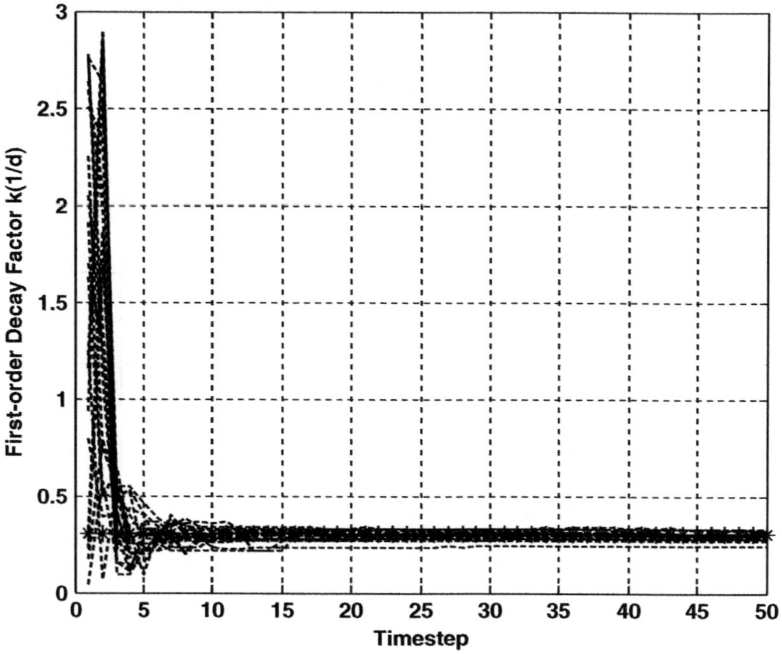

Fig. 2 Estimation of the decay factor k, based on 20 runs,d^{-1}

decay factor k as shown in Fig. 2 indicates that the first guess value for parameter k is far away from true value 0.3. The estimation, however, converges to its true value as the time goes on. Note that Fig. 2 records the parameter estimation for 20 different runs.

To evaluate the performance, the programming was run for 20 times and the RMSE for each run was calculated. Figure 3 shows that for most runs, the difference between estimated BOD concentration and the observation data begins with a larger number and keeps going down through the time steps and finally gets close to zero. But some small fluctuations in RMSE within the first 13 time steps can also be observed. The change in the RMSE indicates that as the assimilation is processed longer, the estimated value for the concentration is getting closer to the observation data, which results in the smaller RMSE along with time. All 20 runs show the clear RMSE decreasing trend only after 13 time steps. It takes the system 11 time steps to reduce RMSE to below 5 mg/l.

Conclusion

Particle Filter approach was used in a two-dimensional subsurface transport model with a decay factor to predict the evolution of the BOD contaminant plume and first-order decay factor. The effectiveness of the proposed method was then evaluated by

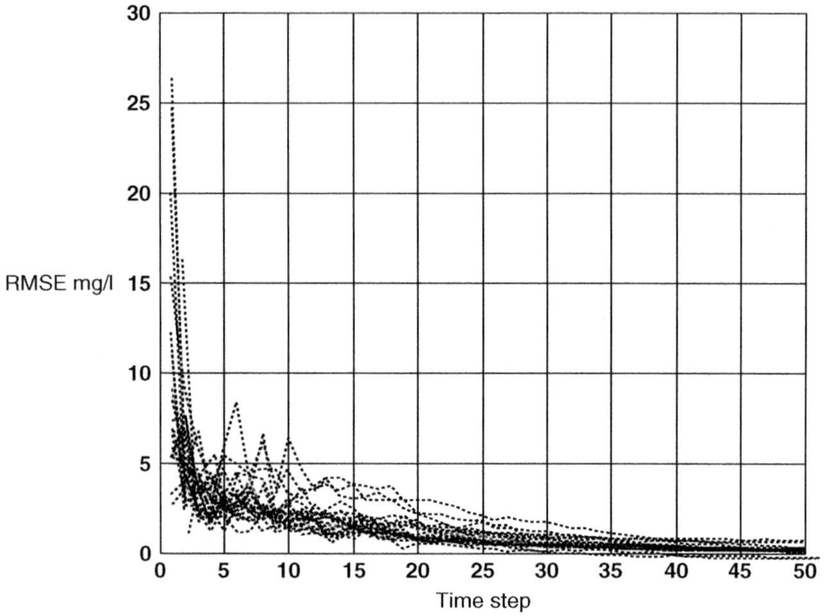

Fig. 3 The difference between simulated observations and Particle Filter estimation based on 20 runs, mg/l

root mean squared error (RMSE). The results show that the curve of RMSE has fluctuations in the beginning of the estimation, but the decreasing trend shows up after 13 time steps when the system gets more "correct" information from observations. The length of the "chaos" time is related on the amount of observation data which the system can "learn" from. With less observation data in assimilation, it will take the system more time steps to show the clear RMSE decreasing trend. These results indicate that the Particle Filter can perform consistently with random numbers processed within the method. The effect of observation error on the estimation was studied by using different observation errors in data assimilation. The results show that assimilation with more reliable observation values will deliver a better estimation. Furthermore, the effect of the decay factor on the estimation was investigated by using different true values of decay factor in observation data simulation. The results indicate that RMSE is more sensitive to k when the true value of k is smaller than 0.10 d^{-1}.

Acknowledgment This work was sponsored by the Department of Energy's Samuel Massie Chair of Excellence Program under grant number DE-FG01-94EW11425. The views and conclusions contained herein are those of the authors and should not be interpreted as necessarily representing the official policies or endorsements, either expressed or implied, of the funding agencies.

References

Chang, S.Y. and Jin, A. (2005). "Kalman filtering with regional noise to improve accuracy of contaminant transport models." *Journal of Environmental Engineering*, 131, 971–982

Chen, T., Morris, J. and Martin, E. (2005). "Particle filters for state and parameter estimation in batch processes." *Journal of Process Control*, 15, 665–673

Ching, J., Beck, J.L. and Porter, K.A. (2006). "Bayesian state and parameter estimation of uncertain dynamical systems." *Probabilistic Engineering Mechanics*, 21(1), 81–96

Doucet, A., Godsill, S. and Andrieu, C. (2000). "On sequential Monte Carolo sampling methods for Bayesian filtering." *Statistics and Computing*, 10, 197–208;

Gordon, N.J., Salmond D.J. and Smith A.F.M. (1993). "Novel approach to nonlinear/non-Gaussian Bayesian state estimation." IEE Proceeding, 140(2), 107–113

Kivman, G.A. (2003). "Sequential parameter estimation for stochastic systems." *Nonlinear Processes in Geophysics*, 10(3), 253–259;

Svetlana, N.L., Kivman, G.A., Schroter, J. and Wenzel, M. (2003). "Sequential weak constraint parameter estimation in an ecosystem model." *Journal of Marine Systems*, 43, 31–49

Yu, Y. and Cheng, Q. (2005). "Particle Filters for maneuvering target tracking problem." *Signal Processing*, 8, 568–580

Zou, S. and Parr, A. (1995). "Optimal estimation of two-dimensional contaminant transport." *Ground Water*, 33(2), 319–325

Modeling the Sorption Behavior of Aromatic Amine Onto Sediment Under Acidic Conditions

Shihua Chen and Marianne C. Nyman

Abstract In this study the sorption behavior of an aromatic amine (benzidine) in the aqueous-sediment systems was investigated. Three sorption mechanisms including hydrophobic partitioning, covalent binding and cation exchange govern the fate of benzidine after its entry into the ecosystems. A sorption equilibrium model was presented to investigate these sorption mechanisms under acidic conditions. Model simulation results indicated that the model fits the experimental data very well. The results also indicated that hydrophobic partitioning and covalent binding contributed more to the overall sorption than cation exchange did with increasing solution pH from 3.1 to near neutral.

Introduction

Aromatic amines are mainly the end products resulting from the reductive processes of nitroaromatics, azo dyes, dinitroherbicides and hydrolytic degradation of various agrochemicals (Weber et al., 2001; Colón et al., 2002). The exposure to aromatic amines poses serious threats to the environment and ecosystems due to their toxic natures (U.S. EPA, 1999). It becomes of great importance and interest to understand how these aromatic amines react with the natural surfaces in the environment. Due to the complex constituents, heterogeneous properties of sediments and pore water (such as mineralogy, cation exchange capacity, grain size distribution, pH, ionic strength, content of organic matter), as well as the aromatic amines containing a specific functional group (e.g., $-NH_2$), at least three mechanisms that occur in the sediment-water systems play important roles during the sorption processes. These three mechanisms include: (I) electrostatic interactions of the cationic organic compounds with negatively charged sediment or soil surfaces (i.e., cation

S. Chen (✉)
Department of Civil and Environmental Engineering, MRC 319, Rensselaer Polytechnic Institute, Troy, NY 12180, USA

G.A. Uzochukwu et al. (eds), *Proceedings of the 2007 National Conference on Environmental Science and Technology*, DOI 10.1007/978-0-387-88483-7_23,

exchange) (Zachara et al., 1986; Brownwell et al., 1990); (II) partitioning of the neutral hydrophobic organic compounds into the sediment or soil organic matter (SOM) and minerals (Spurlock and Biggar, 1994; Huang et al., 1997; Xia and Ball, 1999); and (III) covalent binding of amino functional groups to carbonyl moieties present in soil or sediment matrices (Parris, 1980; Ononye et al., 1989; Thorn et al., 1996; Weber et al., 1996; Weber et al., 2001).

Benzidine (4,4′-diamino-1,1′-biphenyl) was selected from the aromatic amine family as the model compound in this study. Benzidine is not soluble in water, with a water solubility of 520 mg/L at 25°C. It has two p*K*a values of 3.3 and 4.3. Benzidine is a highly toxic compound, and has been identified as a carcinogen by U.S. EPA. To date, only a few studies of benzidine in water-sediment systems can be found in the literature. The objectives of this study were to investigate the sorption mechanisms that govern the fate of benzidine in the water-sediment systems, and to determine the relative significance of each sorption mechanism by some modeling work.

Materials and Methods

Sediments. Sediment samples were obtained from Lake Macatawa located in Holland, MI, USA. Prior to use, the sediment samples were dried in an oven at 100 ± 5°C for 48 h, broken up by a mortar and a pestle, and further homogenized to small grains with a grinder. The OC of sediment sample was 11.2%, and the sediment type was found to be silty-clay with 73% (by weight) of the grain size less than 150 μm.

Batch Sorption Experiment. The various pH conditions of DI-sediment systems were obtained by using different ratios of citric acid to sodium citrate. The buffer components and their corresponding concentrations were as follows: pH 2.8 (0.10 M citric acid/0.01 M sodium citrate), pH 3.2 (0.08 M citric acid/0.04 M sodium citrate), pH 4.6 (0.04 M citric acid/0.07 M sodium citrate), and pH 5.9 (0.01 M citric acid/0.10 M sodium citrate). Benzidine was dissolved in these buffered solutions to make benzidine stock solutions. An aliquot of 5 mL of the buffered benzidine solution was spiked to corresponding tubes containing 0.15 g sediments. The tubes were rotated on an end-over-end rotator at 45 rpm for 30 days and centrifuged at 1350 g for 25 min. After centrifugation, supernatants were removed for HPLC analysis as well as the pH measurement.

Benzidine concentrations were measured with a Shimadzu HPLC equipped with a Shimadzu Premier C18 reversed-phase column (150 × 4.6 mm; 5 μm) and a UV-VIS detector at 263 nm. Mobile phases consisted of acetonitrile and sodium acetate buffer solution (pH range: 4.7–4.9) (1:3 v/v) at a flow rate of 1.0 mL/min.

Sorption Equilibrium Model. The sorption equilibrium model consists of three components: linear hydrophobic partitioning, Langmuir-type covalent binding and quadratic form of cation exchange, as shown in Equation (1).

$$q = q_l + q_{\text{cov}} + q_{ex} = K_d[B] + \frac{S_0 K_{\text{cov}}[B]}{1 + K_{\text{cov}}[B]} + \left(-\frac{K_{ex}}{2D_0}[BH^+] + \sqrt{(\frac{K_{ex}}{2D_0}[BH^+])^2 + \frac{K_{ex}}{D_0}[BH^+]CEC} \right) \quad (1)$$

where q, q_l, q_{cov} and q_{ex} are the totally sorbed concentration, and sorbed concentrations by hydrophobic partitioning, covalent binding and cation exchange (μmol/kg), respectively, K_d is the linear partitioning coefficient (L/kg), [B] is the concentration of neutral benzidine in the solution (μmol/L), K_{cov} is the equilibrium constant (L/μmol), S_0 is the available covalent binding sites for benzidine (μmol/kg), K_{ex} is the exchange equilibrium constant, $[BH^+]$ is the overall cationic benzidine species in the solution (μmol/L), CEC is cation exchange capacity (μmol(-)/kg) of the sediment, and D_0 is the sediment concentration (kg/L), respectively.

Based on the pH-p*K*a relationships, the overall cationic benzidine species in the solution $[BH^+]$ is shown in Equation (2).

$$[BH^+] = \frac{2 + K_{a1} / [H^+]}{1 + K_{a1} / [H^+] + K_{a1}K_{a2} / [H^+]^2} C_e \quad (2)$$

where C_e (μmol/L) is the aqueous benzidine concentration at sampling, k_{a1} and K_{a2} are the acidity constants, respectively.

By fitting the experimental data to Equation (1), the values of fitting parameters of K_d, K_{cov}, S_0, and K_{ex} could be obtained by performing nonlinear least-squares regression with Sigmaplot from Systat Software (Point Richmond, CA, USA).

Results and Discussion

Model Prediction. Sorption isotherms of benzidine in the sediment-water systems under different pH conditions were conducted. The model fitting parameters are listed in Table 1. The fitting curves are plotted in Fig. 1 together with the experimental data.

Table 1 Sorption equilibrium model fitting parameters

	Fitting parameters						Statistics	
pH	K_d	K_{OC}	K_{cov}	K_{ex}	S_0	S_0^*	SEE	r^2
3.1	1.79E-07	1.60E-06	12.76	0.022	3076.27	27	648.76	0.99
3.8	9.75E-08	8.71E-07	15.62	0.015	4227.53	37	550.62	0.99
5.0	462.62	4130	1.05E08	0.012	4153.64	37	1141.43	0.99
6.9	214.43	1914	2.21	5.19E-17	5878.64	52	3381.26	0.90

S_0^* = OC normalized covalent binding sites (mmol/kg OC); SEE = Standard error of estimate; r^2 = coefficient of determination.

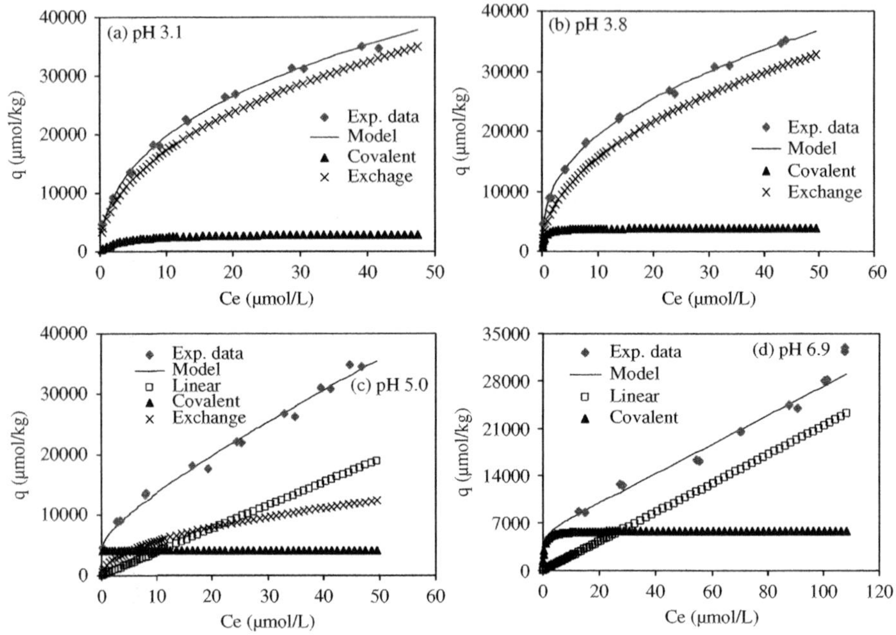

Fig. 1 Model fitting of benzidine sorption at different pH values of (**a**) 3.1, (**b**) 3.8, (**c**) 5.0, and (d) 6.9.

The model simulated the experimental data very well. The goodness-of-fit was above 0.90 for all the data sets, and a value of 0.99 for the goodness-of-fit was observed for some systems. When solution pH was low (e.g., 3.1 or 3.7), the protonated benzidine species were dominant. As a result, cation exchange process dominated the sorption of benzidine. When the slurry pH was increased to 5.0, both neutral and cationic benzidine species existed in relatively significant quantities. All three sorption processes occurred, while covalent binding and hydrophobic partitioning existed in increasing contributions. In contrast, the contribution from cation exchange to the overall sorption of benzidine decreased. When slurry pH increased up to near neutral (i.e., 6.9), over 99% of benzidine molecules were neutral, and were sorbed via hydrophobic partitioning and covalent binding mechanisms. The contribution of cation exchange became negligible.

The fitted OC-normalized distribution coefficient K_{OC} values were 1914 (pH 6.9) and 4130 (pH 5.1) L/kg. Means and Wijayaratne (1989) investigated the sorption of benzidine on estuarine colloidal organic matter and found the K_{OC} to be 3430 L/kg at pH 7.9. According to the authors, their reported value was somewhat higher than corresponding K_{OC} data obtained in sediment systems mainly because aromatic amines are more strongly bound to estuarine colloids. In another study, the same researchers calculated the K_{OC} to be 1116 L/kg using Equation 3 (Wijayaratne and Means, 1984).

$$\log K_{OC} = -0.693 \log S(\mu g/mL) + 4.851 \quad (3)$$

Similarly, Sabljić and co-workers (1995) employed the Quantitative Structure-Activity Relationship (QSAR) model to quantify the relationships between the soil sorption coefficients and the n-octanol/water partition coefficients (log K_{OC} vs. log K_{OW}) based on a reliable set of experimental log Kow data. Those researchers proposed a model for anilinium compounds under neutral conditions, as illustrated in Equation 4.

$$\log K_{OC} = 0.85 + 0.62 \log K_{OW} \quad (4)$$

Applying the reported K_{OW} value of 1.34 for benzidine to Equation (4) gives a K_{OC} value of 2884 L/kg, which is also in good agreement with the predicted value (1914 L/kg at neutral pH of 6.9). As for the covalent binding mechanism portion, the fitted parameters suggested that the covalent binding sites in the applied sediment matrices were about 27–52 mmol/kg OC under different pH conditions.

Model Sensitivity Analysis. To demonstrate the effects of different processes (i.e., partitioning, cation exchange, and covalent binding) and fitting parameters on the shape of the fitting curves for the experimental data, model calculations were carried out using the data set at pH 5.0 as a basic (optimal) set. Each model fitting parameters were assigned a value of 150% or 50% of its basic value, while all the other parameters remained unchanged. This calculation procedure was repeated until all the fitting parameters were evaluated. The resulting curves are presented in Fig. 2.

From the results, when K_d, K_{cov} and S_0 were increased to 150%, all the predicted sorbed concentrations q_e (μmol/kg) increased. In comparison, when these fitting

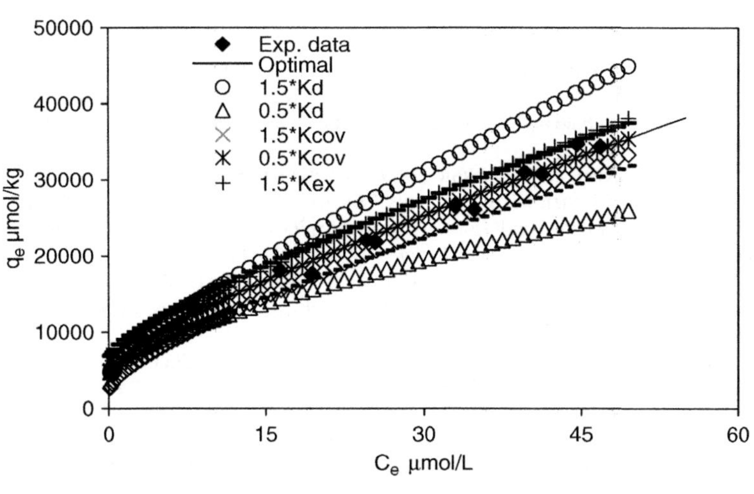

Fig. 2 Sensitivity analysis for model fitting parameters. Model calculations were carried out using the data set at pH 5.0 as an optimal set

parameters were reduced by 50%, the predicted q_e were smaller than the optimal value. As for K_{cov}, neither change had appreciable effect on the predicted q_e values. It may be concluded that the model is least sensitive to changes in K_{cov} values, but most sensitive to changes in K values. In addition, influence of K_{ex} or S_0 on the model sensitivity is between K_d and K_{cov}.

Conclusions

A sorption mechanism based equilibrium model was developed in this study. The fitted model parameter of K_{OC} (1914–4130 L/kg) was in accordance with published values. The covalent binding sites were found to be 27–52 mmol/kg OC in the applied pH range of 3.1–6.9. Model sensitivity analysis revealed that the predicted model was the least sensitive to changes in K_{cov} values, but the most sensitive to changes in K_d values. This model strategy may provide one approach to predict the complex sorption behavior of aromatic amines in the ecosystems.

Acknowledgments The authors thank the U.S. National Science Foundation for financial support in the form of a CAREER award (BES-0093191) to MC Nyman.

References

Brownwell, B.J., H. Chen, J.M. Collier, and J.C. Westall. 1990. "Adsorption of Organic Cations to Natural Materials." *Environ. Sci. Technol.* 24(8): 1234–1241.

Colón, D., E.J. Weber, and G.L. Baughman. 2002. "Sediment-Associated Reaction of Aromatic Amines. 2. QSAR Development." *Environ. Sci. Technol.* 36(11): 2443–2450.

Huang, W., T.M. Young, M.A. Schlautman, H. Yu, and W.J. Weber, Jr. 1997. "A Distributed Reactivity Model for Sorption by Soils and Sediments. 9. General Isotherm Nonlinearity and Applicability of the Dual Reactive Domain Model." *Environ. Sci. Technol.* 31(6): 1703–1710.

Means, J.C., and R.D. Wijayaratne. 1989. "Sorption of Benzidine, Toluidine, and Azobenzene on Colloidal Organic Matter." In I.H. Suffet and P. MacCarthy (Eds.), Aquatic Humic Substances: Influence on Fate and Treatment of Pollutants, pp. 209–222. American Chemical Society, Washington, DC.

Ononye, A.I., J.G. Graveel, and J.D. Wolt. 1989. "Kinetic and Spectroscopic Investigations of the Covalent Binding of Benzidine to Quinones." *Environ. Toxicol. Chem.* 8(4): 303–308.

Parris, G.E. 1980. "Covalent Binding of Aromatic Amines to Humates. 1. Reactions with Carbonyls and Quinones." *Environ. Sci. Technol.* 14(9): 1099–1106.

Sabljić, A., H. Gusten, H. Verhaar, and J. Hermens. 1995. "QSAR Modeling of Soil Sorption. Improvements and Systematics of LogK_{oc}Vs. LogK_{ow} Correlations." *Chemosphere* 31(12): 4489–4514.

Spurlock, F.C., and J.W. Biggar. 1994. "Thermodynamics of Organic Chemical Partition in Soils. 2. Nonlinear Partition of Substituted Phenylureas from Aqueous Solution." *Environ. Sci. Technol.* 28(6): 996–1002.

Thorn, K.A., P.J. Pettigrew, and W.S. Goldenberg. 1996. "Covalent Binding of Aniline to Humic Substances. 2. ^{15}N NMR Studies of Nucleophilic Addition Reactions." *Environ. Sci. Technol.* 30(9): 2764–2775.

U.S. EPA. 1999. Integrated Risk Information System (IRIS) on Benzidine. National Center for Environmental Assessment, Office of Research and Development, Washington, DC.

Weber, E.J., D. Coln, and G.L. Baughman. 2001. "Sediment-Associated Reactions of Aromatic Amines. 1. Elucidation of Sorption Mechanisms." *Environ. Sci. Technol.* 35(12): 2470–2475.
Weber, E.J., D.L. Spidle, and K.A. Thorn. 1996. "Covalent Binding of Aniline to Humic Substances. 1. Kinetic Studies." *Environ. Sci. Technol.* 30(9): 2755–2763.
Wijayaratne, R.D., and J.C. Means. 1984. "Sorption of Polycyclic Aromatic Hydrocarbons by Natural Estuarine Colloids." *Mar. Environ. Res.* 11(2): 77–89.
Xia, G., and W.P. Ball. 1999. "Adsorption-Partitioning Uptake of Nine Low-Polarity Organic Chemicals on a Natural Sorbent." *Environ. Sci. Technol.* 33(2): 262–269.
Zachara, J.M., C.C. Ajinsworth, L.J. Felice, and C.T. Resch. 1986. "Quinoline Sorption to Subsurface Materials: Role of pH and Retention of the Organic Cation." *Environ. Sci. Technol.* 20(6): 620–627.

Radioactive Contaminanant Transport in Subsurface Porous Environment

An Jin and Shoou-Yuh Chang

Abstract The deterministic model is developed based on the discrete advection-dispersion-adsorption equations with radioactive decay mechanism. The data assimilation scheme is designed to utilize the Kalman Filter (KF) to incorporate the knowledge of uncertainties in both the model and the measurement. Spatially correlated regional noise structures are proposed and integrated into a data assimilation scheme. This model demonstrates that the data assimilation scheme reduces the uncertainty and predicts more accurately than a deterministic model. Through absorbing information from observation, the predictive plumes of radioactive contamination from the assimilation system can follow the change of a randomized irregular plume shape in the real world more closely than a non-assimilation deterministic model.

Introduction

Radioactive agent has its decay chain. To simulate the complete transport process of the radioactive agent and the corresponding decay products in the subsurface environment, this study introduced the retardation factor to simplify the interactive effects between the porous materials and the dispersion, advection and decay, leading no explicit absorption term in the transport equation. Subsurface properties fluctuate in space and time (Heuvelink and Webster, 2001). The subsurface environment is more heterogeneous and anisotropic compared to the atmosphere and surface water environment. Due to natural intrinsic uncertainties as well as the infeasible measurability, the prediction is inevitably uncertain and some random variables must be introduced to represent the unpredictable variation. Therefore, stochastic data assimilation such as Kalman filter (KF) can be integrated with deterministic models to give better predictions. Spatially independent Gaussian noise is easy to obtain and is widely assumed in many areas for examining KF properties. However,

A. Jin (✉)
North Carolina A&T State University, Greensboro, North Carolina 27411, USA

G.A. Uzochukwu et al. (eds), *Proceedings of the 2007 National Conference on Environmental Science and Technology*, DOI 10.1007/978-0-387-88483-7_24,

for subsurface modeling, it is more reasonable to consider the data and noise as regional i.e. spatially correlated instead of spatially independent Gaussian (Conwell et al., 1997; Goovaerts, 1999; Brooker, 2001; Ferreyra et al., 2002). In this study, a radioactive contaminant transport model with KF is constructed. Then a random field simulation scheme is utilized to examine the impact of spatially correlated noise on the effectiveness of the data assimilation results.

Approach

Radioactive Contaminant Plume in Porous Subsurface Environment

For radioactive contaminants transport in subsurface environment, contributions by nuclear actions of the radioactive agent and sorption caused by porous medium must be taken into considerations. This radioactive decay is a first-order kinetic process with decay rate k. By adding absorption and decay mechanisms to the general advection-dispersion equation, we obtained

$$\frac{\partial C}{\partial t} = D\left(\frac{\partial^2 C}{\partial x^2}\right) - V\left(\frac{\partial C}{\partial x}\right) - \frac{\rho}{\eta}\left(\frac{\partial S}{\partial t}\right) - kC$$

where C is the concentration of contaminant (mass of solute per unit mass of solution), D_x and D_y are the longitudinal and transverse dispersion coefficients, respectively, V is the average linear pore liquid velocity, ρ is bulk density of the porous medium, η is the porosity, S is the contaminant concentration in the sorbed phase and k is the rate of the radioactive decay. In practice, we can simplify the equation by assuming linear partitioning between the C and S and introducing retardation factor R:

$$\frac{\partial C}{\partial t} = \frac{D_x}{R}\left(\frac{\partial^2 C}{\partial x^2}\right) + \frac{D_y}{R}\left(\frac{\partial^2 C}{\partial y^2}\right) - \frac{V}{R}\left(\frac{\partial C}{\partial x}\right) - \frac{0.693}{RL_{half}}C \tag{1}$$

where L_{half} is the half-life of the radioactive decay and R is the retardation factor

$$R = 1 + \rho\left(\frac{K_P}{\eta}\right)$$

where we assumed K_p is linear partition coefficient.

Stochastic Markov Chain

Stochastic Markov chain is used for simulating the complicated real-world transport processes which usually have unknown or uncertain sources, inaccurate transport parameters or numeric errors. The stochastic dynamics can be represented as deterministic dynamics of transport and a random noise term

$$\mathbf{x}_{k+1} = \mathbf{A}\mathbf{x}_k + \mathbf{p}_k, \quad \mathbf{k} = \mathbf{0}, \mathbf{1}, \mathbf{2}, \mathbf{3}, \ldots \tag{2}$$

where $\mathbf{x}_k$ and $\mathbf{x}_{k+1}$ are the subsurface state variables at time k and k+1, respectively. $\mathbf{A}$ is the finite-difference operator that advances the $\mathbf{x}_k$ by one time step. $\mathbf{p}_k$ is the model system error, the difference between the model prediction and the optimal estimate of the true value. The model error vector, $\mathbf{p}_k$, is assumed to have covariance matrix $\mathbf{Q}_k$ and zero mean.

Observation $\mathbf{z}_k$ is described by data conversion matrix $\mathbf{H}$ and error $\mathbf{o}_k$. Similarly, $\mathbf{o}_k$ is assumed to be the error of observations that has covariance matrix $\mathbf{R}_k$ and zero mean:

$$\mathbf{z}_k = \mathbf{H}\mathbf{x}^T{}_k + \mathbf{o}_k \tag{3}$$

Optimal Estimation and Data Assimilation with KF

Considering a stochastic process defined by Equations (2) and (3) above, the Kalman optimal estimator (Kalman and Bucy, 1961) is

$$\mathbf{x}_{k+1}(+) = \mathbf{x}_{k+1}(-) + \mathbf{K}_{k+1}(\mathbf{z}_{k+1} - \mathbf{H}\mathbf{x}_{k+1}(-)) \tag{4}$$

where $\mathbf{x}_{k+1}(+)$ indicates the estimated value after the KF adjustment, and $\mathbf{x}_{k+1}(-)$ the value before the KF adjustment, i.e. the predicted value from the model. The matrix $\mathbf{K}_{k+1}$ is determined by

$$\mathbf{K}_{k+1} = \mathbf{P}_{k+1}(-)\mathbf{H}^T(\mathbf{H}\mathbf{P}_{k+1}(-)\mathbf{H}^T + \mathbf{R}_{k+1})^{-1} \tag{5}$$

where $\mathbf{P}_{k+1}$ is the optimal estimate error covariance matrix, $\mathbf{M}^T$ and $\mathbf{M}^{-1}$ denote the transpose and inverse of matrix $\mathbf{M}$, respectively. $\mathbf{P}_{k+1}$ can be obtained by recursive calculating

$$\mathbf{P}_{k+1}(+) = \mathbf{P}_{k+1}(-) - \mathbf{P}_{k+1}(-)\,\mathbf{H}^T[\mathbf{H}\,\mathbf{P}_{k+1}(-)\,\mathbf{H}^T + \mathbf{R}]^{-1}\mathbf{H}\,\mathbf{P}_{k+1}(-) \tag{6}$$

and

$$\mathbf{P}_{k+1}(-) = \mathbf{A}\mathbf{P}_k(+)\mathbf{A}^T + \mathbf{Q}\mathbf{k} \tag{7}$$

Regional Noise Field Simulation

System noise is usually denoted as a vector $\mathbf{p_k}$ where $\mathbf{p_k}$ is q-by-1 column vector. The statistical structure of Gaussian noise with time independence is usually described as

$$E\{p_k\} = 0 \tag{8}$$

and

$$E\{p_k\, p_l'\} = Q_r \delta_{kl}, \quad \delta_{kl} = 0 \; if\, k \neq l, \; \delta_{kl} = 1 \; if\, k = l \tag{9}$$

where Q is q-by-q symmetric and positive semi-definite. **Q** is a diagonal matrix if noise is spatially independent Gaussian. Since the noise will be integrated into KF, the most convenient way is to create the noise that satisfies given error covariance structures or given semi-variogram structures.

Let $\mathbf{p_k}$(t) be a 2-D random noise state vector, $t \in G_s \subseteq \Re^2$, with given covariance matrix Q(h) that represents a certain spatial dependence of the $\mathbf{p_k}$(t)

$$Q_{ij} = Q(t_j - t_i) = Q(h_{ij}) \tag{10}$$

where t is space location index, G_s is grid of system noise injection, h_{ij} is distance between the space location t_i and t_j.

Assuming that the spatially-dependent $\mathbf{p_k}$(t) has a reasonable decomposition with components of a spatially-independent unit Gaussian vector $\mathbf{g_k}$(t) and a linear transformation **L**

$$p_k\,(t) = L g_k\,(t) \tag{11}$$

with $E(gg^T) = I$ (Chang and Jin, 2005), the linear transformation matrix **L** can be solved as the lower triangular of Cholesky decomposition of Q if Q is positive definite. i.e.

$$Q = LL^T \tag{12}$$

Experiments

The model grid G_m is defined on a 2-D plane domain. G_m has n_x grid points in x direction and n_y grid points in y direction. In the numerical scheme, we set n_x = 20, n_y = 11, dx = 1.5 m, dy = 1.5 m, dt = .20 day. We assumed that horizontal dispersion D_x = 1.5 m^2/day, D_y = .25 m^2/day, porosity = .40, velocity = 1.5 m/day along the x direction and retardation = 1.525. The initial condition is an instantaneous contaminant source injected into a location with coordinates (5,5). Initial contamination background is set to 10 ppm.

Since the transport process and its predictions are actually occurring in a geographically specific subsurface area, it is obvious that to utilize a regional error frame to simulate the uncertainty is more reasonable than to use a geo-independent error frame.

Root Mean Squared Error (RMSE) was chosen as the error parameter to evaluate the KF effectiveness on the whole domain. *RMSE*(*k*) is defined as

$$RMSE\,(k) = \sqrt{\frac{1}{N_{ij}-1}\sum_{i,j}[C^E(i,j,k) - C(i,j,k)]^2} \tag{13}$$

Results

Prediction with Half-life 216 Day and Inaccurate Hydrologic Parameters

In practice, the sampling places and times are restricted and description of the hydrologic parameters can never be considered complete and accurate as required by subsurface transport models. More effort in continuously pursuing completeness and accuracy in the labs and sites may reduce the errors caused by these hydrological parameters, but this is usually very costly due to highly heterogeneous subsurface environment. Applying KF data assimilation scheme can provide another cost-effective way to improve accuracy by merging finite observations. In this experiment, we simulate the scenario that has one inaccurate hydrologic parameter in the prediction model and test the impact of regional noise **R**. For simplicity, the average linear pore liquid velocity, V, is assumed inaccurate in the prediction model. V is increased by 20% when the true fields were simulated. However, a deterministic model with the uncertain input still employs the given V in its prediction.

To reflect the regional nature of the pollutant in the subsurface environment, observation error matrix **R** is assumed to satisfy $\mathbf{R}[i,j] = E\{\mathbf{o}[i]\mathbf{o}[j]\} = c^2 \exp(-h_{ij}/a)$, where h_{ij} is grid distance between grid point i and j, c and a are positive parameters that need to be specified in a particular model. The regional noise field simulation scheme has been developed and examined by verifying experimental semivariogram of the generated noise fields.

The prediction from the numerical model that has no data assimilation scheme is shown in Fig. 1. Note that the pollutant plume in the prediction field moves behind the plume of the true field due to the inaccurate V used in prediction model.

Both Figs. 2a and 2b give KF filtered results of contaminant plume at 6.4 day. With the data assimilation scheme, the qualities of model predictions are improved. However, by comparing the two filtered results, the KF using regional noise structure (Fig. 2b) is able to follow the irregular shapes of true plume more accurately due to a more effective data assimilation process. In Fig. 2b, the prediction

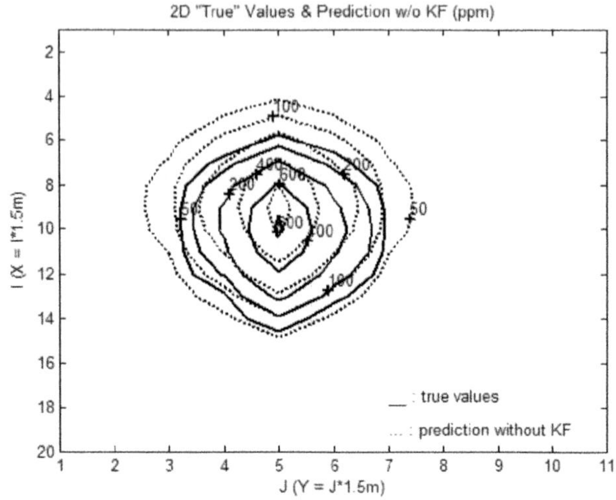

Fig. 1 Prediction without KF Data Assimilation (Half-life = 216 day)

contours in the significant pollution areas (with pollutant concentration greater than 100 ppm) follow the true contours more closely than what has shown in Fig. 2a, KF using Spatially Independence Gaussian(SIG) noise structure.

The correction effect of different KFs can be seen more clearly in Fig. 3, in which the comparison of the average prediction errors (RMSE) for the three schemes, the deterministic model scheme without the KF filter, the model with the KF filter, and the KF model with the regional error structure are calculated at different time steps. As shown in Fig. 3, by applying the regional noise KF data assimilation scheme,

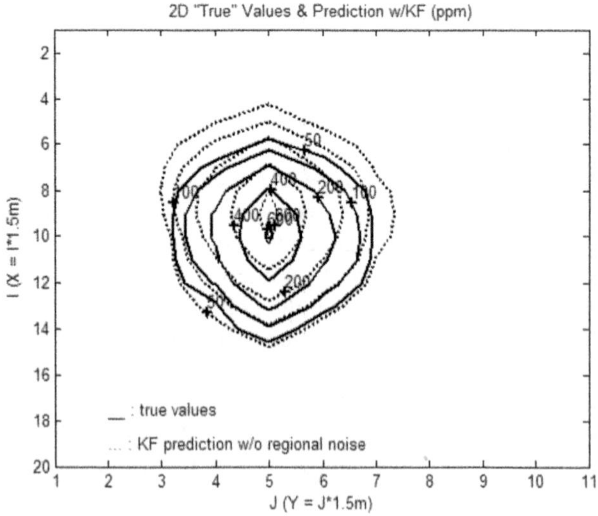

Fig. 2a Predictions improved by KF with SIG covariance **R**

Fig. 2b Predictions by KF with regional covariance **R**

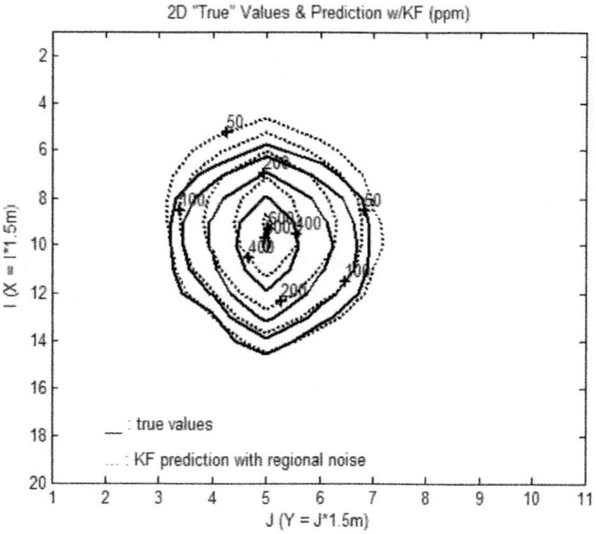

RMSE is converged to 25 ppm at 6.4 day. This is a 37% decrease compared to the estimate from the model without KF and about 29% less compared to the model with KF using SIG noise structure. This correction process of KF involved in prediction is very effective in correcting the error that is caused by inaccurate hydrologic parameter V. Even without knowing which parameter is inaccurate, this correction process occurred by sampling from only nine sites and applying appropriate obser-

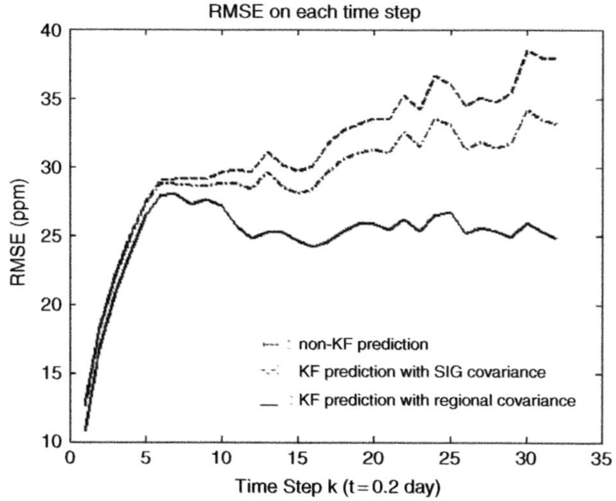

Fig. 3 Prediction Error (RMSE) (halftime = 216 day)

vation error covariance matrix **R** in KF. As a comparison, Fig. 3 suggests that the stochastic model with KF data assimilation is more accurate than the non-KF deterministic model; moreover, the accuracy can be greatly improved by selecting the correct noise statistic structure in the KF scheme.

Conclusions

A hydrodynamic model is derived to simulate the radioactive contaminant in a subsurface porous environment. Errors arising from numerical models and from the uncertain geo-physical parameters can result in unavoidable prediction deviations. Test cases have examined the adaptive performance in the cases that having an inaccurate hydrologic parameter.

The results indicate that the prediction error (RMSE) of the KF scheme using a regional noise covariance matrix is smaller than that from the model's prediction without KF or that from the KF using a SIG noise matrix. The results from test cases also show the potential applications of a regional noise structure in environmental transport modeling. Applying a regional noise structure in KF data assimilation improved the KF correction process for inaccurate hydrologic parameters.

References

Brooker, P.I. (2001) "Modelling spatial variability using soil profiles in the Riverland of South Australia." *Environmental International*, 27(2001), 121–126.

Chang, S.-Y. and Jin, A. (2005). "Kalman filtering with regional noise to improve accuracy of contaminant transport models." *Journal of Environmental Engineering*, 131(6), 971–982.

Conwell, P.M., Silliman, S.E. and Zheng, L. (1997). "Design of a piezometer network for estimation of the variogram of the hydraulic gradient: The role of the instrument." *Water Resources Research*, 33(11), 2489–2492.

Ferreyra, R.A., Apezteguia, H.P., Sereno, R. and Jones, J.W. (2002) "Reduction of soil sampling density using scaled semivariograms and simulated annealing." *Geoderma*, 110(2002), 265–289.

Goovaerts, P. (1999) "Geostatistics in soil science: state-of-the-art and perspectives." *Geoderma*, 89(1999), 1–45.

Heuvelink, G.B.M. and Webster, R. (2001) "Modelling soil variation: past, present, and future." *Geoderma*, 100(2001), 269–301.

Kalman, R.E., and Bucy, R.S. (1961) "New results in linear filtering and prediction theory." Trans. ASME, Series D, *Journal of Basic Engineering*, 83, 95–108.

Transport and Degradation of a Trichloroethylene Plume Within a Stream Hyporheic Zone

John B. Williams, Gary Mills, Daniel Barnhurst, Sherry Southern, and Noelle Garvin

Abstract In predicting natural attenuation rates for contaminant plumes, it is vital to determine flow patterns, suitability of chemical and microbial conditions, and seasonality. Savannah River Site's CMP Pits operated from 1971 until 1979; receiving chemicals, metals, and pesticides. Now a Superfund site, monitoring-wells indicated perchloroethylene (PCE) and trichloroethylene (TCE) had seeped beneath the vadose zone. It was unknown how the plume was entering Pen Branch valley below and whether natural attenuation was degrading the contaminant load. Our study focused on plume transport and exchange within the critical hyporheic zone beneath Pen Branch and helped to ground-truth model the plume borders. We also determined reductive dechlorination of PCE and TCE into dichloroethylene and vinyl chloride. Over forty sampling holes were augered into the hyporheic zone and adjacent floodplain along with 12 stream stations. Chemical conditions linked to natural attenuation (e.g. H_2S, Fe^{+2}, and NH_3) were monitored to identify reductive dechlorination suitability along with temperature, pH, redox, and dissolved oxygen. Plume flow displayed complex entry patterns, but natural attenuation was documented by higher levels of cis-dichloroethylene (cis-DCE) (61.5 μg/l) compared to PCE or TCE. High means of hyporheic PCE (26.5 μg/l) and TCE (6.7 μg/l) compared to overlying stream water PCE (0.5 μg/l) and TCE (0.2 μg/l) raise new transport pathway questions.

Introduction

Groundwater contamination from perchloroethylene (PCE) and trichloroethylene (TCE) is a common pollution problem worldwide. In the United States alone, TCE was present at 852 of the 1,416 sites proposed for the Environmental Protection Agency (EPA) National Priorities List (Board on Environmental Studies

J.B. Williams (✉)
South Carolina State University, Orangeburg, SC, USA

G.A. Uzochukwu et al. (eds), *Proceedings of the 2007 National Conference on Environmental Science and Technology*, DOI 10.1007/978-0-387-88483-7_25,

and Toxicology, 2006). Recent incidents of this contaminant continue to surface as noted for Camp Lejeune marine base USA where a PCE plume potentially impacted thousands of individuals using a residential water supply from 1968 to 1985 (Maslia et al., 2007). The Camp Lejeune PCE plume was traced to a single point source, dry-cleaning business, but had spread as a plume into the drinking supply.

At the U.S. Department of Energy's (DOE) Savannah River Site (SRS), PCE and TCE were used in large quantities during its construction and operations. Disposal of much PCE and TCE was placed into unlined pits near a hill top above Pen Branch stream along with other chemicals, metals, and pesticides (SRS Citizen's Advisory Board, 1999). This disposal site, CMP Pits, operated from 1971 to 1979 when it was closed and backfilled. Plumes, resulting from seepage have been documented flowing down-slope beneath CMP Pits (SRS Citizen's Advisory Board, 1999). Remediation actions near the plume source include soil vapor extraction of volatile organic compounds (VOC) from the vadose zone and soil excavation and removal. Natural attenuation offers a possible remediation action for the plume fringe and our study is evaluating its effectiveness. Of particular interest is degradation activity in the dynamic hyporheic zone beneath Pen Branch where groundwater and surface water interactions occur. The objectives of our study were to detect plume fringe boundaries (helping ground-truth hydrological models) and to determine the concentrations of PCE and TCE degradation products in hyporheic zone sediments and water. We are also determining exchanges between the hyporheic zone and adjacent surface waters and floodplain soils. Results will help SRS determine the relative effectiveness of natural attenuation in reducing contaminant loads at Pen Branch.

Materials and Methods

Pen Branch, a tributary within the Savannah River drainage, is located in the valley below CMP Pits. Hydrological models developed at SRS mapped TCE/PCE plume flow beneath wetlands along Pen Branch and wetlands sampling wells verified the plume presence. We augered a total of 39 sample holes into the streambed and banks along Pen Brach. These stations were dispersed along the Pen Branch reach bordering the plume. Several of the holes were clustered in sections near known plume "hot spots" and six holes were augered into the floodplain on the opposite side of Pen Branch from the plume. Seven surface water stations were also sampled.

Hyporheic zone cores were collected by first driving a 90.0 cm length of 4.0 in (10.2 cm) open diameter PVC pipe into the stream bottom with a sledge hammer and pounding board. This pipe functioned like a caisson; keeping new stream waters out. After stream waters were removed by bailing, holes were hand-augered using a 3 in (7.6 cm) auger bucket to depths of up to 1.5 m. Most holes reached a clay-like confining layer, but some holes were shallower due to obstructions or varying confining layer depths. Two replicate sediment samples for VOC analysis were collected from each core by stainless steel spatula and placed into 10 ml septa-seal vials and crimped. Water samples were pumped from hole bottoms using a Masterflex^{TM} E/S Portable Sampler peristaltic pump with No. 15 Tygon^{TM} tubing and then replicate 5.0 ml samples were placed into 10 ml septa-seal vials and crimped. All VOC samples were stored on ice until transport to the lab refrigerator. Sample blanks were included using coded station numbers to determine background levels due to analytical procedures.

In addition to direct VOC analysis of hole water and sediments, passive diffusion bag (PDB) samplers were placed into most holes to cumulatively sample VOC's for intervals up to three weeks duration. PDB's were constructed similar to Harte, et al. (2000) with ca. 30 ml of deionized, contaminant-free water enclosed in polyethylene bags. Two replicate samples for VOC analysis were collected from each PDB.

Water chemistry data were also collected at each hole using a Hydrolab^{TM} multiprobe meter including: temperature, dissolved oxygen (DO), pH, redox, and conductivity. Where hole water was not too turbid, additional chemistry measures of hydrogen sulfide, ferrous iron, nitrate, and ammonia were made using Hach^{TM} Company field kits that involved chemical color reactions and colorimetric comparisons.

Analyses of VOC were performed using an Agilent^{TM} 6890 Series II gas chromatograph with an electron capture detector and interfaced to an Agilent^{TM} 7694 automated headspace sampler programmed to equilibrate samples at 50°C for 10 min. Separation of PCE, TCE and DCE isomers was achieved using a DB-VRX 60 m $\times$ 0.25 mm $\times$ 1.4 μm capillary column.

Results and Discussion

Significant degradation of plume TCE and PCE into cis-DCE had occurred as the plume moved down-slope to the Pen Branch hyporheic zone. At nearly all stations, cis-DCE was much higher in concentration than TCE or PCE (Fig. 1). Natural attenuation accomplished VOC degradation for a large portion of the original plume TCE

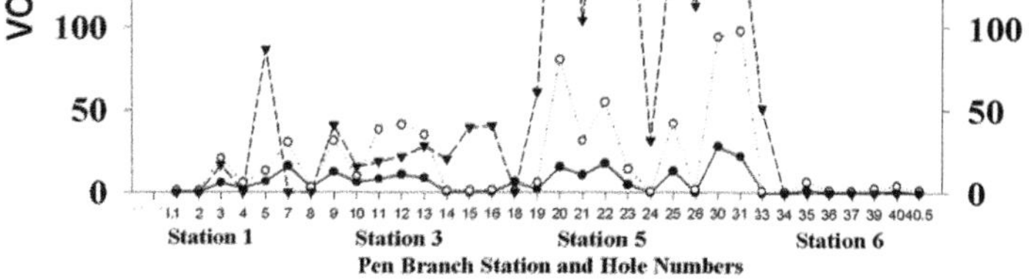

Fig. 1 Mean hole water TCE vs PCE vs cis-DCE in Pen Branch Hyporheic Zone during June–August 2007

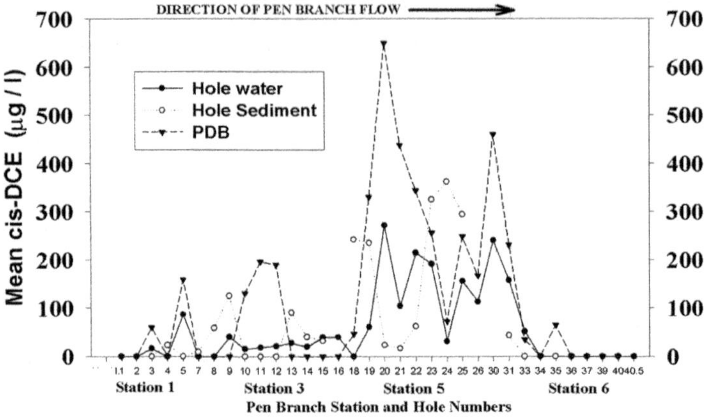

Fig. 2 Mean cis-DCE determined from hole water, sediments, and PDB's in Pen Branch hyporheic zone during June–August 2007

and PCE. VOC concentrations were highest in the zones around Station 5 (Fig. 1). This pattern was consistent with modeled plume flow and validated the model accuracy.

A comparison of cis-DCE levels captured by our different sampling techniques yielded varying results (Fig. 2). In general, highest concentrations were found from PDB samples. Comparisons between sediment versus hole water cis-DCE did not vary consistently (Fig. 2). At some stations, cis-DCE was higher from sediment samples than from hole water, while at other stations, the reverse condition was true. Future sampling should evaluate the relative precision of a cumulative sample from PDB's versus the time-specific grab-samples from sediment or hole water. Precipitation and daily hydrological flow variations might impact grab sample precision more.

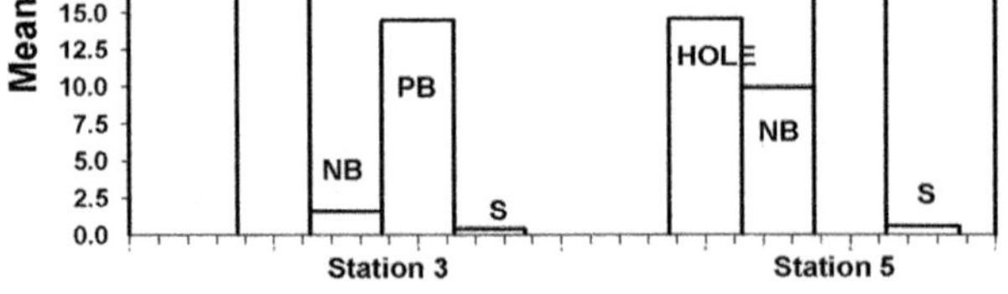

Fig. 3 Mean PCE for different zones at Stations 3 and 5 in Pen Branch, June–August 2007

Mean PCE for water samples from Pen Branch hyporheic zone holes and bank holes on the plume-side at Stations 3 and 5 were much higher than stream surface waters or bank holes on the shore across from the plume (Fig. 3). It was not clear from our results why so little VOC showed up in surface stream samples (Fig. 3). Less PCE was detected from holes on the opposite Pen Branch bank from the plume. However, the relatively high PCE levels in the opposite bank at Station 5 seem to indicate that the down-slope hydrological flows from CMP Pits are strong enough to transect the Pen Branch hyporheic zone.

Conclusions

As groundwater approaches a stream system, a seepage interface develops between the stream channel and subsurface phreatic zones generating geochemical and redox gradients that will affect natural attenuation processes. This strongly suggests that understanding this final interface between ground and surface waters is critical to predicting the flux of TCE/PCE contaminants into regulated receiving waters. Our results indicate significant concentrations of PCE/TCE are present in the shallow hyporheic sediments within the stream bed. Concentrations of cis and trans DCE were variable but generally increased then subsequently decreased or disappeared in the stream surface waters and is consistent with microbial degradation in the hyporheic zone. For other natural attenuation variables including pH, temperature, ferrous iron, hydrogen sulfide, and redox, hyporheic conditions were consistent with near-optimal natural attenuation conditions listed by Wiedemeier, et al. (1998).

Our results also showed concentrations in the stream hyporheic zone were often much higher than the values measured in the adjoining floodplain groundwater; indicating there may be significant migration of contaminants or that they are transported in preferential flow paths. This also suggests careful monitoring of the system is needed to determine if the system has reached a stable, steady-state or is progressing towards a more contaminated state. All current natural attenuation assessments of PCE/TCE remediation on the SRS are predicated on upland and floodplain studies and consequently, degradation processes in hyporheic sediments are not considered. Our results indicate that the attenuation capacity of the floodplain ecosystem for the CMP plume has already been exceeded and contaminants are within cm of the stream interface. Clearly, assessing the natural attenuation capacity of the hyporheic zone is critical for predicting further migration of contaminants and evaluating the longer term feasibility of natural attenuation.

Acknowledgments The efforts of South Carolina State University student interns: Chris Carter, Stephanie Roach, Landis Chambers, Kendall Simmons, Ta'mara Brown, Lanita Peterson, and Eric Foxworth were greatly appreciated. University and SRS planning assistance were also given generously by Dr. Judith D. Salley, Larry Anderson, and Janelle Jansen.

References

Board on Environmental Studies and Toxicology. 2006. *Assessing the Human Health Risks of Trichloroethylene.* National Research Council of The National Academies, The National Academies Press, Washington, DC, 448 p.

Harte, P.T., M.J. Brayton, and W. Ives. 2000. *Use of Passive Diffusion Samplers for Monitoring Volatile Organic Compounds in Ground Water.* USGS Fact Sheet 088-00 4p.

Maslia, M L., J.B. Sautner, R.E. Faye, R.J. Suárez-Soto, M.M. Aral, W.M. Grayman, W.J.J. Wang, F.J. Bove, P.Z. Ruckart, C. Valenzuela, J.W. Green, Jr., and A.L. Krueger. 2007. *Analyses of Groundwater Flow, Contaminant Fate and Transport,and Distribution of Drinking Water at Tarawa Terrace and Vicinity,U.S. Marine Corps Base Camp Lejeune, North Carolina: Historical Reconstruction and Present-Day Conditions.* Agency for Toxic Substances and Disease Registry, U.S. Department of Health and Human Services, Atlanta, GA, 116 p.

SRS Citizen's Advisory Board, 1999. *Chemicals, Metals, And Pesticides (CMP) Pits.* SRS Citizen's Advisory Board Publ. Recommendation No. 83, 2p.

Wiedemeier, T.H., M.A. Swanson, D.E. Moutoux, E.K. Gordon, J.T. Wilson, B.H. Wilson, D.H. Kampbell, P.E. Haas, R.N. Miller, J.E. Hansen, and F.H. Chapelle. 1998. *Technical Protocol for Evaluating Natural Attenuation of Chlorinated Solvents in Ground Water.* United States Environmental Protection Agency, Office of Research and Development, Washington DC, EPA/600/R-98/128.

Part V
Global Climate Change

Using a Spoken Diary and Heart Rate Monitor in Modeling Human Exposure to Airborne Pollutants for EPA'S Consolidated Human Activity Database

Curry I. Guinn and Daniel J. Rayburn Reeves

Abstract In this paper, we describe technology used to collect fine-grained data of individual daily human activity including information related to time, exertion, specific activity, and location. This data uses the Consolidated Human Activity Database developed by the EPA's National Exposure Research Laboratory and is used to build models of human exposure to airborne pollutants. This paper will look specifically at subject compliance and quality in a small pilot study and the automatic classification of activity and location based on a spoken language diary collected from a digital recorder.

Introduction

The purpose of this study was to develop and test a method that would assist the Environmental Protection Agency, and the scientific community at large, in the generation of an activity/location/time/energy expenditure database of sufficient detail to accurately predict human exposures and dose. Collecting daily location/activity data along with measures of pulmonary dynamics to assess human exposure to airborne pollutants has proven to be expensive, to be burdensome to the participant, and to produce unreliable data. This study tested the feasibility of modifying existing technology to reduce burden on respondent, cost and to improve the reliability of data collection.

One goal of this study was to develop a protocol with highly detailed reliable information with low subject burden, two features which are generally inversely correlated. Our protocol tested the use of a digital voice recorder to collect activity/location diary data assuming it to be a less burdensome and a more reliable method than using paper and pencil diaries or hand-held computers. We expected the data to be more complete and reliable than retrospective reports (diaries filled

C.I. Guinn (✉)
University of North Carolina Wilmington, USA

G.A. Uzochukwu et al. (eds), *Proceedings of the 2007 National Conference on Environmental Science and Technology*, DOI 10.1007/978-0-387-88483-7_26,

out at the end of day) because the recorders are easy to use, the diary entries are made as the events occur, and we expected that participants would be more likely to complete the study because of the reduced burden. The data collection plan was also expected to show that the cost of the transcription of the diary can be reduced substantially by using speech and language processing to translate the digital diaries into the EPA's Comprehensive Human Activity Database (CHAD).

Background and Previous Research

Capturing Environmental Exposure Data. The definition of environmental exposure provided by Ott (Ott, 1982) and later adapted by others (NAS 1991; USEPA, 1992) is "an event that occurs when there is direct contact at a boundary between a human and the environment with a contaminant of a specific concentration for an interval of time." This definition implies that four variables must be measured to accurately characterize exposure: location (L), time (T), activity (A), and concentration (C). The EPA developed a system to store and systematically analyze the available data, the Consolidated Human Activity Database (CHAD) (McCurdy et al., 2000).

Paper-based diaries, electronic diaries, voice-recorded diaries, and observational techniques have all been used to collect data about temporal activities, spatial locations, product use, and dietary consumption of research participants (Johnson et al., 2001). Diary methods relying on recall are not highly reliable and have a relatively high respondent burden, which negatively impacts participant compliance. Observational techniques are extremely costly and burdensome. Post-study processing of diary entries is labor intensive unless simplified reporting protocols are employed and automatic processing systems are developed.

In our study, participants are trained to record their current activity and location whenever the participant changes either of those conditions (RTI, 2001). They are trained to use simple English sentences such as "I am on the bus on my way to South Square Mall." The coder's task is to take spoken language utterances like "I am on the bus on my way to South Square Mall" and select the appropriate CHAD location code (31140: Travel by bus) and activity code (18400: Travel for goods and services).

In order to reduce coder burden, one of our goals was to use the computer to assist in taking the participant statements and returning the appropriate CHAD codes. A common approach to text abstraction is to use statistical NLP techniques that analyze a training corpus to build probabilities that can be used to choose the most likely semantic categories for an utterance. In this study, a selection of the total corpus was set aside as training data. From this training set, human coders selected CHAD activity and location codes for each utterance. Using Bayesian statistics, unigram, bigram, and trigram probabilities were generated for each set of words for each activity and location code.

Experimental Platform

Two physical devices were employed – one to capture the heart rate data; the other to capture the voice diary.

Heart Rate Monitoring. Each subject wore a custom-built heart rate monitor with three EKG electrodes. This device recorded heart rate at 2 min intervals. Further, if the subject's heart rate changed by more than 15 beats per minute from the previous measurement, the device would beep – a signal that the subject should make a voice diary entry.

Voice Diaries. Subjects carried a holstered digital recorder (Sony ICD-MS1) to make their diary entries. Because of cumbersome headsets and unreliable wireless microphones, subjects were required to use the built-in microphone on the device.

During training, participants were instructed to make a voice diary entry whenever they changed location, changed activity, or if the heart rate monitor beeped.

Experimental Results

Nine participants were involved in this study for a duration of one week in addition to the training period. A summary of some of the characteristics of these participants is given in Table 1. Across the entire data collection period for the nine participants, the average daily experiment period was 8.56 h.

Recordings per day. There were 1350 diary entries across the nine participants. The average number of recordings per participant was 29 per day. Given the average monitoring time of 8.56 h, participants were making 3.39 recordings per hour. An interesting trend over the course of the field trial was a precipitous decline in the number of recordings as the week progressed. During the first three days of the trial, participants averaged 34.81 recordings a day. During the last 2 days of the trial,

Table 1 Subject characteristics

ID	Sex	Occupation	Age	Education
1	F	Manages Internet Company	52	Some College
2	F	Grocery Deli Worker	18	Some College
3	M	Construction Worker	35	High School
4	F	Database Coordinator	29	Graduate Degree
5	F	Coordinator for Non-profit	56	Some College
6	M	Unemployed	50	High School
7	M	Retired	76	High School
8	M	Disabled	62	High School
9	M	Environment Technician	56	Graduate Degree

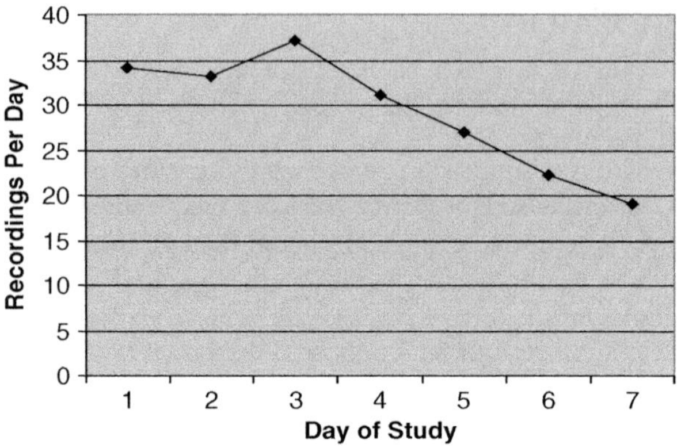

Fig. 1 Average recordings per day across time

participants averaged 20.67 recordings a day. This suggests that participant fatigue of the recording process is significant. Figure 1 shows the diary reporting trend for participants over 7 days of the trial.

Prompting based on heart rate change. In an attempt to encourage participants to report their change in activities, the heart rate monitor would prompt the participant with a tone if the participant's heart rate changed by 15 beats per minute between two recording intervals. This tone was fairly unobtrusive. During training, participants were told to record their current location and activity whenever they heard the tone. Our trial study indicates that this tone encouraged diary entries with only limited success. By analyzing the heart rate data, we can determine when a tone was issued by the monitor. By cross-referencing those times with the times of diary entries, we can determine whether an entry was made around the time of the tone. For this analysis, we assumed the participant was following protocol if they made a diary entry at any point in a 3-min interval surrounding the time of the tone. Table 2 provides the average number of prompting tones issued per day and the percentage of times a subject made an entry corresponding to a tone.

Diary entry quality. *Of the 1350 diary entries, 133 (9.85%) contained no encodable data. The subject's inadvertent start of recording or failure to stop a previous recording was the overwhelming cause of these false entries. One disadvantage of the voice diary is that participants are given no prompts (other than from the heart rate monitor) or guidelines interactively while making recordings. This problem was manifest in the data collected. In 90 entries (6.6%) the subject's activity could not be coded by a human listener. In 20 entries (1.5%) a human encoder could not determine the location. Examples of this sort of deficient diary entry might only include the location ("I'm in the kitchen") or only the activity ("I'm putting on my clothes"). It may be possible to infer the location or the activity in some cases, particularly if the previous diary entries were made in close time proximity.*

Table 2 Heart rate change indicator tones and subject compliance

Subject	Number of Tones	% of Times Subject Made a Diary Entry
1	22.1	45%
2	41.8	29%
3	32.5	36%
4	33.0	55%
5	33.3	36%
6	15.6	40%
7	32.5	37%
8	26.0	22%
9	22.7	31%

Natural Language Processing Methodology

The natural language processing task is to take the words returned from the speech recognizer and determine the appropriate CHAD encoding for location and activity. As an example, "my bedroom making the bed" would be encoding as **30125 (Bedroom)** and **11200 (Indoor Chores)**. The strategy employed was to build statistical language models of the domain.

The valid diary entries were divided into a training set (925 entries) and a test set (279 diary entries). A human coder went through each set and determined the appropriate CHAD activity and location codes. Then uniqram, bigram, and trigram probabilities were determined for each entry in the training set. As each element of the test set is processed, these unigram, bigram and trigram scores are used to select maximally likely category (Guinn, 2006).

Statistical Processing Results. *The training set was first processed using a human listener to perform the speech recognition to obtain raw transcripts of the voice diaries. The statistical techniques described above were applied to the training set. Then, using these statistics, the most likely activity and location was determined for the test set. We experimented with different thresholds by having the system not make a guess if the score was too low or if the relative scores of the top two choices were close. Table 3 shows the precision and recall given a threshold of 0.3 while Fig. 2 shows the relative tradeoff between precision and recall given different threshold values.*

Discussion and Future Work

One drawback of the system described above is that data is captured passively by the system with no immediate feedback to the user. In practice, the only time the system communicated to the subject was when the system issued a beep if the subject's

Table 3 Statistical processing accuracy of hand-transcribed Data with Threshold of 0.3

Subject	**Act. Prec.**	**Act. Recall**	**Loc. Prec.**	**Loc. Recall**
1	93.3	66.6	76.5	61.9
2	54.5	27.9	62.5	46.5
3	55.6	35.7	62.5	47.6
4	82.5	51.5	74.1	60.6
5	87.5	66.0	78.0	73.6
6	87.9	39.9	67.3	48.6
7	68.1	64.0	55.9	80.6
8	60.4	40.4	56.0	68.4
9	89.5	72.3	93.0	85.1
Total	75.5%	51.6%	69.5%	63.7%

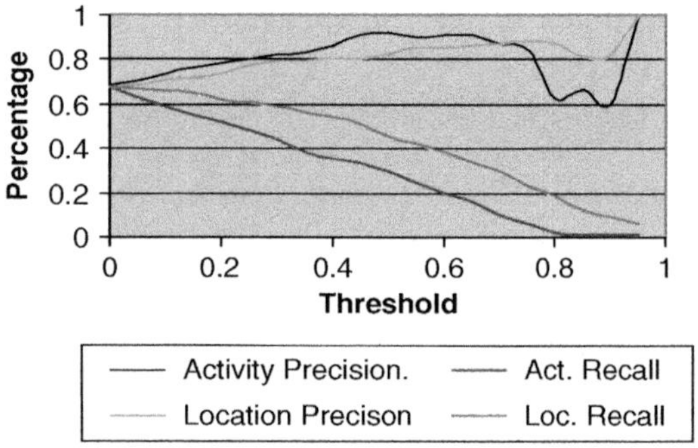

Fig. 2 Threshold values affect the precision and recall: The higher the threshold, the greater the precision but the lower the recall

heart rate changed significantly. This beep was a signal to the subject to make a diary entry.

A better solution is to create a dialog between the subject and the system with more elaborate prompting and immediate error-correcting dialogs when the system is unable to process the user's diary entry. To create this system, we are currently building an input device with the PocketPC technology using a wireless mic/headset, built-in GPS, a pen-based diary, and peripherals for exertion and motion data gathering (Fig. 2).

This system will also fuse data from multiple input sources and use discrepancies to trigger dialogs with the user. For instance, the system will have constant access to Global Positioning System data which includes not just coordinates but also speed.

If the voice diary entry indicates that the person is at home but the GPS reports the position is away from home and traveling at 55 mph, then the system can prompt the user to re-enter the voice diary entry. In some situations, the system could prompt with very specific questions with expected Yes/No answers – a response which has a high likelihood of correct speech recognition.

Acknowledgments This work has been funded by the Environmental Protection Agency under contract 68-D-99-012 and as a subcontract to RTI International.

References

Guinn, C., Crist, D., and Werth, H. A. 2006. "Comparison of Hand-Crafted Semantic Grammars Versus Statistical Natural Language Parsing in Domain-Specific Voice Transcription," *Proceedings of Computational Intelligence*, B. Kovalerchuk (Ed.), San Francisco, CA, pp. 490–495.

Johnson, T., Long, T., and Ollison, W. 2001. *A survey of prospective technologies for collecting human activity data.* Contractor report for the American Chemistry Council.

McCurdy, T., Glen, G., Smith,L., and Lakkadi,Y. 2000. "The National Exposure Research Laboratory's Consolidated Human Activity Database". *Journal of Exposure Analysis Environment Epidem,* 10:566–578.

National Academy of Sciences (NAS). 1991. *Human exposure assessment for airborne pollutants; advances and opportunities.* National Academy of Sciences, Washington, DC.

Ott, W.R. "Concepts of human exposure to air pollution." 1982. Environment International, 7: 179–196.

RTI, *Longitudinal activity data for selected susceptible population subgroups.* 2001. Final Report for EPA Contract No. 68-D-99-012 (Task Order 0012).

U.S. Environmental Protection Agency (USEPA). 1992. "Guidelines for exposure assessment". *Federal Register*, 57(104):22888–22938.

Proposal for Focus Shift in Addressing Climate Change and Environmental Assesment

Gennadyi Gurman

Introduction: Establish Need for a Universally Acceptable Quantifiable Constant for Measuring Environmental Quality Objectively

In chemistry and physics there exist universal units that represent constants or constant intervals that can be used to express various qualities of our world in an unchanging manner. Celsius can be used to express our world in terms of temperature, atmospheres measure pressure, newtons measure force, and meters can measure distance and space.

The quest of this train of thought is to start working in a direction that will create a unit that will be used to measure or quantify environmental systems in order to objectively assess their suitability for life.

Today's greatest problem with assessing environmental degradation is our lack of an effective universal unit that can quantify our results and present them objectively. Balmford et al. notes that problem in a recent article on biodiversity loss (Balmford et al., 2005). Several attempts have been undertaken to deal with that problem. Inhaber proposed an Environmental Quality Index (EQI) that would take into account air quality, water quality, noise, wildlife, pesticides, radiation, and many other aspects of the environment (Inhaber, 1974). United States Environmental Protection Agency (EPA) put together a report on the environment that lists 181 separate indicators (EPA, www.epa.gov/indicators/). Hundreds of scientists all over the world present findings that point to changes in a multitude of ecosystems. Different scientists working in different ecosystems use different organisms as their indicator species, the proverbial canaries in a mineshaft. Jaquish et al. measures environmental and genetic changes using litter size in callitrichid primates (Jaquish et al., 1996), Roger Payne's "Toxic Odyssey" uses levels of bioaccumulated toxins

G. Gurman (✉)
Queens Botanical Garden, Flushing, NewYork, USA

G.A. Uzochukwu et al. (eds), *Proceedings of the 2007 National Conference on Environmental Science and Technology*, DOI 10.1007/978-0-387-88483-7_27,

in whales to gauge the health of oceanic systems (Payne, 2004), Weber uses chloride ion concentrations in carbonate rocks as environmental indicators (Weber, 1964), Atienzak et al. uses fitness and changes in random amplified polymorphic DNA to observes effects of copper toxicity from the environment (Atienzar et al., 2001), Romanek and Grossman look at stable isotopes in *Tridacna* clams as environmental indicators (Romanek and Grossman, 1989), and Tinner calculates changes in Central European vegetation in response to climate changes (Tinner and Lotter, 2001).

Even more recently the journal EcoHealth ran a special section on wetland health indicators. In the section Wardrop et al. proposes a new taxonomy of ecological indicators and is supported by several case studies on the subject (Brooks et al., 2007; Wardropw et al., 2007; Hershner et al., 2007; Hanowski et al., 2007; Reiss and Brown, 2007; Lane, 2007). Though this approach is certainly useful in gaining a glimpse at the health of a specific ecosystem in a given time, it once again addresses only specific systems (wetlands in this case) and does not answer the question of optimal conditions.

Many of the presented findings seem to point out relationships between environmental factors and certain species or ecosystems as well as claim that some of these changes are negative changes in the environment. However, how does one qualify an environmental change as "good" or "bad" without a constant quantifier that is applicable across the board?

Establish DNA as Qualifier of Life

Whenever we put a qualifier nametag on an environmental change we are looking at that change from the perspective of what it means to our species. Hernroth et al. uses blue mussels (*Mytilus edulus*) as an indicator of environmental conditions, but only because they can be used as proxies for human pathogens (Hernroth et al., 2002). The rationale brought forth by opponents of rainforest deforestation is frequently the destruction of known and – as of yet – unknown resources that could be used by humans. In 2004, the New York Academy of science dedicated an issue of its annals to the impact of ecological changes on animal health and disease control (Daszak et al., 2004). The same rationale is used for the defense of oceans, disruption in natural nutrient cycles, climate change and pollution.

When measuring environmental change one has to have a scale against which to compare it. Humans, numerous as they may be, represent only one of the many estimated millions, if not billions of life forms on this planet. Measuring environmental changes against human interest is not only unethical on a global level but is also contrary to the scientific nature of universal quantification. Just as Newton's laws don't only govern Earth, environmental quantification should not be applicable only to one species but all life, on all levels of development – from the most primitive of prokaryotes to highest of eukaryotes. Of course once the general rules have been established, specific cases may be investigated. Just as the Earth's gravitational pull measured at is 9.8 newtons, there may a specific environmental enumeration that will signify an optimal human environment.

Taxonomy is looked at for the appropriate unit against which to measure. The attempt here is to find a correlation of "environmental pressures" to "living things." Since this equation must have a Y for every X, where X is "living thing" and Y is "environmental pressures" there must be a unifying characteristic that is present in all life and is indicative of life and only life that stands in for X. In taxonomy organisms are grouped on their evolutionary trees using a dichotomous approach, where closely related groups share many similar characteristics, whereas groups that are more distant from each other may share fewer and fewer characteristics. Generally, characteristics that arose earlier in the evolutionary process are more widely dispersed – "having eukaryotic cells" is an example – and later innovations may be constricted to a single genus or even species, opposable thumbs and binocular vision are such examples (Fig. 27.1).

What is the oldest structure that is present in all living organisms and is indicative of the presence of life?

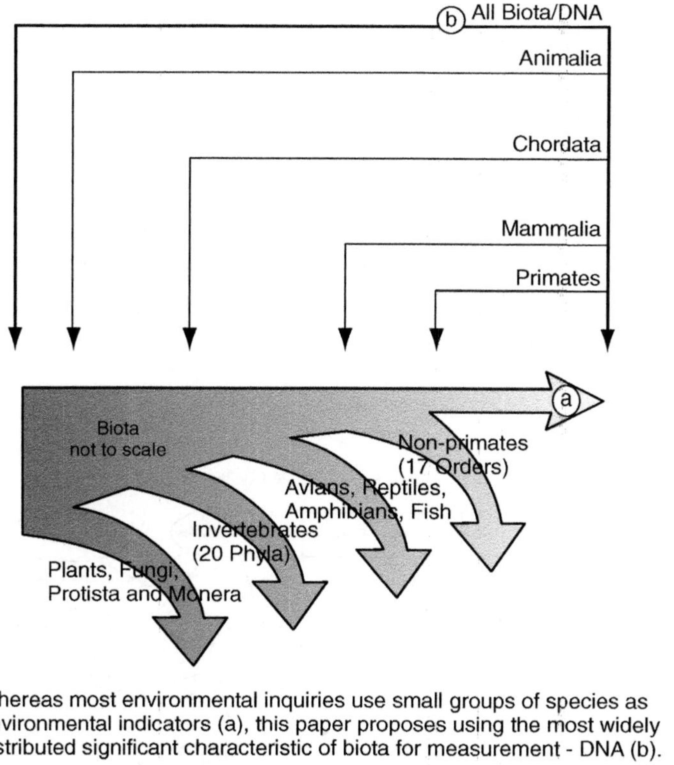

Fig. 27.1 Taxonomic approach to DNA as an indicator

The only structure that is so widely dispersed and is mandatory to living things is DNA.

DNA as a Universal Indicator – 4 Functions and 3 Resources

In order to quantify environmental influence on an organism this paper presents an organization model. DNA is used as the point of influence due to its prevalence throughout all biota and importance in both physical and physiological attributes of an organism thus allowing the effects to be visible across populations, species and kingdoms. Consideration is given to DNA's primary functions (replication, maintenance, building instructions, and processes instructions) and the resources necessary to maintain those functions (energy, materials, and medium). Systematic overview then becomes necessary to isolate the combinations of environmental pressures affecting the functions and resources needed to maintain life (Fig. 27.2).

The functions of DNA can be summed up in the following series:

1. Replication: includes both mitosis-like multiplication (copying) and reproductive meiosis-like behavior. RNA replication is a possible candidate for inclusion.
2. Maintenance: the maintenance of a particular DNA strand, correction of mutations or shutdown of mutated portions and self-destruction.
3. Building instructions: coded instructions for the sequencing of peptides, both building blocks and enzymes.
4. Processes instructions: coded instructions for order of operations.

The resources necessary for the maintenance of DNA functions that can be affected by environmental pressures can be expressed as follows:

1. Energy: there must be a source of energy driving the reactions – occurring in and near DNA – to completion.

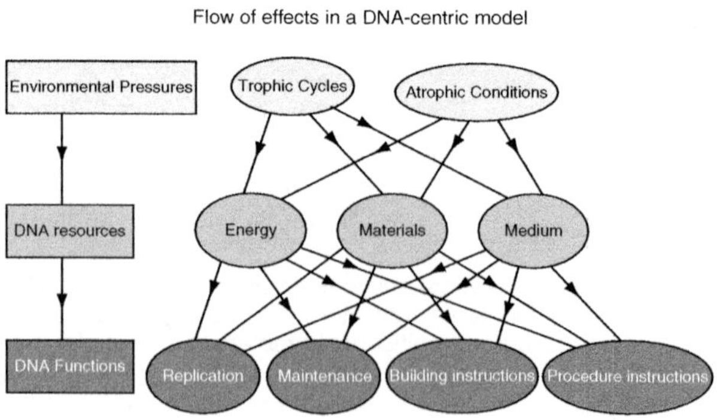

Fig. 27.2 Flow of effects in a DNA-centric model

2. Materials: there must be a source of materials (building blocks) both for internal (DNA maintenance and replication) and external (organic structures and bodily functions) composition processes.
3. Medium: there must be a stable medium in which the DNA molecule can perform its functions. This is a mainly chemical analysis of the environment, best expressed in ratios of organically vital chemical compounds and elements.

The factors that control the access to these resources can be expressed as trophic cycles and atrophic conditions. The trophic cycles are the cycles that are moved by the participation of food chains and include but aren't limited to carbon, nitrogen, water, and oxygen cycles. Atrophic conditions are the combination of geological, chemical and climactic conditions that can alter the accessibility and purity of the energy, materials, and medium.

Through using this approach in combination with data on historical environmental changes and changes in quality and quantity of DNA, a formula may be obtained that will quantify environmental impacts without anthropological bias.

Application of Environmental Quantification

The use of environmental indicators is already widely applied in many fields in order to test the quality of ecosystems. These indicators provide an excellent measurement of current conditions, and if available over time, can provide a look at a progression of events. The system proposed in this paper is meant to look at the optimal conditions that can exist in any given ecosystem. Combined with information derived from similar environmental indicators on the current state of the ecosystem this system can provide a quantified deviation from optimum conditions. Such an unbiased look can allow for a quantification and qualification of the effects of climate change upon various ecosystems. If the indicators are narrowed in range to the needs of a specific species, *Homo sapiens* for example, the optimal climate conditions can be identified for that organism.

A secondary application can be seen in sustainable urban planning and architecture. Optimal environmental conditions can become the guiding principle in sustainable design and in meshing of ecological concepts with everyday human needs.

References

F.A. Atienzar, V.V. Cheung, A.N. Jha, and M.H. Depledge, 2001 "Fitness Parameters and DNA Effects Are Sensitive Indicators of Copper-Induced Toxicity in *Daphnia magna*" Toxicologica Sciences, 59, 241–250

A. Balmford, L. Bennun, B. ten Brink, D. Cooper, I.M. Côté, P. Crane, A. Dobson,* N. Dudley, I. Dutton, R.E. Green, R.D. Gregory, J. Harrison, E.T. Kennedy, C. Kremen, N. Leader-Williams, T.E. Lovejoy, G. Mace, R. May, P. Mayaux, P. Morling, J. Phillips,

K. Redford, T.H. Ricketts, J.P. Rodríguez, M. Sanjayan, P.J. Schei, A.S. van Jaarsveld and B.A. Walther, 14 January 2005 "The Convention on Biological Diversity's 2010 Target" Science, 307(5707), 212–213

R.P. Brooks, G.P. Patil, S. Fei, A.I. Gitelman, W.L. Myers, and E.D. Reavie, June 2007 "Next Generation of Ecological Indicators in Wetland Condition" EcoHealth; 4(2), 176–178

P. Daszak, G.M. Tabor, A.M. Kilpatrick, J. Epstein, and R. Plowright, October 2004 "Impact of Ecological Changes on Tropical Animal Health and Disease Control" Science, 1026, 1–11

EPA The Draft Report on the Environment www.epa.gov/indicators/

J. Hanowski, N. Danz, R. Howe, G. Niemi, R. Regal, June 2007 "Consideration of Geography and Wetland Geomorphic Type in the Development of Great Lakes Coastal Wetland Bird Indicators" EcoHealth, 4(2), 194–205

B.E. Hernroth, A.-C. Coden-Hansson, A.-S. Rehnstam-Holm, R. Girones, and A.K. Allard, September 2002 "Environmental Factors Influencing Human Viral Pathogens and Their Potential Indicator Organisms in the Blue Mussel, *Mytilus edulis*: the First Scandinavian Report" Applied and Environmental Microbiology, 68, 4523–4533

C. Hershner, K. Havens, D.M. Bilkovic, D. Wardop, June 2007 "Assessment of Chesapeake Bay Program Selection and Use of Indicators" EcoHealth, 4(2), 187–193

H. Inhaber. 1974. "Environmental Quality: Outline for a National Index for Canada" Science, 186, 798–805

C.E. Jaquish, J.M. Cheverud, and S.D. Tardif, 1996 "Genetic and Environmental Impacts on Litter Size and Early Infant Survival in Three Species of Callitrichids" The Journal of Heredity, 87(1), 74–77

C.R. Lane, June 2007 "Assessment of Isolated Wetland Conditions in Florida Using Epiphytic Diatoms at Genus, Species, and Subspecies Taxonomic Resolution" EcoHealth, 4(2), 219–230

R. Payne, 11 June 2004 "A Toxic Odyssey" Science, 304, 1584–1586

K.C. Reiss, and M.T. Brown, June 2007 "Evaluation of Florida Palustrine Wetlands: Application of USEPA Levels 1, 2, and 3 Assessment Methods" EcoHealth, 4(2), 206–218

C.S. Romanek, and E.L. Grossman, October 1989 "Stable Isotope Profiles of *Tridacna Maxima* as Environmental Indicators" PALAIOS, 4,(5), 402–413

W. Tinner, and A.F. Lotter, June 2001 "Central European Vegetation Response to Abrupt Climate Change at 8.2 ka" Geology, 29(6), 551–554

D.H. Wardropw, C. Hershner, K. Havens, K. Thornton, D.M. Bilkovic, June 2007 "Developing and Communicating a Taxonomy of Ecological Indicators: A Case Study from the Mid-Atlantic" EcoHealth, 4(2), 179–186

J.N. Weber, September 1964 "Chloride Ion Concentration in Liquid Inclusions of Carbonate Rock as Possible Environmental Indicator" Journal of Sedimentary Research, 34(3), 677–680

Sensor Webs in Digital Earth: Monitoring Climate Change Impacts

Matt Heavner, Rob Fatland, Eran Hood, Cathy Connor, Tracy Lee Hansen, Mary Sue Schultz, Tom LeFebvre, and Albert Esterline

Abstract The University of Alaska Southeast is currently implementing a sensor web identified as the SouthEast Alaska MOnitoring Network for Science, Telecommunications, Education, and Research (SEAMONSTER). From power systems and instrumentation through data management, visualization, education, and public outreach, SEAMONSTER is designed with modularity in mind. We are utilizing virtual earth infrastructures to enhance both sensor web management and data access. We will describe how the design philosophy of using open, modular components contributes to the exploration of different virtual earth environments. We will also describe the sensor web physical implementation and how the many components have corresponding virtual earth representations. This presentation will provide an example of the integration of sensor webs into a digital earth. We suggest that sensor networks and sensor webs should integrate into digital earth systems and provide a resource easily accessible to both scientists and the public. The initial scientific application of the SEAMONSTER sensor web is to monitor climate change impacts of glaciated watersheds in Southeast Alaska. Melting glaciers are dominating the biogeochemistry of watersheds and as the glaciers disappear, this influence will diminish. By monitoring these watersheds using a sensor web, we are improving knowledge regarding impacts of climate change.

Introduction

Impacts of climate change are especially pronounced in the polar regions of the planet (ACIA, 2005). Because these areas are less accessible to the majority of the population, the regions are historically less well studied. Additionally, public awareness of the impacts of climate change in the polar-regions is often lacking. One important aspect of climate change impacts is on the cryosphere, or frozen water

M. Heavner (✉)
University of Alaska Southeast, Juneau, Alaska, USA

G.A. Uzochukwu et al. (eds), *Proceedings of the 2007 National Conference on Environmental Science and Technology*, DOI 10.1007/978-0-387-88483-7_28,

regions of the world. By understanding the watershed scale impacts associated with changes in the glaciers, we provide insight into the potential responses of the large masses of ice (*e.g.* Greenland, Antarctica)

Deploying a large number of sensors or utilizing satellite remote sensing is the present technology for environmental remote sensing over a large geographical region. NASA and other groups are exploring the concept of a sensor web. A NASA Earth Science Technology Office Advanced Information Systems Technology workshop on Sensor Webs developed the definition:

> A sensor web is a coherent set of heterogeneous, loosely-coupled, distributed nodes, interconnected by a communications fabric that can collectively behave as a single dynamically adaptive and reconfigurable observing system. The Nodes in a sensor web interoperate with common standards and services. Sensor webs can be layered or linked together. (AIST, 2007)

The critical difference between a sensor web and a sensor network is the communication and semi-autonomous operation (and reconfiguration) of the heterogenous nodes. We report on the SEAMONSTER project which implements sensor web technology in Southeast Alaska and provides a specific example of the difference between a sensor network and a sensor web. The communication and reconfiguration aspects of a sensor web are described in more detail by AIST (2007) and Delin (2004).

Motivation

The University of Alaska Southeast campuses are located within the diverse ecosystems of Juneau, Sitka, and Ketchikan. The campuses are contained within the 17 million acre Tongass National Forest and border the Juneau Icefield, which contains 38 major glaciers covering 1,500 square miles, and a glacial fjord system containing thousands of islands.

The South East Alaska MOnitoring Network for Science, Telecommunications, Education, and Research (SEAMONSTER) is a NASA-sponsored smart sensor web project designed to support collaborative environmental science with near-real-time recovery of large volumes of environmental data. Southeastern Alaska is a challenging research environment because of access issues. Due to the mountainous terrain and the island nature of southeast Alaska, access to research sites is unpredictable and expensive and often a primary constraint to research. As a result, the development of a smart sensor web in southeast Alaska will be of tremendous benefit to ongoing earth science and ecological research by UAS and state and federal agency researchers. The UAS campus has access to diverse environments with both intertidal ocean and glaciers within 10 km of campus. The initial geographic focus is the Lemon Creek watershed near Juneau Alaska with expansion planned for subsequent years up into the Juneau Icefield and into the coastal marine environment of the Alexander Archipelago and the Tongass National Forest.

The overarching concept of SEAMONSTER is that the first generation of sensor web technologies currently exist. SEAMONSTER is an implementation of those technologies intended to act as an educational tool, a public resource, a scientific resource, and a sensor web testbed.

Seamonster Study Sites

The Lemon Creek watershed illustrated in Fig. 1 is accessible from Juneau while offering the extreme conditions found in Southeast Alaska. Micro-sensor clusters (motes) are indicated as red dots and Microserver (more capable computer) nodes are orange triangles in this example installation. The Lemon Creek watershed is relatively compact (~31 square kilometers) and begins on Lemon Glacier with two supra-glacial lakes at its head. These lakes have periodic outburst drainages that enter Lemon Creek proper via sub-glacial drainage channels. Below this glacier Lemon Creek flows through steep wilderness terrain, fed by several tributaries, before reaching a region of active mining and industrially zoned use, finally flowing past residential housing in Juneau and emptying into the ocean.

UAS has been involved in monitoring the outburst activity of supra-glacial lakes using in-situ pressure transducers with data-loggers. Instrumentation of the lakes,

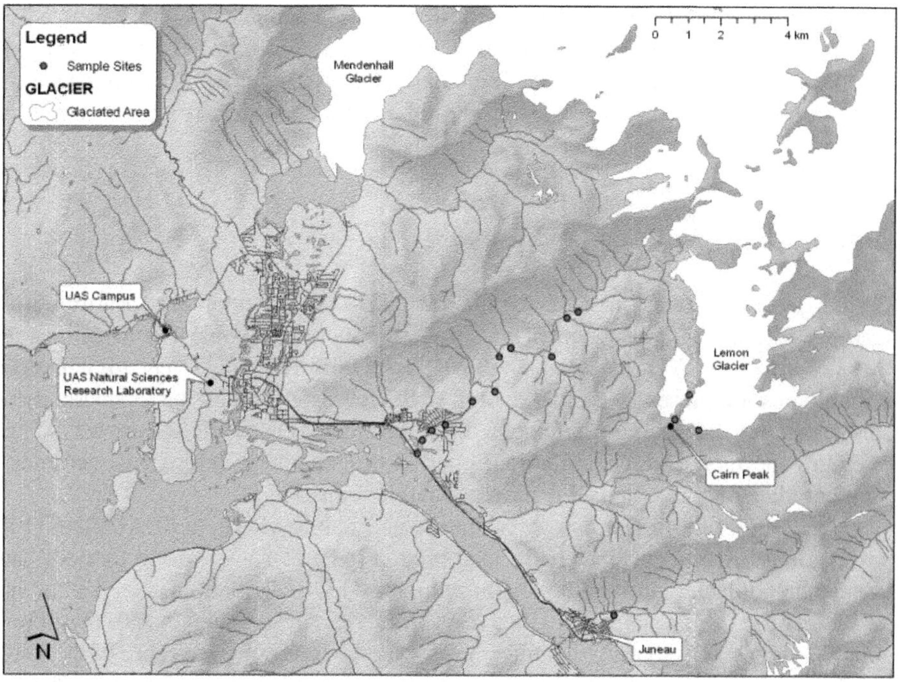

Fig. 1 The Lemon Creek watershed, showing motes, Microserver nodes, and glaciated areas

the Lemon Glacier surface (to study any glacial surge or reaction to lake drainages, via geophones and GPS observations), and throughout the watershed provide water quality monitoring, meteorological parameter monitoring (for autonomous prediction of possible drainage events and the resultant change in sensor-web state for enhanced data acquisition), and hazard monitoring. The Lemon Creek watershed is instrumented via a sensor web to monitor turbidity, flow, temperature, and meteorological parameters. The catastrophic drainage of the lakes and the resulting impact on the watershed is a transient event that requires sensor web technology to provide autonomous adaptation to current environmental conditions to correctly observe the impacts of the lake drainage on the Lemon creek Watershed.

The sensor platforms of SEAMONSTER operate via battery banks and solar panel recharging. The power constraint limits the sampling frequency and operation time of the computers on the sensor platforms. Therefore long sleep cycles are required for power management. When a sudden lake drainage is detected (either by a sudden drop in the pressure transducer readings in the lake, or a drastic change in the water temperature at the terminus of the glacier), all platforms can receive the alert signal and increase the sample rate to observe the lake drainage and associated watershed impacts.

A second SEAMONSTER study site is the Mendenhall Glacier, portrayed in Fig. 2. Changes in glacier volume and extent are key indicators of climate change. The Mendenhall Glacier is located 10 km northeast of the UAS campus. SEAMONSTER is used to monitor the mass balance (and retreat) of Mendenhall Glacier in partnership with the US Forest Service. The maintenance of a consistent mass balance program on the Mendenhall will help fill an approximately 3,000 km gap in glacier monitoring along the Pacific Coast of North America. The observations are communicated to the public using digital earth technologies to increase public access.

Digital Earth and Sensor Webs

Digital Earth is a phrase to describe the virtual 3-D representation of Earth that is spatially referenced and hyperlinked. Examples include Google Earth, Microsoft Virtual Earth, NASA Worldwind, and ESRI's ArcGIS framework. The digital earth technologies are presently rapidly evolving. Standards efforts by the Open Geospatial Consortium are being implemented for the SEAMONSTER sensor web. This leverages the increasingly available digital earth technology for sensor web management, data analysis, and education and public outreach.

Using Digital Earth technology, scientific and public access to SEAMONSTER observations of climate change is increased. Figure 2 shows the retreat of the Mendenhall Glacier during the 2007 summer. This information is given to the US Forest Service Mendenhall Glacier Visitors Center and shared with the about 300,000 visitors. By sharing this information via the Internet, even more of the public can be made aware of these observations (Connor and Prakash, 2008). A second example of the use of digital earth and sensor webs is the display of water quality

Created 9-11-07 by Logan Berner (berner.logan@gmail.com) of the University of Alaska SE Natural Science Department as part of the jointly funded NASA-NOAA SEAMONSTER project

Fig. 2 Mendenhall glacier terminus recession, 2007

parameters in Google Earth via kml as shown in Fig. 3. In the Lemon Creek Watershed, this plot shows water temperature by color (4.5–9.0°C, seen in legend on left). One can explore the drainage and see that the main channel is cooler (light blue) and dominated by glacial melt. The non-glacial tributaries (orange, red, yellow) are warmer before joining the main channel.

Additional education and public outreach activities include involving K-12 teachers participating in the NSF funded EDGE program in understanding sensor web applications for glacier research. Each teacher participating sends two K-12 students to a summer UAS workshop. The students and teachers are able to maintain involvement with the SEAMONSTER project through the digital earth files we distribute.

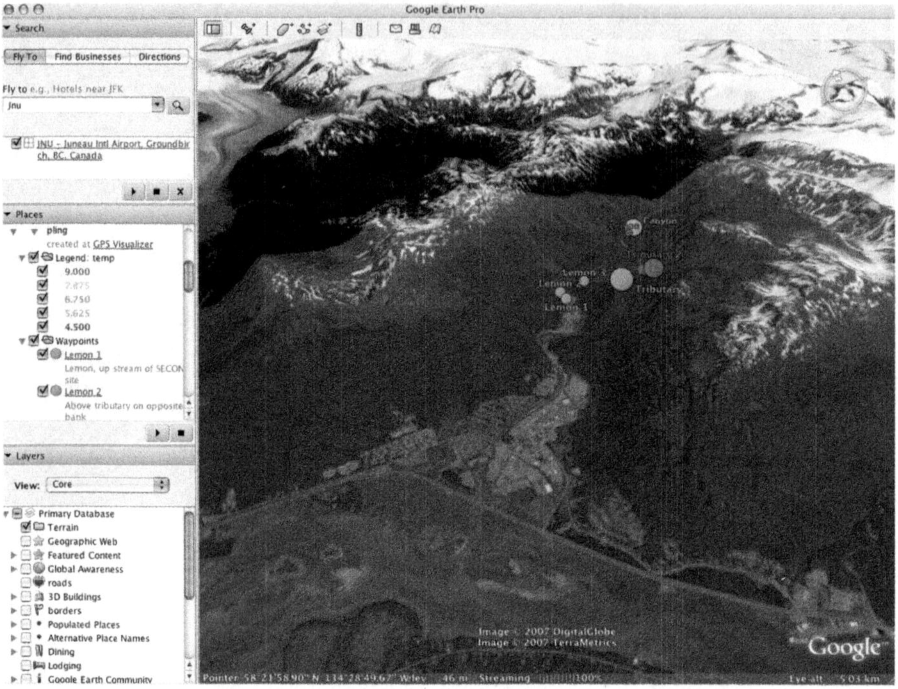

Fig. 3 Display of water quality parameters in Google Earth via kml

Finally, Lemon Glacier was studied as part of the 1957–1958 International Geophysical Year. Lemon Glacier studies for the 2007–2008 International Polar Year are creating a legacy of next generation scientists prepared for sensor web use.

Conclusions

Sensor webs represent the next technological step in environmental monitoring of both long-term climate trends and transient events. Incorporating sensor web and digital earth technologies improves scientific studies, educational efforts, and public outreach all associated with climate change. Plans for the immediate future include establishing SEAMONSTER study sites in other locations of interest in southeast Alaska, enhancing Web access to SEAMONSTER data to exploit more fully the notion of digital earth, and investigating agent technology (Wooldridge, 1997) as a way to realize collaboration among nodes in a sensor web.

Acknowledgments Funding for SEAMONSTER is provided by NASA Earth Science Technology Office grant AIST-05-0105, NOAA Education Partnership Panel Interdisciplinary Scientific Environmental Technology (ISET) Cooperative Science Center Grant, and NSF Research Experience for Undergraduates Grant No. 0553000. Logan Berner, Marijke Habermann, Erica Halford, Josh Jones, Edwin Knuth, Nick Korzen, Holly Moeller, David Sauer, Shannon Siefert, Suzie Teerlink, Larry Brewster and James Miller have provided SEAMONSTER support.

References

Arctic Climate Impact Assessment Scientific Report (ACIA), 2005, Cambridge University Press, Cambridge.

Report from the Earth Science Technology Office (ESTO) Advanced Information Systems Technology (AIST) Sensor Web Technology Meeting, February 13–14, 2007 http://esto.nasa.gov/sensorwebmeeting/files/AIST_Sensor_Web_Meeting_Report_2007.pdf

Delin, K.A. 2004. "Sensor Webs in the Wild." In N. Bulusu and S. Jha (Eds.), Wireless Sensor Networks: A Systems Perspective, Artech House, London.

Wooldridge, M. 1997. "Agent-based Software Engineering." IEEE Proc. on Software Engineering. 144: 26–37.

Connor, C.L. and A. Prakash. 2008. "Experiential Discoveries in Geoscience Education: The EDGE Program in Alaska." Journal of Geoscience Educ. 56: 179–186.

Thermal Characterization of Biodegradable Poly (Lactic Acid)/Clay Nanocomposites

Parakalan Krishnamachari, Jian Zhang, Jizhong Yan, Abolghasem Shahbazi, Leonard Uitenham, and Jianzhong Lou

Abstract Aliphatic polyesters such as polylactides have various applications due to their biodegradable and/or biocompatible character. Poly (L-lactide) is produced from L-lactic acid, which is derived itself from the fermentation of corn or sugar beet. PLA with a glass transition temperature around 55 °C is a relatively stiff and brittle polymer with low deformation at break. Polymer nanocomposites have been a subject of a large number of research studies in the last decade. However, to commercialize polymer nanocomposites (especially bio-degradable polymer nanocomposite) there are still a number of technical barriers. Polymer nanocomposites based on initially biodegradable Poly (lactic acid) (PLA) and organically modified layered silicates were prepared by melt processing using a Brabender twin screw mixer. Several organically modified montmorillonite (Nanoclay) were incorporated at 1% loading level into the PLA. However, a completely exfoliated morphology has been evidenced by X-ray diffraction analysis for the combination of PLA and the natural montmorillonite modified with a quaternary ammonium salt, Cloisite® 30B, this is the most hydrophilic among the organically modified montmorillonite. Further studies were done on the same combination with loading levels of 1%, 2%, 3%, 4% and 5%. Thermal stability of the Nanocomposites was studied using TGA. An increase in thermal stability with the clay content is observed by TGA with a maximum obtained for a loading of 3 wt% of 30B nanoclay. Glass transition and melting point data were collected and analyzed using DSC. The onset of Tg has been increased by the addition of the 30B nanoclay.

Introduction

Most biodegradable polymers have excellent properties comparable to many petroleum-based plastics and may soon be competing with commodity plastics. So, biodegradable polymers have great commercial potential for bio-plastic, but

---

J. Lou (✉)
Department of Chemical Engineering, North Carolina A&T State University, Greensboro, North Carolina 27411, USA

G.A. Uzochukwu et al. (eds), *Proceedings of the 2007 National Conference on Environmental Science and Technology*, DOI 10.1007/978-0-387-88483-7_29,

Table 1 Nanoclay commercial names and surface treatments

Nanoclay	CEC (meq/100 g clay)	Surfactants
Cloisite® Na+	92.6	None
Cloisite® 10A	125	$PhCH_2N^+(HT)(CH3)_2$
Cloisite® 15A	125	$(HT)_2N^+ (CH3)_2$
Cloisite® 20A	95	$(HT)_2N^+ (CH3)_2$
Cloisite® 25A	95	$(CH3)_2(C8H17)N^+(HT)$
Cloisite® 30B	90	$TN^+(CH3)(C2H4OH)_2$
Cloisite® 93A	90	$(HT)_2HN^+(CH3)$

HT: Hydrogenated Tallow Amine; T: Tallow Amine

some of the properties such as brittleness, low heat distortion temperature, high gas permeability, and low melt viscosity for further processing restrict their use in a wide-range of applications (Ray et al., 2003). Therefore, modification of the biodegradable polymers through innovative technology is a formidable task for materials scientists. On the other hand, nanoreinforcement of pristine polymers to prepare nanocomposite has already proven to be an effective way to improve these properties concurrently. So, preparation to processing of biodegradable polymer-based nanocomposites, that is, green nanocomposites are the wave of the future and considered as the next generation materials.

Materials: The Polymer – Poly lactic acid under the commercial name (Poly-D/L- lactide (PDLLA) – PLA 4060D) was supplied by NatureWorks. The resin was provided in the form of pellets. The various organoclay, under commercial names of Cloisite 30B, 15A, 20A, 25A, 93A and Na+ were supplied by Southern Clay Products Inc. (Gonzalez, TX) (see Table 1).

Nanocomposite Preparation: The mixtures were prepared by mixing the PLA polymer pellets with various nanoclays in a C.W. Brabender Twin Screw mixer. The sample was hand-mixed at room temperature before feeding into the Twin Screw. The temperature was set at 190°C. The rotation speed was set at 20 rpm. Brabender twin Screw Mixer – contains two 42 mm (1–5/8) diameter, counter-rotating (toward each other) intermeshing screws. For electrical heating the barrel is equipped with two heater collar zones of 2300 watts each.

Characterization

TA Instruments DSC Q100 was used to obtain the glass transition data. The samples were characterized using a temperature modulated DSC operated in a modulated mode at a heating rate of 5°C/min under an inert atmosphere. Perkin Elmer TGA 7 series was used to obtain the thermo gravimetric analysis. The test was conducted at heating rate of 5°C/min under inert conditions from 40°C to 600°C. The interlay spacing d_{001} was examined by D8 Discover Series 2 X-ray diffractometer manufactured by Bruker Axs. The measurements were carried out using reflection

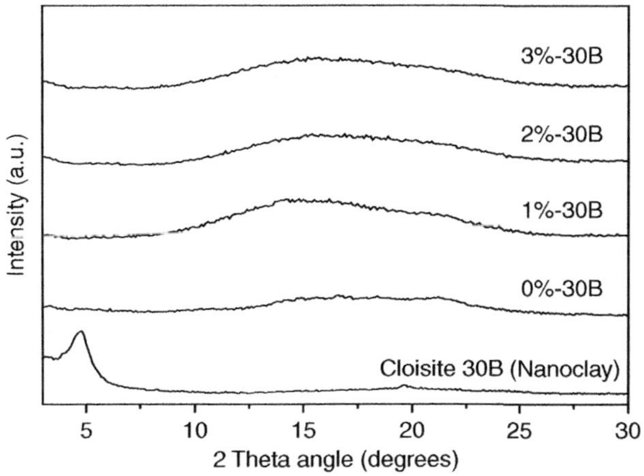

Fig. 1 XRD Diffractograms: Cloisite 30B and PLA with 0%, 1%, 2% and 3% (wt %) of Cloisite 30B

geometry and CuKα radiation (wavelength $\lambda = 0.154$ nm) operated at 40 kV and 100 mA.

Exfoliation from melt mixing: The structure of the hybrids formed as a result of melt mixing in the Brabender Twin Screw Mixer was analyzed using XRD. It should be noted that the Twin Screw was run at a very low rpm of 20 to enable more residence time for the melt mixing and also to avoid the molecular degradation of the PLA resin matrix due to excessive shear.

In Fig. 1, the XRD patterns of the Organoclay 30B and PLA/30B composites with 1%, 2% and 3% of 30B were compared. The primary silicate reflection at $2\theta = 4.9°$ ($d_{001} = 1.8$ nm) was recorded for the 30B powder. The unfilled PLA (0%-30B) was characterized by background scattering with low intensity. For the PLA/30B nanocomposites, there were no noticeable XRD peaks of 30B observed at low-angle range, confirming that the exfoliation of silicate layers of 30B in PLA matrix was obtained as a result of melt mixing. This lack of intergallery clay diffraction is due to the disorderly and random distribution of the clay platelets within the PLA matrix.

In contrast to the PLA/30B hybrids, the other PLA/Nanoclay hybrids were found to have a peak corresponding to the nanoclay in the diffraction patterns of the composites (see Fig. 2). While the peak was weak in some cases; it was very prominent in other cases. From this observation it could be concluded that apart from PLA/30B hybrids none of the other resin matrix/nanoclay combination have been exfoliated. It should be noted that the 30B is the most hydrophilic in nature among the various nanoclays.

Following this exfoliation data, DSC and TGA studies were done only on the PLA/30B nanocomposites with varying weight fractions of the nanoclay, Cloisite 30B.

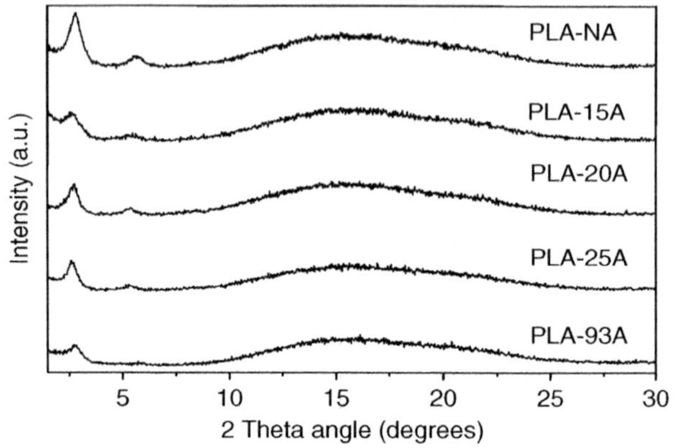

Fig. 2 XRD diffractograms: PLA with 1% wt of NA, 15A, 20A, 25A, 93A

DSC: The DSC Thermograms of the PLA-30B nanocomposites can be seen in Fig. 3 and the values for the glass transition temperatures (Tg) in Table 2. It can be clearly seen that the onset of Tg has been increased by the addition of the 30B nanoclay.

The Tg of the unfilled PLA (0% 30B in the Fig. 3) is 47.4°C. An increase of 6°C in the Tg can be observed by the addition of 1% 30B. A further 2°C increase is seen by the addition of 2% 30B. However at a loading of 3%, the Tg is 51.4°C which is

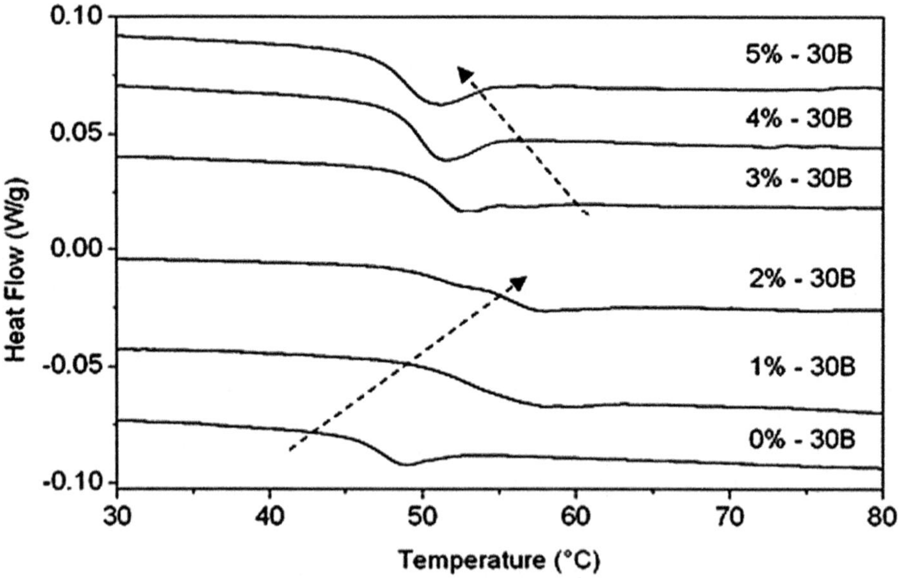

Fig. 3 DSC – Glass Transition temperature: PLA with 0%, 1%, 2%, 3%, 4% and 5% (wt %) of Cloisite 30B. The arrows indicate the trend of the Tg

Table 2 Glass transition temperatures of various composition of 30B nanoclay

Nanocomposite System (PLA-30B)	Glass transition temperature (°C)
0% – 30B (Unfilled Polymer)	47.39
1% 30B	53.22
2% 30B	55.22
3% 30B	51.34
4% 30B	49.93
5% 30B	48.96

still higher than the unfilled PLA but lower than that of the Tg of 2% filler loading. The Tg continues to go down at 4% and 5% loading levels of 30B.

The glass transition temperature is a complex phenomenon depending on various factors like chain flexibility, molecular weight, branching/cross linking, intermolecular attraction and steric effects (Cowie, 1991; Burton and Folkes, 1986). The increase in Tg that comes with the lower loading levels of the nanoclay 30B could be attributed to the decrease in free volume in the polymer nanocomposite due to cross-linking caused by the intermolecular attraction between the diols of the organic modifier (Cloisite 30B) with the C=O bonds present in the PLA.

The lowering of Tg by the addition of the Nanoclay has also been observed by (Lee et al., 2003). At higher loading levels the nanoclay particles may provide steric factors that play a role in increasing the chain flexibility of the PLA resin system. This increased flexibility and mobility of the PLA backbones leads to a decrease in the glass transition temperatures. It is possible that the plasticizing effects of clay surfactants are more pronounced at higher loading levels.

Thermogravimetric Analysis: An increase in thermal stability with the clay content is observed by TGA, with a maximum obtained for a loading of 3 wt% in 30B nanoclay (see Fig. 4). This increase in thermal stability could be attributed to

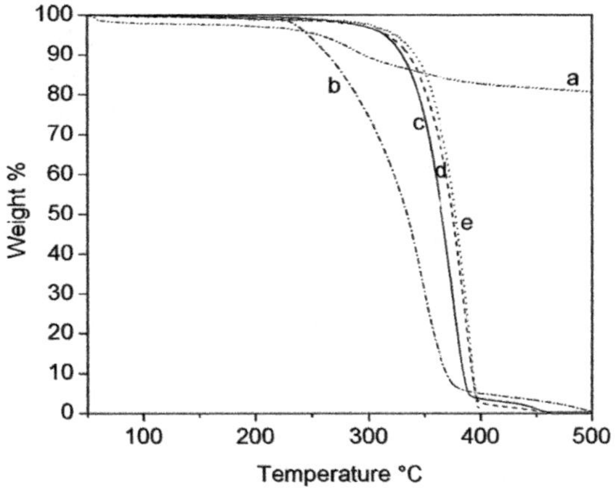

Fig. 4 TGA analysis of Cloisite 30B (a) and PLA with 0% (b), 1% (c), 2% (d) and 3% (e) of Cloisite 30B

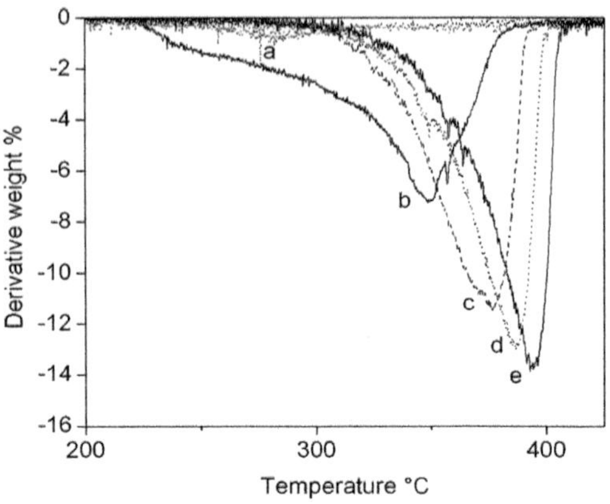

Fig. 5 TGA – Derivative Weight loss – Cloisite 30B (a) and PLA with 0% (b), 1% (c), 2% (d), and 3% (e) of Cloisite 30B

Table 3 Temperature at which maximum weight loss occurs for the PLA-30B nanocomposite system (as determined from the Derivate Weight loss curves)

Nanocomposite System (PLA-30B)	Temperature at max weight loss (°C)
Cloisite 30B	281.36
0% – 30B (Unfilled Polymer)	349.29
1% 30B	376.70
2% 30B	386.30
3% 30B	393.75

an ablative reassembling of the silicate layers which may occur on the surface of the nanocomposites creating a protective physical barrier. In addition to that the volatization might also be delayed by the labyrinth effect of the silicate layers exfoliated in the nanocomposites (Gilnian et al., 1998). The d-TGA curves (derivative weight loss curves) are a fair indication of the temperature at which the maximum weight loss is triggered (see Fig. 5 and Table 3). It could be clearly observed that there is a shift towards the positive direction with increasing loading levels of the nanoclay filler 30B. The increase in the thermal stability has also been attributed to the high thermal stability of clay and to the interaction between the clay particles and the polymer matrix. (Wen and Wikes, 1996; Fischer et al., 1999; Petrovic et al., 2000)

Conclusion

Polymer nanocomposites based on initially biodegradable Poly (lactic acid) (PLA) and organically modified layered silicates were prepared by melt processing using a Brabender twin screw mixer. Several organically modified montmorillonite (Nanoclay) were incorporated at 1% loading level into the Poly (Lactic Acid). However a completely exfoliated morphology has been evidenced by X-ray diffraction analysis only for the combination of the PLA and the natural montmorillonite modified with a quaternary ammonium salt namely Cloisite® 30B, the most hydrophilic among the organically modified montmorillonite.

Further studies were done on the same combination with loading levels of 1%, 2% and 3%. Thermal stability of the nanocomposites was studied using the TGA. An increase in thermal stability with the clay content is observed by TGA with a maximum obtained for a loading of 3 wt% of Cloisite 30B nanoclay. Glass transition and melting point data were collected and analyzed using the DSC. The onset of Tg has been increased by the addition of the 30B nanoclay.

References

Burton, R.H., Folkes, M.J. 1986. In: Clegg, D.W., Collyer, A.A. (ed.), Mechanical properties of reinforced thermoplastics. London: Elsevier.

Cowie, J.M.G. 1991 Polymers: chemistry & physics of modern materials, Chapman & Hall, New York.

Fischer, H.R., Gielgens, L.H., Koster, T.P. 1999. M Acta Polym, 50, 122.

Gilnian, J.W., Kashiavagi, T.C.L., Giannelis, E.P., Manias, E., Lomakin, S., Lichtenhan J.D. et al., 1998. In: Le Bras, M., Caniino, G., Bourbigot, S., Delobel, R. (eds.), Fire retardancy of polymers, The Royal Society of Chemistry, Cambridge.

Lee, J.H., Park, T.G., Park, H.S., Lee, D.S., Lee, Y.K., Yoon, S.C., Nam, J.D. 2003. Thermal and mechanical characteristics of poly(-lactic acid) nanocomposite scaffold, Biomaterials, 24(16), 2773–2778.

Petrovic, X.S., Javni, L., Waddong, A., Banhegyi, G.J. 2000. J Appl Polym Sci, 76, 133.

Ray, S.S., Yamada, K., Okamoto, M., Fujimoto, Y., Ogami, A. and Ueda, K. 2003 New polylactide/layered silicate nanocomposites. Polymer, 44(21), 6633–6646.

Wen, J., Wikes, G.L. 1996. Chem Mater, 8, 1667

Tools for Carbon Management: Potential Carbon Footprint Reduction Through Fuel Switching

Parikhit Sinha and David M. Cass

Abstract As institutions of higher education try to meet the American College and University Presidents' Climate Commitment for moving toward a carbon neutral campus, they can utilize a number of carbon management tools, including greenhouse gas accounting protocols, carbon-friendly technologies, and carbon markets. However, carbon management strategies may have substantial costs, and it may be prudent to take a cost/benefit approach to evaluating them. In the case of fuel switching as a greenhouse gas emission reduction strategy, the costs and benefits of switching from fossil fuels to renewable fuels were evaluated for industrial boilers (annual heat input of 450,000 MMBtu/year) servicing a large (5 million gross square feet) campus. Switching from #6 fuel oil to B100 biodiesel led to a substantial (90%) carbon footprint reduction, but at a considerable annual incremental cost ($7.56 million). Switching from #6 fuel oil to a more efficient fossil fuel type (natural gas) resulted in a moderate (31%) carbon footprint reduction at a more reasonable annual incremental cost ($1.3 million).

Introduction

In the United States, debate over climate change science has shifted to debate over climate change action. This shift has come about due to the increased scientific consensus on the causes, trends, and potential impacts of anthropogenic climate change, combined with heightened public awareness, state-led climate action, and understood links between climate change and energy security. For institutions and businesses, climate change presents risks and opportunities. Risks include climate risks (e.g. extreme weather events, sea level rise, changes in water supply, altered biological patterns, etc.) and regulatory risks (stroke-of-the-pen risks that result in new rules for environmental compliance). Along with risks, opportunities include enhanced demand for green technology and sustainable business practices. Institutions and businesses that take leadership roles in managing greenhouse gas

P. Sinha (✉)
O'Brien & Gere, Blue Bell, Pennyslvania, USA

G.A. Uzochukwu et al. (eds), *Proceedings of the 2007 National Conference on Environmental Science and Technology*, DOI 10.1007/978-0-387-88483-7_30,

(GHG) emissions will be in a better position to adapt to a future carbon-constrained world.

A number of carbon management tools are available for managing GHG emissions, including greenhouse gas accounting protocols, carbon-friendly technologies, and carbon markets. Greenhouse gas accounting protocols guide the development of a carbon footprint through establishment of organizational and operational boundaries for direct and indirect energy use, emission factors for greenhouse gases, rules to avoid double-counting, and tools for tracking emissions over time. Carbon-friendly technologies are those that aim to maximize energy efficiency, reuse waste fuel or energy, make use of renewable fuels, and/or sequester greenhouse gases, in an effort to minimize greenhouse gas emissions per energy provided. When combined with demand-side energy conservation practices, carbon-friendly technologies can be used to meet an institution's carbon footprint reduction goals. Carbon markets (e.g., European Climate Exchange, Chicago Climate Exchange) are market mechanisms used to establish the price of carbon and allow it to be traded in order to manage GHG emissions with minimum economic harm. Carbon can be traded when a verified GHG emission reduction project is used to generate a carbon credit, or when a market member's carbon footprint reduction goals are exceeded and generate carbon credits.

In the area of higher education, the American College and University Presidents' Climate Commitment (ACUPCC) is a high-visibility effort to address global warming by garnering institutional commitments to neutralize GHG emissions, and to accelerate the research and educational efforts of higher education to equip society to stabilize the earth's climate. As institutions of higher education, universities have inherent leadership roles in the scientific community and may also serve as role models for sustainability. Universities that take on the ACUPCC charge can initiate short and long-term strategies toward achieving carbon neutrality. Specifically, upon signing the ACUPCC, universities agree to create institutional structures to guide the development and implementation of a carbon management plan, to create a comprehensive inventory of annual GHG emissions, and to develop an institutional action plan to achieve climate neutrality.

In developing an overall strategy for carbon management, the costs and benefits of potential strategies may be considered prior to choosing a course of action. The costs can be quantified in terms of investment required per unit of carbon dioxide equivalent emissions reduced, and benefits can be quantified in terms of percentage reduction in carbon footprint. The costs and benefits of fuel switching, a fundamental carbon management strategy, are presented below for a large university campus serviced by an industrial boiler plant.

Methods

Greenhouse gas emissions are closely tied to combustion of fossil fuels. Therefore, switching from fossil fuels to renewable fuels is a key carbon management strategy. A feasibility analysis of fuel switching was performed for an institution of higher education equipped with industrial boilers providing steam and heat to campus buildings (5 million gross square feet [GSF]). GHG emissions and fuel costs based on the U.S. Department of Energy (EIA, 2007) were estimated for five types of fuel: #6 fuel oil, #2 fuel oil, natural gas, B20 biodiesel (20% biodiesel and 80% #2 fuel oil), and B100 biodiesel (100% biodiesel).

Annual volume throughput of all fuels were normalized to yield 450,000 MMBtu of heat content. GHG emissions for each fuel type were calculated in pounds of CO_2 equivalent per year (lb CO_2E/yr) using the following equation:

$$Emission = \sum_{GHG} AT \times FD \times FHC \times EF \times GWP \quad (1)$$

with input parameters from WRI and WBCSD (2005) summarized in Table 1.

Table 1 Emission parameters

Fuel	Annual throughput (AT)	Fuel density (FD)	Fuel heating content (FHC)	GHG emission factor (EF)	Global Warming Potential (GWP)
#6 Oil	3,036,437 gal	7.8 lb/gal	19 MBtu/lb	174.5 lb CO_2/MMBtu, 0.024 lb CH_4/MMBtu, 0.0015 lb N_2O/MMBtu	CO_2=1, CH_4=21, N_2O=296
#2 Oil	3,169,014 gal	7.1 lb/gal	20 MBtu/lb	161.3 lb CO_2/MMBtu, 0.024 lb CH_4/MMBtu, 0.0015 lb N_2O/MMBtu	CO_2=1, CH_4=21, N_2O=296
Natural Gas	475,687,103; cubic feet (cf)	0.043 lb/cf	22 MBtu/lb	121.0 lb CO_2/MMBtu, 0.013 lb CH_4/MMBtu, 0.0003 lb N_2O/MMBtu	CO_2=1, CH_4=21, N_2O=296
B20 Biodiesel	3,267,024 gal	7.1 lb/gal	19.4 MBtu/lb	165 lb CO_2/MMBtu, 0.734 lb CH_4/MMBtu, 0.010 lb N_2O/MMBtu	CO_2=1, CH_4=21, N_2O=296
B100 Biodiesel	3,781,512 gal	7.0 lb/gal	17 MBtu/lb	179.7 lb CO_2/MMBtu, 0.734 lb CH_4/MMBtu, 0.010 lb N_2O/MMBtu	CO_2=1, CH_4=21, N_2O=296

Since biodiesel is a renewable fuel, emissions of CO_2 from combustion of biodiesel were not included in the GHG inventory. However, since combustion of biodiesel emits other GHGs (CH_4 and N_2O), which are not sequestered during the growth of biodiesel feedstock, emissions of CH_4 and N_2O from biodiesel combustion were included in the GHG inventory. For B20 biodiesel, only 20% of the CO_2 emissions were discounted, whereas for B100 biodiesel, 100% of the CO_2 emissions were discounted. Annual GHG emissions were reported for each fuel and normalized with respect to GSF. Annual fuel costs were calculated and also normalized with respect to GSF.

Results and Discussion

Annual greenhouse gas emissions from industrial boilers (annual heat input of 450,000 MMBtu/year) servicing a 5 million gross square foot campus are shown in Table 2 for five fuels. Overall, #6 fuel oil is the least expensive of the five fuels and has the highest GHG emissions; B100 biodiesel is the most expensive of the five fuels and has the lowest GHG emissions. Natural gas is the second least expensive fuel and also has the second lowest GHG emissions, emitting less than B20 biodiesel.

In order to discuss the feasibility of switching from #6 Fuel Oil to other fuels for industrial boilers, costs and benefits of fuel switching are presented in Table 3. Due to the substantial additional annual cost ($7.56 million) of switching from #6 fuel oil to B100 biodiesel, the 90% reductions in GHG emissions from B100 biodiesel may be cost-prohibitive. However, switching from #6 fuel oil to natural gas can result in significant (31%) reductions in GHG emissions at a lesser incremental annual cost

Table 2 Fuel switching comparison

Fuel type	GHG emissions (lb CO_2E/yr)[a]	GSF served	GHG lbs per GSF served	Cost of fuel type per unit[e]	Cost per GSF (USD)
#6 Fuel Oil	78,966,671	5,000,000	15.8	1.138 USD/gal	0.69
#2 Fuel Oil	73,026,676	5,000,000	14.6	2.162 USD/gal	1.37
Natural Gas	54,606,402	5,000,000	10.9	1.001 cents/cf	0.95
B20 Biodiesel[b]	60,074,987	5,000,000	12.0	2.362 USD/gal	1.54
B100 Biodiesel[c,d]	8,268,299	5,000,000	1.7	3.162 USD/gal	2.39

[a] – Emissions calculated using Equation 1 and Emission Parameters table (WRI and WBCSD, 2005).
[b] – B20 Biodiesel costs an average of $0.20 more than #2 Fuel Oil. (cost per unit B20 biodiesel = cost per unit #2 Fuel Oil + $0.20)
[c] – B100 Biodiesel costs an average of $1.00 more than #2 Fuel Oil. (cost per unit B100 biodiesel = cost per unit #2 Fuel Oil + $1.00)
[d] – Cost of B100 does not include cost to convert boilers to accommodate B100 fuel characteristics
[e] – Fuel costs taken from the U.S. Department of Energy (EIA, 2007).

Table 3 Feasibility of fuel switching

Fuel switching (from #6 fuel oil)	Additional annual cost ($ relative to #6 fuel oil)	Potential emission reductions (lb CO_2E/yr relative to #6 fuel oil)	Potential emission reductions relative to #6 fuel oil (% of existing footprint)
#2 Fuel Oil	$3,395,943	5,939,996	8%
Natural Gas	$1,306,163	24,360,269	31%
B20 Biodiesel	$4,342,921	18,891,684	24%
B100 Biodiesel	$7,556,298	70,698,373	90%

($1.3 million). Another way to compare costs and benefits between the fuel types is to evaluate the cost of each fuel per percentage point of GHG footprint reduction achieved. For example, natural gas costs $42,340.81 per percentage point of GHG reduction, whereas B100 Biodiesel costs $84,400.20 per percentage point of GHG reduction. B100 is approximately twice as expensive as natural gas in achieving a set GHG reduction goal. However, it is capable of achieving a much larger overall reduction, if fully implemented.

Conclusions

Institutions of higher education can serve as role models for sustainability by developing carbon management plans with the eventual aim of achieving a carbon neutral campus. However, the costs associated with implementation of GHG emission reduction strategies can be substantial, and it may be prudent to take a cost/benefit approach to evaluating different strategies. In the case of fuel switching as a GHG emission reduction strategy, the costs and benefits of switching from fossil fuels to renewable fuels was evaluated for industrial boilers (annual heat input of 450,000 MMBtu/year) servicing a large (5 million GSF) campus. Switching from #6 fuel oil to B100 biodiesel led to a substantial (90%) carbon footprint reduction, but at a considerable annual incremental cost ($7.56 million). Switching from #6 fuel oil to a more efficient fossil fuel type (natural gas) resulted in a moderate (31%) carbon footprint reduction at a more reasonable annual incremental cost ($1.3 million).

Acknowledgments We would like to acknowledge O'Brien & Gere for their encouragement of original technical research.

References

Energy Information Administration (EIA). 2007. Gasoline and Diesel Fuel Update. United States Department of Energy, Washington, DC. (Available at: http://tonto.eia.doe.gov/oog/info/gdu/gasdiesel.asp).

World Resources Institute (WRI) and World Business Council for Sustainable Development (WBCSD). 2005. Calculation Tool for Direct Emissions from Stationary Combustion Calculation Worksheets, Version 3.0. (Available at: http://www.ghgprotocol.org).

Coral Disease and Global Climate Change

Garriet W. Smith and Marie-Ange Smith

Abstract The health of the world's coral reefs have steadily declined over the past two decades. The primary diver of this decline is as an increase in shallow ocean temperatures. Increased temperatures stress the normal metabolism of corals and change the normal microbiota associated with healthy colonies. Often the result is the development of diseased colonies. We tested the susceptibility of the scleractinian coral *Montastrea faveolata* to infection by a number of *Vibrio spp.* which were previously isolated from diseased corals at *in situ*temperatures (28°C) and increased temperatures (31°C) in temperature controlled seawater aquaria. Both inoculated and control corals in aquaria kept at *in situ*temperatures showed no disease signs after two weeks. Inoculated corals in aquaria kept at increased temperatures developed disease signs after three days, although control corals at increased temperatures remained disease free. We suggest that the inoculated, increased temperature corals developed disease signs due to (1) decreased resistance under increased temperatures, (2) increased growth rates of the pathogens, (3) initiation of the transcription of toxin genes, or (4) a combination of the first three. We suggest that the increase of global coral disease is due, in part, to increased global temperatures.

Introduction

Global climate change has affected ecosystems in a number of ways including; range extensions of various species, phase-shifts of dominant organisms (Harvell et al., 2002), changes in parasitic loads (Torchin et al., 2003) and an increase in emerging infectious diseases (Daszak et al., 2000). This has been well documented in the marine environment for both vertebrate and invertebrate hosts (Harvell et al., 1999) and coral animals are no exception (Harvell et al., 2007).

G.W. Smith (✉)
University of SC, Aiken, SC, USA

G.A. Uzochukwu et al. (eds), *Proceedings of the 2007 National Conference on Environmental Science and Technology*, DOI 10.1007/978-0-387-88483-7_31,

Increased seawater temperatures affect corals and lead to a disease state in a number of ways. Healthy coral animals are composed of the coral tissue, symbiotic diatoms (zooxanthellae) and a normal microbiota (Gil-Agudelo et al., 2007). This holobiont normally exists near it's maximum temperature range. Increases in temperatures can disrupt this symbiotic association in a number of ways including; changing the normal microbiota (Gil-Agudelo et al., 2007), changing the normal clade distribution of zooxanthellae (Baker, 2001) or decreasing normal resistance mechanisms of the host (Harvell et al., 2007). Coral diseases occurring after exposure to increased seawater temperatures have been reported for both gorgonian (Harvell et al., 2001) and scleractinian (Rosenberg and Ben-Haim, 2002; Jones et al., 2004) corals.

Bacteria in the genus *Vibrio* cause coral bleaching (a type of disease) and adhesion of the bacterium to the coral surface is temperature dependent (Toren et al., 1998). *Vibrio charchariae* was identified as the cause of white-band disease (Gil-Agudelo et al., 2006), also associated with increased temperatures. Cervino et al. (2004) isolated *Vibrio* species from corals with yellow-band disease. Aquaria inoculations with these bacterial strains resulted in signs of yellow-band disease and elevated temperature increased the rate of spread of the disease and coral mortality. Yellow-band disease was shown to be a disease primarily affecting the zooxanthellae (Cervino et al., 2004). Previous host-pathogen studies were conducted in the Florida Keys although yellow-band disease is found throughout the Caribbean. To determine if the Florida Keys isolates could cause yellow-band disease in corals at the other extreme of the Caribbean (Mexico), and if temperature played a role, we inoculated coral fragments from Mexico under two temperature regimes and observed the result.

Materials and Methods

Experiments were conducted at the Marine Station of the Universidad Nacional Autonoma de Mexico in Puerto Morelas, Yucatan Peninsula. Coral fragments (4 × 4 cm) were removed from *Montastraea annularis*, growing on reefs near the marine station. These were placed on PVC ring stands filled with modeling clay. Four rings, each containing a coral fragment, were placed in a one-liter beaker and 500 ml of seawater was added. Four beakers (for each temperature treatment) were randomly placed into flow through temperature-controlled aquaria (28°C and 31°C). All coral fragments were allowed to equilibrate at temperature for 48 h before inoculation.

Inoculations were made by adding approximately 10^8 cells to a 10 ml solution of sterile seawater and 0.5 g sterile carbonate sand. The slurry was allowed to incubate overnight so that cells could absorb to the sand particles. Yellow-band strains used in this study were previously isolated by Cervino et al. (2004) and the designations used were M23 (YBM 23, GenBank accession number AY770832), M3121 (FLG2A, AY770830) and FL3122 (YBFL 3122, AY770831). As a control, a slurry of *Escherichia coli* K12 was also prepared. After acclimation, inoculation was performed by withdrawing 0.5 ml of the slurry (containing approximately 10 sand grains) and delivering the slurry directly to the surface of the coral fragments with a pipette.

All fragments were observed twice a day for signs of yellowing. After three days, 0.1 ml of mucus was removed from the surface of each fragment near the inoculation site and plated on TCBS (Difco). The carbon source utilization patterns (BioLog) of yellow colonies arising on the medium, were compared with the inoculum strains. At the end of the experiment, treatment water (in the beakers) was treated with a hypochlorite solution and disposed of on land.

Results and Discussion

One fragment inoculated with M23 and incubated at 31°C showed yellowing after one day. After two days (Fig. 1), two (of the four) fragments inoculated with M23, two fragments inoculated with M3121 and one inoculated with FL3122, all incubated at 31°C, showed yellowing (Table 1). By day four, all fragments inoculated with yellow-band isolates and incubated at 31°C showed yellowing. None of the

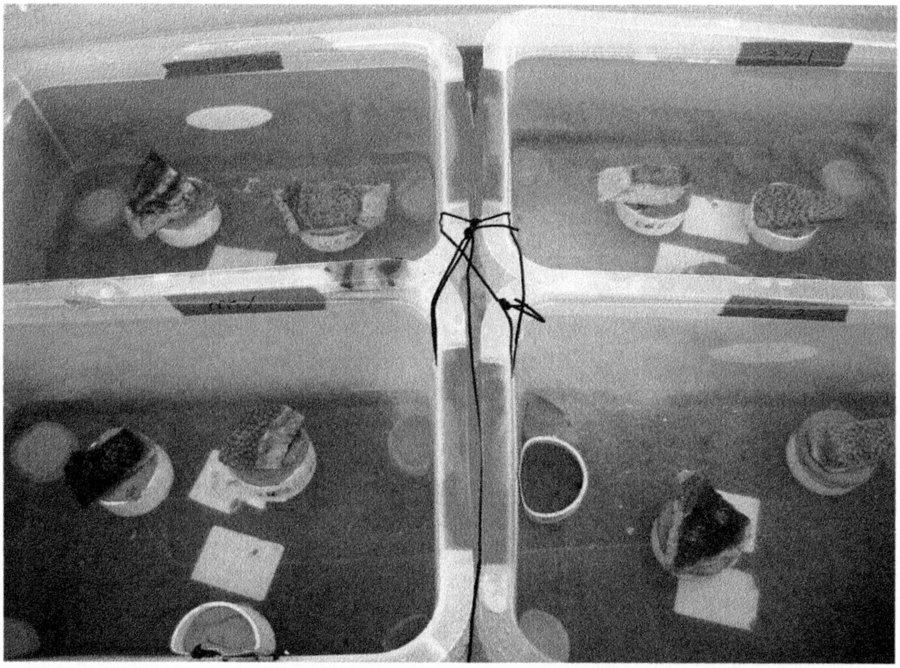

Fig. 1 Examples of coral fragments removed from experimental beakers for photographing after two days. *Top left*: FL3122. Fragment on *right* is yellowing (*lighter color*). *Top right*: M3121. Both are yellowing *(lighter shade)*. Both *Bottom* containers have M23 inoculated fragments on the *right* side and controls on the *left* side

Table 1 Progression of yellowing among coral fragments

Temperature	28°C				31°C			
Strain	Control	M23	M3121	FL3122	Control	M23	M3121	FL3122
Day	1,2,3	1,2,3	1,2,3	1,2,3	1,2,3	1,2,3	1,2,3	1,2,3
# Yellowed	0,0,0	0,0,0	0,0,0	0,0,0	0,0,0	1,2,4	0,2,4	0,1,4

*E. coli*inoculated fragments showed yellowing areas and none of the samples incubated at 28°C showed yellowing after three days. The *E. coli* inoculated fragments and fragments left at 28°C were observed for another two and a half weeks and still showed no signs of yellow-band disease.

Mucus samples obtained after three days and plated on TCBS, showed at least some *Vibrio* on all samples. Analysis of carbon source utilization patterns showed that although a number of *Vibrio* strains were present in all samples, those that corresponded with the inocula (M23, M3121 and FL3122) were only found in mucus that had been inoculated with each specific strain. In other words, this indicates there was no cross contaminations among the treatments. In addition, we were unable to reisolate *E. coli* from fragments inoculated with *E. coli*.

Conclusions

A statistical analysis of this experiment was unnecessary since all of the putative pathogens caused yellowing in all of the 31°C replicates, but none of the 28°C fragments or controls. This is not surprising because the strains used were known yellow-band associates. Nevertheless, the experiment showed that all strains could cause yellowing in corals other than those in the Florida Keys and only under increased temperature regimes. The mechanism for temperature-induced pathogenesis on corals is only known for one pathosystem. That is for bleaching of *Oculina patagonica* by *Vibrio shiloi*. Adhesion of the bacterium to the coral host only occurs at increased temperatures. Additionally, only after adhesion and penetration of the bacterium does the production of a proline rich toxin occur. It is interesting that both *Vibrio shiloi* and the yellow-band vibrios are zooxanthellae pathogens. Presently it is not known if the yellow-band strains also produce toxins that lead to disease, but this is not uncommon among this genus. This paper also reinforces the concern for coral reef health in an environment affected by global warming.

Acknowledgments This work was supported by a grant from the World Bank Targeted Research Program through the Global Environmental Fund to the Coral Disease Working Group. Photographs were taken by Dr. Jessica Ward, SIO, San Diego, CA.

References

Baker, A.C. 2001. "Bleaching of reef corals promotes rapid response to environmental change." Nature 411:765–766.

Cervino, J.M., R.L. Hayes, T.J. Goreau, S.W. Polson, S.C. Polson, R.J. Martinez and G.W. Smith. 2004. "Vibrio spp. Infection and elevated temperatures as links to Yellow Blotch/Band disease in the Caribbean." Applied Environmental Microbiology 40(11):6855–6864.

Daszak, P., A.A. Cunningham and A. D. Hyatt. 2000. "Emerging infectious diseases of wildlife-threats to biodiversity and human health." Science 287:443–449.

Gil-Agudelo D.L., C. Myers, G.W. Smith and K. Kim. 2006. "Changes in the microbial communities associated with *Gorgonia ventalina*during Aspergillosis infection." Diseases of Aquatic Organisms. 69:89–94.

Gil-Agudelo D.L., D.P. Fonseca, E. Weil, J. Garzón-Ferreira and G.W. Smith. 2007. "Bacterial communities associated with the mucopolysaccharide layers of three coral species affected and unaffected with dark spots disease." Canadian Journal of Microbiology 53:465–471.

Harvell, C.D., K. Kim, J.M. Burkholder, R.R. Colwell, P.R. Epstein, J. Grimes, E.E. Hofmann, E. Lipp, A.D.M.E. Osterhaus, R. Overstreet, J.W. Porter, G.W. Smith and G. Vasta. 1999. "Emerging marine diseases: Climate links and Anthropogenic factors." Science 285:1505–1510.

Harvell, C.D., K. Kim, C. Quirolo, J. Weir and G.W. Smith. 2001. "Coral bleaching and disease: contributors to 1998 mass mortality in *Briareum asbestinum*(Gorgonacea)." Hydrobiologia 460:97–104.

Harvell, C.D., C.E. Mitchell, J.R. Ward, S. Altizer, A.P. Dobson, R.S. Ostfeld, and M.D. Samuel. 2002. "Climate warming and disease risks for terrestrial and marine biota." Science 296:2, 158–162.

Harvell C.D., S. Merkel, E. Jordan-Dahlgren, E. Rosenberg, L. Raymundo, G. Smith, E. Weil and B. Willis. 2007. "Coral Diseases: the balance between coral and microbial associates." Oceanography 20:58–81.

Jones, R.J., J. Bowyer, O. Hoegh-Guldberg and L.L. Blackall. 2004. "Dynamics of a temperature-related coral disease outbreak." Marine Ecology Progress Series 281:63–77.

Rosenberg, E. and Y. Ben-Haim. 2002. "Microbial diseases of corals and global warming." Environmental Microbiology 4:318–326.

Torchin, M.E., K.D. Lafferty, A.P. Dobson, V.J. McKenzie and A.M. Kuris. 2003. "Introduced species and their missing parasites." Nature 421:628–630.

Toren, A., L. Landau, A. Kushmaro, Y. Loya and E. Rosenberg. 1998. "Effect of temperature on adhesion of *Vibrio* strain AK-1 to *Oculina patagonica* and on coral bleaching." Applied and Environmental Microbiology 64:1379–1384.

Climate Change, Drought, and Wetland Vegetation

Brant W. Touchette and Sara E. Steudler

Abstract Many climate change models forecast global increases in drought severity and duration along mid-latitudes. Areas that would typically receive ample precipitation allowing for the maintenance of complex vegetative assemblages may undergo substantial declines in species numbers as available surface and subsurface waters diminish. Few studies have considered climate-associated changes on aquatic and wetland systems. This is of particular concern, as wetlands are among the most threatened ecosystems globally. Therefore the purpose of this study was to understand how low-water availability influences the growth and productivity of an ecologically-important wetland plant, water willow (*Justicia americana* L.), in southeastern United States. In this study we used experimental microcosms maintained under controlled greenhouse conditions. Plants were allowed to acclimate in the microcosms for six weeks prior to initiating experimental treatments which included completely emerged (control), simulated 1-in-5 year precipitation low (for 2-, 4-, and 6-wks), and 1-in-25 year drought (for 2-, 4-, 6-wks). The results indicate that *J. americana* is relatively resistant to short periods of water deficit, even as relative water content (θ) in tissues fell below 40 percent of full capacity. Extended periods of severe drought (greater than 2 wks for 1-in-25 yr precipitation), however, resulted in substantial reductions in plant growth, productivity, and survival. Thus, while *J. americana* can tolerate short durations of water sacristy, long-term water deprivation as predicted in some climate change models could adversely affect the overall productivity and survival of this ecologically-important wetland plant species.

B.W. Touchette (✉)
Elon University, Elon, NC, USA

G.A. Uzochukwu et al. (eds), *Proceedings of the 2007 National Conference on Environmental Science and Technology*, DOI 10.1007/978-0-387-88483-7_32,

Introduction

Climate change models vary considerably with respect to possible scenarios for regional temperatures and precipitation. The Hadley model (HadCM2), for example, predicts minimal warming along with significant increases in precipitation (~22%) in southeastern United States by the end of the 21st century. In contrast, the Australian Bureau of Meteorology Research Center (BMRC) and the Canadian Global Coupled Model (CGCM) forecast somewhat dryer conditions resulting in a decrease in precipitation between 3 and 22% (Rosenberg and Edmonds 2005). Clearly the predicated outcomes of various models for the Southeast, including the amount and direction of future precipitation, are subjected to considerable uncertainty. Nevertheless, one consequence of elevated temperatures is lower soil moisture, as higher temperatures promote greater evapotranspiration. Therefore, any projected increase in temperature, even with slight increases in precipitation, will likely influence vegetative assemblages in the Southeast and possibly favor plants that are more tolerant to both warmer temperatures and dryer soils.

In addition to the expected increases in evapotranspiration, extended periods of little to no rainfall would likely worsen already compromised plant-water relations. Drought as defined as the absence of appreciable precipitation over extended periods allowing for the depletion of soil moisture while promoting injury to plants (Kramer 1983), is largely dependent on annual and seasonal climate conditions. Regions that receive comparatively high annual precipitation but with minimal rain events during the growing season, will likely be more susceptible to drought in comparison to areas that receive more temporally distributed rainfall. Therefore, basing precipitation entirely on an annual rainfall could discount damaging drought events – especially those that often occur in mid- to late-summers in the Southeast. The response of vegetation to decreased water availability from projected drought conditions is of great concern in most climate change scenarios (Wigley et al. 1984; Hanson and Weltzin 2000). While there are considerable differences in the projected direction and magnitude of future precipitation scenarios (Hanson and Weltzin 2000), it has generally been accepted that future climate conditions will include increases in both drought severity and duration (Gregory et al. 1997; Hanson and Weltzin 2000). Indeed, based on the CGCM, BMRC, and HadCM2 scenarios, droughts in North America will likely intensify in the Southeast, southern Rocky Mountains, and parts of the Northwest by the end of the century.

There is growing interest in defining drought-associated changes in natural vegetation; however, most of the studies to date are limited to terrestrial ecosystems such as grasslands and forests (e.g., Mangan et al. 2004; Breshears et al. 2005). Considerably fewer studies have addressed climate-associated changes in water availability on freshwater systems (Sorenson et al. 1998; Dawson et al. 2003). Therefore, the purpose of this study was to evaluate growth, biomass allocations, and water relations in an herbaceous wetland plant (*Justicia americana* L. Vahl.) to simulated drought conditions (up to 6 wks of a 1-in-5 and 1-in-25 year precipitation low). *J. americana* was selected as a model species for this study because of its ecological importance, including its role as a dominant littoral plant within many lakes and reservoirs of the Southeastern United States – a region that may experience appreciable declines in soil moisture by the end of the century.

Materials and Methods

Newly emerged shoots of *Justicia americana* were collected in early spring from natural populations in the central Piedmont of North Carolina (Badin Lake).Shoots were transplanted into greenhouse maintained microcosms (15 L) containing natural lake sediment (~12 cm; sandy clay soil texture) and water (~ 5 cm above soil level) at plant densities observed in natural field populations (~200 shoots m^{-2}). Plants were allowed to acclimate to above conditions for 6 weeks to ensure proper plant establishment as indicated by new leaf development and stem elongation. Following the acclimation period, simulated drought conditions were initiated by removing surface water, over a one week period, until water levels reached the upper portion of the substratum. During this period, drought-treated plants (n = 5 for each treatment) received watering of foliage and sediment comparable to a 1-in-5 (1.1 cm water wk^{-1}) or a 1-in-25 (0.6 cm water wk^{-1}) year low precipitation rate and periodicity (based on summer precipitation data from the central Piedmont region of North Carolina over a 75 year period). Control microcosms (n = 5) were watered at mean summer rates and periodicity (2.9 cm wk^{-1}) while maintaining water levels at 5 cm above the substratum. Drought-treated microcosms were exposed to 2-, 4-, or 6-weeks of simulated drought prior to returning to pre-drought conditions (i.e., mean summer precipitation rates and periodicity, and 5 cm of standing water).

Leaf relative water content (θ) was evaluated weekly according to Barr and Weatherley (1962) as modified by Touchette (2006) using the following equation:

$$\theta = \frac{(W_f - W_d)}{(W_f - W_d)} \times 100$$

where W_f was the fresh weight recorded during collection, W_t was the turgid weight, and W_d was the oven dry weight (60°C, until constant weight). Turgid weight was determined by placing leaf samples in 50 mL sealed vials containing deionized water, and allowing the samples to reach full turgor in darkness overnight. Soil water content was evaluated using the gravimetric technique described by Black (1965), and expressed as mass of water present per dry weight of sample. A small soil core (1 cm in diameter) was used to collect the sample through the 12 cm soil profile. Water mass was determined by weighing the soil sample before and after drying (60°C until constant weight).

Phytomass was determined by carefully removing plants from microcosms at the end of 8 weeks, which included the experimental drought period (between 2- and 6-wks) and a recovery period (up to 6 wks). The plants were sorted between above- (stems and leaves) and belowground tissues (roots and rhizomes), and dried (60°C) to constant weight.

Repeated measure analysis of variance (ANOVA; general linear model [GLM] procedure) were performed using SAS statistical software (SAS Institute Inc., Cary, NC) on tissue water content and soil moisture data. A pairwise multiple comparison procedure (Holm-Sidak) was used for post-hoc evaluations. Biomass data for control and drought treated plants were compared using a one-way ANOVA followed by a Holm-Sidak post-hoc analysis. All comparisons were considered significant at an $\alpha = 0.05$.

Results and Discussion

The experimental microcosm system simulated drought conditions with a concomitant water table decline typical of lacustrine wetlands (Touchette et al. 2007). For the 1-in-5 year low, soil moisture levels decreased by approximately 35% by week-2, and subsequently maintained a 60% decline in moisture during weeks-4 and -6. In contrast, the 1-in-25 year drought lost approximately 50% of its soil moisture by week-2, 80% by week-4, and 90% by week-6 (Table 1).

The restricted water availabilities resulted in significant decreases in θ by week-4 for both low and high drought treatments ($p<0.01$; Fig. 1). In contrast, no significant tissue water loss was observed in treated plants during the first two weeks

Table 1 Soil moisture levels % of full saturation), plant biomass (aboveground, belowground, and total), and plant survival (num.m^{-1}; initial planting contained ~200 plants m^{-2}) observed in experimental microcosms. Soil moisture content was recorded on day-14, -28, and -42 prior to weekly watering. Plant biomass and survival were recorded on week-8 (between2-and 6-wks post-drought). Significant differences from the control for biomass and survival are indicated by an asterisk ($p < 0.05$). Date is presented as mean ±1 SE

Parameter/Sample Day	Control	2-Week	4-Week	6-Week
Soil Moisture (% of max.);1-in-5 yr low				
Day-14	Saturated	60 ± 4	65 ± 5	65 ± 3
Day-28	Saturated	Saturated	39 ± 3	46 ± 6
Day-42	Saturated	Saturated	Saturated	42± 4
Soil Moisture (% of max.): 1-in-25 yr low				
Day-14	Saturated	48 ± 4	51 ± 5	49 ± 4
Day-28	Saturated	Saturated	22 ± 31	6 ± 3
Day-42	Saturated	Saturated	Saturated	11 ± 4
Biomass (g dry wt.m^{-2}); 1-in-5 yr low				
Aboveground	487 ± 108	380 ± 38	158 ± 70*	167 ± 45*
Belowground	449 ± 141	300 ± 49	104 ± 36*	159 ± 47*
Total	936 ± 240	680 ± 83	262 ± 101*	326 ± 91*
Biomass (g dry wt. m^{-2}); 1-in 5 yr low				
Aboveground	487 ± 108	455 ± 40	90 ± 35*	9 ± 5*
Belowground	449 ± 141	318 ± 39	65 ±11 *	17 ± 8*
Total	936 ± 240	773 ± 68	155 ± 47*	26± 13*
Survival (shoots m^{-2}); 1 in-5 yr low	221 ± 5	218 ± 7	130 ± 21*	137 ± 17*
Survival (shoots m^{-2}); 1 in-25 yr low	221 ± 5	219 ± 8	118 ± 19 *	27 ± 9*

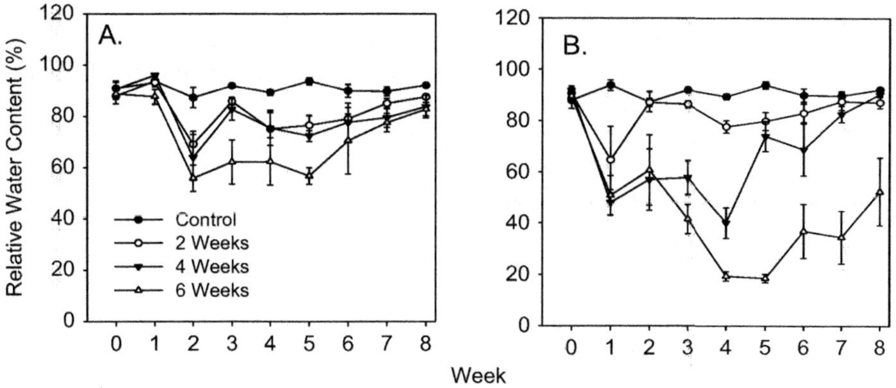

Fig. 1 Relative water content (θ) in leaf tissues exposed to a 1-in-5 year (left panel) and a 1-in-25 year (right panel) drought for two (open circle), four (filled inverted triangle), and six (open triangle) weeks. Drought was initiated at week-0 and continued for 2-, 4-, or 6-weeks, respectively. Data is presented as means ± 1 SE

of simulated drought (p = 0.05 and 0.21, for the low and high drought treatments, respectively). Interestingly, plants that experienced diminished θ due to limited water availabilities were slow to recover following water repletion. For example, plants that lost approximately 40% of their tissue water required an additional 3–4 weeks of saturated conditions before water content return to pre-drought levels (Fig. 1). This suggests that drought induced plant stresses may continue far beyond the actual drought period and that even short periods of water deficit can promote long-term water stress in these plants. This response is likely attributed to xylem cavitations, wherein restricted xylem flow prevented optimal water transport, thereby extending the plant's physiological water deficit.

The apparent water stress in the 4- and 6-week drought treated plants resulted in significant reductions in plant biomass ($p<0.001$ for both above- and belowground tissues; Table 1). In the 1-in-5 year precipitation low, above- and belowground biomass was between 20 and 35% of the control plants. Similarly, in the 1-in-25 year drought, above- and belowground biomass was reduced to less than 18% of the controls. This reduction in plant biomass was attributed, in part, to declines in plant survival, as mortality in treated plants varied from approximately 55% (4- and 6-weeks of 1-in-5 year low, and 4-week 1-in-25 year low) and 12% (6-weeks, 1-in-25 year low; Table 1).

This study illustrates the potential sensitivity of wetland vegetation to climate-induced drought. Short drought periods (as little as 4 weeks) may adversely affect the growth and survival of these ecologically important plants. As wetlands provide a number of important ecological/environmental functions (e.g., habitat, water clarification, sediment stabilization, nutrient/pollution removal), any disturbance of these systems, including a decline in productivity and survival, could deleteriously influence these functions. Possible outcomes to elevated temperatures with simultaneous declines in soil moisture may include: (*i*) a decline in total wetland coverage, resulting in lower overall ecological function, (*ii*) an acceleration of eutrophication of inland waters (rivers, lakes, and reservoirs) by minimizing the amount of nutrient uptake and retention, (*iii*) a decrease in wetland plant diversity – possibly selecting plants that maintain greater hydrological distributions, (*iv*) the removal of important key-stone species (plant or animal), and (*v*) the establishment and expansion of invasive and non-indigenous plant species.

Acknowledgments We are grateful for the research assistance provided by A. Jennings, N. Painter, N. Wright, and K. Chuchra-Zbytniuk. This study was partially sponsored through a grant-in-aid from Elon University.

References

Barr, H. D., and P. E. Weatherley. 1962. "A Re-examination of the Relative Turgidity Technique of Estimating Water Deficit in Leaves." Australian Journal of Biological Sciences. 15:413–428.

Black, C. A. 1965. Methods of Soil Analysis: Part I. Physical and Mineralogical Properties. American Society of Agronomy, Madison, Wisconsin, USA.

Breshears, D. D., N. S. Cobb, P. M. Rich, K. P. Price, C. D. Allen, R. G. Balice, W. H. Romme, J. H. Kastens, M. L. Floyd, J. Belnap, J. J. Anderson, O. B. Myers, and C. W. Meye. 2005. "Regional Vegetation Die-off in Response to Global-Change-Type Drought." Proceedings of the National Academy of Sciences. 102:15144–15148.

Dawson, T. P., P. M. Berry, and E. Kampa. 2003. "Climate Change Impacts on Freshwater Wetland Habitats." Journal of Nature Conservation 11:25–30.

Gregory, J. M., J. F. B. Mitchell, and J. A. Brady. 1997. "Summer Drought in Northern Mid-latitudes in a Time-Dependent CO_2 Climate Experiment." Journal ofClimate 10:662–686.

Hanson, P. J., and J. F. Weltzin. 2000. "Drought Disturbance from Climate Change: Response of United States Forests." The Science of the Total Environment 262:205–220.

Kramer, P. J. 1983. Water Relations of Plants. Academic Pres, Inc. New York.

Mangan, J. M., J. T. Overpeck, R. S. Webb, C. Wessman, and A. F. H. Goetz. 2004. "Response of Nebraska Sand Hills Natural Vegetation to Drought, Fire, Grazing, and Plant Functional Type Shifts as Simulated by the Century Model." Climate Change 63:49–90.

Rosenberg, N. J., and J. A. Edmounds. 2005. "Climate Change Impacts for the Coterminous USA: An Integrated Assessment: From Mink to the 'Lower 48'." Climate Change 69:1–6.

Sorenson, L. G., R. Goldberg, T. L. Root, and M. G. Anderson. 1998. "Potential Effects of Global Warming on Waterfowl Populations Breeding in the Northern Great Plains." Climatic Change 40:343–369.

Touchette, B. W. 2006. "Salt Tolerance in a *Juncus roemerianus* Brackish Marsh: Spatial Variations in Plant Water Relations." Journal of Experimental Marine Biology and Ecology 337:1–12.

Touchette, B. W., L. R. Iannacone, G. E. Turner, and A. R. Frank. 2007. "Drought Tolerance Versus Drought Avoidance: A Comparison of Plant-Water Relations in Herbaceous Wetland Plants Subjected to water withdrawal and repletion." Wetlands 27:656–667.

Wigley, T. M. L., K. R. Briffa, and P. D. Jones. 1984. "Predicting Plant Productivity and Water Resources." Nature 312:102–103.

Dynamics of Soil Organic Nitrogen and its Functions in Soil Carbon Stabilization Processes

Xianzhi Song

Abstract Soil organic nitrogen (SON) plays an important role in long-term ecosystem productivity and the global C cycle. But estimation and prediction involved with SON for soil C sequestration are not completely accurate due to insufficient knowledge of the dynamics of SON and its chemical composition. We hypothesized that the dynamics of SON has a correlated relationship with SON thermal properties. This hypothesis was examined through measuring the optimal pyrolysis temperatures of SON in soils and humic and fulvic acids via pyrolysis gas chromatographic atomic emission detection method (GC/AED). It appeared that SON optimal pyrolysis temperature shifted from 500°C for the surface soil to 400°C for the 25–38 cm depth soil. The low shift of optimal pyrolysis temperatures with increasing humification was also observed from the independent fulvic and humic acids (from 700°C to 600°C), suggesting a relationship of SON thermal property with its dynamic changes. The high thermal stability of SON in combination with the major non-heterocyclic N components coupled with polynuclear C illustrates the contributions of SON to the biological and chemical stabilities of SOM.

Introduction

Soils contain the largest near-surface reservoir of terrestrial C, three times as much C as terrestrial vegetation does (Neff et al. 2002). Although the mechanisms that control long-term N and C cycling are difficult to discern because of complex nature of biological, chemical, and physical interactions among a diversity of microorganisms, organic matter and its functional groups, and mineral surfaces, it has been well known that the stability and accumulation of SON has a positive correlation with the stability and accumulation of SOM (Knops and Tilman 2000). Up to date, however,

X. Song (✉)
Western Carolina University, Cullowhee, NC, USA

G.A. Uzochukwu et al. (eds), *Proceedings of the 2007 National Conference on Environmental Science and Technology*, DOI 10.1007/978-0-387-88483-7_33,

the dynamics of SON and its function in SOM stabilization has not been well understood because of insufficient knowledge of the composition and structures of SON.

In current models of soil C and N cycling, N dynamics are mainly based on external inputs, internal N cycling with the plant-soil system, and N losses (Ogle et al. 2007). Most of those N measurements are only related to inorganic N that accounts for less than 1% of the total N in soils. In contrast, SON, which accounts for more than 90% (Vairavamurthy and Wang, 2002), is treated as a black box. Knowledge of SON dynamics and composition is essential for understanding the changing global C cycle because SON is one of the controlling factors in soil carbon storage size and turnover (Knops and Tilman 2000).

The thermal stability of organic compounds in soils is determined by their chemical structures and binding mineral particles. Thermal analysis measurement has been used to evaluate a relationship between "quality" and thermodynamic stability or biodegradation of soil organic matter and humic substance (Siewert 2001; Francioso et al. 2005; Kuzyakov et al. 2006; Dorodnikov et al. 2007). It also has been used as a tool to differentiate SOM with different humification states (Kuzyakov et al. 2006).

Objective. The primary objective of this study was to test a hypothesis, the dynamics of SON in association with SOM humification and stabilization processes has a correlated relationship with its thermal property or optimal pyrolysis temperatures. The secondary objective is an attempt to identify the function of SON in SOM stabilization mechanisms. An analytical pyrolysis coupled with gas chromatograph/atomic emission detection and cryogenic system (Py-GC/AED), which was developed for characterizing N components of humic and fulvic acids (Song and Farwell, 2004), is employed in this study.

Material and Method

The soils studied were taken from a forest land located at 44° 05′50"N, 103°47′57"W, with an annual mean air temperature of 3.2°C and rainfall of 41.1 cm. The soil texture is channery loam at the surface, and pale brown channery clay loam and channery loam at the depth of 25–38 cm. The soil samples collected at the surface of 0–12 cm and depth of 25–38 cm, and the humic and fulvic acid that were bought from the International Humic Substance Society were analyzed for the total C and N with a Vario Max CNS elemental analyzer.

The Py-GC/AED system mainly consists of a pyrolyzer, a cryogenic preconcentrator, an Agilent 6890A series GC, and an HP G2350A atomic emission detector. Pyrolysis temperature was programmed from 200 to 900°C with a constant increment of 100°C. GC separation was carried out at 35°C, with an initial hold for 10 min and then a ramp to 230°C at a rate of 5°C min^{-1}, where the final temperature was held for 7 min. More detailed information about the Py-GC/AED and its operational conditions can be found in Song and Farwell (2004).

Results and Discussion

Unlike traditional thermal property measurement with TG-DSC in which weight loss is measured accompanied by simultaneous measurement of energy release or consumption, N loss at a temperature was measured in this study by the total integrated peak area at N emission wavelength of 174 nm. The total N emitted from the surface and depth of 25–38 cm soils at different pyrolysis temperatures are recorded and shown in Fig. 1.

The starting recording temperature of the samples was 200°C. The initial weight loss of the samples before 190–200°C mainly came from bound water (Kuzyakov et al. 2006). For the surface soil, the N loss increased with increasing temperature and reached its maximum at 500°C. In contrast, for the deep soil, the maximal N loss was found at 400°C.

With increasing soil depth, C/N ratio decreased as shown in Table 1, accompanied by decreased plant residues, increased microbial origin, increased degree of humification and a longer turnover time (Six et al. 2001). The low shift of the optimal pyrolysis temperature with depth suggests a decrease of thermal stability of SON with increased age and humification. Dorodnikov et al. (2007) suggested that the biological availability of thermal labile SOM pools (less than 480°C) was higher than that of the thermal stable pool decomposed above 480°C. It was noted that N losses from decaying leaves and grasses were distributed in a wide temperature range, from 400°C to 800°C (figures were not shown here). Possibly, at the early decomposition stages of SOM, thermal labile N components were easily incorporated into organism tissue, while, thermal stable plant – derived N components were decomposed.

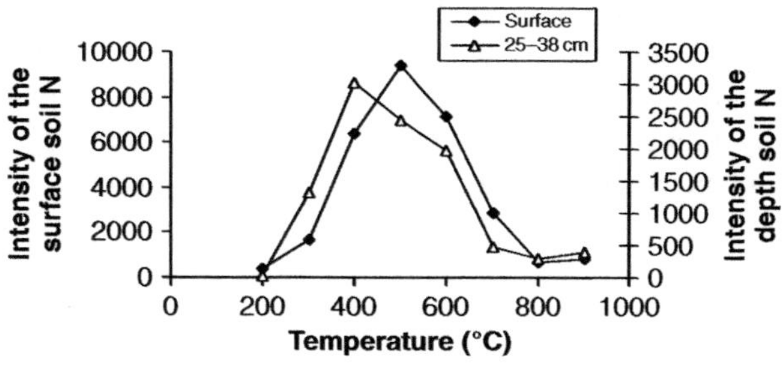

Fig. 1 Optimal pyrolysis temperatures of the SON in the surface and depth of 25–38 cm soils.

Table 1 C and N composition and C/N ratio of the samples

Sample name	Humic acid	Fulvic acid	Surface soil	25–38 cm soil
C %	58.13	50.12	8.054	1.156
N %	4.14	3.75	0.415	0.065
C/N	14.04	13.36	19.39	17.75

The optimal pyrolysis temperature of SON was not only determined by its internal structure or associated energy, it was also affected by associated minerals. The shifts of the optimal pyrolysis temperature of SON may have been contributed by both associated minerals and internal structural changes of SON.

Humic and fulvic acid, two major fractions of humic substances that were the most abundant and important components of SOM, were dissociated from minerals by acid and base extraction and tested for the optimal pyrolysis temperature of SON (Fig. 2). A clear shift of the optimal pyrolysis temperature from 700°C for the fulvic acid to 600°C for the humic acid was observed. The turnover time of humic acid was 10 times longer than that of fulvic acid and the ratio of humic acid to fulvic acid increased with increased humification and time (Huang and Bollag 1998). It is noteworthy to point out that the humic acid and fulvic acid are independent from the soils examined in this study.

Comparisons of the SON between the soils and humic and fulvic acids in terms of N loss verses temperature show some similarities and differences. The observation of the low shift of the optimal pyrolysis temperature with increased humification from both the soils and humic and fulvic acids suggests the existence of a relationship between thermal property of SON and its dynamic changes in SOM humification processes. Further research is needed to confirm and formulize this relationship so that it may be able to apply to C and N models for predicting the dynamics of SON and the stability of SOM pools.

Higher optimal pyrolysis temperature of the humic and fulvic acids than that of the soils may have resulted from their different parental materials. The soil, from which the humic and fulvic acids were extracted, were obtained from very deep, somewhat poorly drained soils on moraines and till plains. Soil organic nitrogen with different origins and humifcation degrees possess different thermal properties (Francioso et al. 2005). This further illustrates a relationship between the dynamics of SON and its thermal property.

In comparison with SOM, of which about 68–79% are decomposed at temperatures between 190 and 390°C (Kuzyakov et al. 2006), based on the data from this study, SON has higher thermal stability, which is consistent with the result reported by Francioso et al. (2005). High thermal stability of SON may suggest its high chemical and biological stability (Siewert 2001).

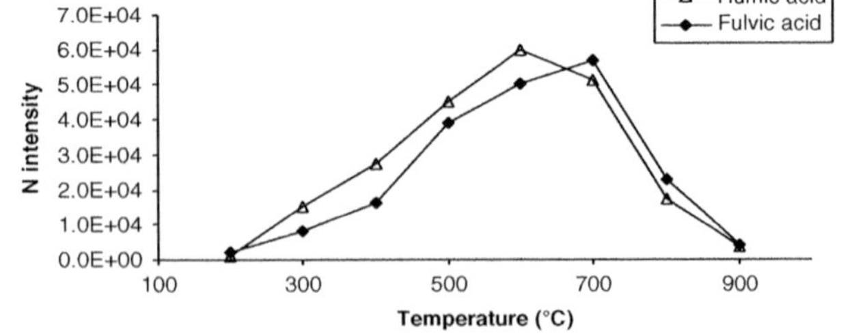

Fig. 2 Optimal pyrolysis temperatures of the SON in the humic and fulvic acids

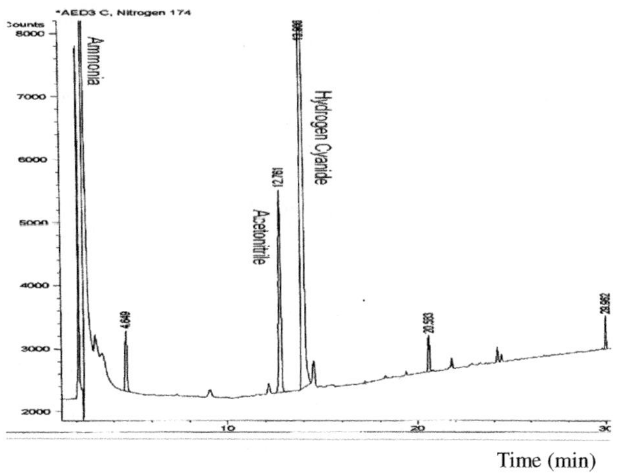

Fig. 3 Py-GC/AED N chromatogram of the humic acid at 500°C

If intense decomposition reaction at temperatures higher than 450C is due to the cracking of high molecular weight polynuclear systems (Francioso et al. 2005), N may exist in heterocyclic forms. GC/AED analyses of the N collected from the decomposition of the humic acid (Fig. 3) show that the major N species are non-heterocyclic N. Heterocyclic N including pyridine and pyrrole with retention time at 20.583 and 29.962 min, respectively, accounts for only a very small portion of the total N. Combination of major non-heterocyclic N with polynuclear aromatic C plus high thermal stability of SON further suggests functions of SON in biological and chemical stabilities of SOM.

In summary, our data showed that SON thermal property, optimal pyrolysis temperature, was correlated with its dynamic changes in SOM stabilization and humification processes in soils. Relatively high thermal stability of SON in combination with major non-heterocyclic N components coupled with polynuclear C suggests contributions of SON to the stabilities of SOM. Further research is needed from a wide variety of soils and humic substances to confirm and formulize the relationship between the dynamics of SON and its thermal property before it can be applied to any models of C and N cycles.

References

Dorodnikov, M., A. Fangmeier, and Y. Kuzyakov. 2007. "Thermal Stability of Soil Organic Matter Pools and Their δ^{13}C Values after C_3–C_4 Vegetation Change." Soil Biology and Biochemistry. 39:1173–1180

Francioso, O., D. Montecchio, P. Gioacchini, and C. Ciavatta. 2005. "Thermal Analysis (TG/DTA) and Isotopic Characterization (^{13}C – ^{15}N) of Humic Acids From Different Origins." Applied Geochemistry. 20:537–544.

Huang, P. M., and J. M. Bollag. 1998. "Minerals-organics-microorganisms interactions in the soil environment." In P.M. Huang, N. Senesi and J. Buffle (Eds.), Structure and Surface Reactions of

Soil Particles, IUPAC Series on Analytical and Physical Chemistry of Environmental Systems. Volume 4. John Wiley and Sons. Chichester, England, pp. 3–39.
Knops, M. H., and D. Tilman. 2000. "Dynamics of Soil Nitrogen and Carbon Accumulation for 61 Years After Agricultural Abandonment." Ecology. 81:88–98.
Kuzyakov, Y., A. Mitusov, and K. Schneckenberger. 2006. "Effect of C_3–C_4 Vegetation Change on $\delta^{13}C$ and $\delta^{15}N$ Values of Soil Organic Matter Fractions Separated by Thermal Stability." Plant and Soil. 283:229–238.
Neff, J.C., A.R. Townsend, G. Glexiner, S.J. Lehman, J. Turnbull, and W.D. Browman. 2002. "Variable Effects of Nitrogen Additions on the Stability and Turnover of Soil Carbon." Nature. 419:915–917.
Ogle, S.M., F.J. Breidt, M. Easter, S. Williams, and K. Paustian. 2007. "An Empirically Based Approach for Estimating Uncertainty Associated With Modelling Carbon Sequestration in Soils." Ecological Modelling. 205(3–4):453–463.
Siewert, C. 2001. "Investigation of the Thermal and Biological Stability of Soil Organic Matter." Ph. D. Thesis. Shaker-Verlag, Aachen.
Six, J., G. Guggenberger, K. Paustian, L. Haumaier, E.T. Elliott, and W. Zech. 2001. "Sources and Composition of Soil Organic Matter Fractions Between and Within Soil Aggregates." European Journal of Soil Science. 52:607–618.
Song, X., and S.O. Farwell. 2004. "Pyrolysis Gas Chromatography Atomic Emission Detection Method for Determination of N-containing Components of Humic and Fulvic Acids." Journal of Analytical and Applied Pyrolysis. 71:901–915.
Vairavamurthy, A., and S. Wang. 2002. "Organic Nitrogen in Geomacromolecules: Insights on Speciation and Transformation with K-edge SANES Spectroscopy." Environmental Science and Technology. 36:3050–3056.

Greenhouse Gas Mitigation Potential of Corn Ethanol: Accounting for Corn Acreage Expansion

Lyubov A. Kurkalova, Silvia Secchi, and Phillip W. Gassman

Abstract Combustion of corn-based ethanol substantially offsets net Greenhouse Gas (GHG) emissions by the recycling of carbon dioxide (CO_2) stored as biomass. However, the GHG mitigation effect may be partially countered by the growth of CO_2 emissions from the increased corn production resulting from the increased demand for corn and higher corn prices. This study assesses the effect for a major U.S. crop production region, the Boone River watershed in Iowa. We use economic models on GIS-based land use and soils data for some 8,275 fields in the watershed to estimate the effect of an increase in corn prices on crop rotations and tillage intensity. The associated changes in soil carbon are estimated using the field-scale Environmental Policy Impact Climate simulation model by comparing soil organic carbon content under the baseline with that under the simulated cropping patterns. A corn price increase from \$2 to \$4 per bushel (\$78.74 to \$157.47 per t) is predicted to lead to the switch from the corn-soybean rotation to continuous corn and from the mulch to conventional tillage on the most of the watershed. The associated CO_2 emissions are estimated at 63,711 tons (57,799 t) per year higher than under the baseline.

Introduction

When burned for energy production, biofuels are releasing carbon dioxide (CO_2) that has been removed from the atmosphere through photosynthesis and stored as biomass, thus "recycling" this potent GHG. While this feature of biofuels production and use has been among the chief arguments in hailing the expansion of corn-based ethanol production, several recent discussions (see e.g., Sibbing, 2007) caution against overstating the carbon recycling effect of corn ethanol. Growth in

L.A. Kurkalova (✉)
North Carolina A&T State University, Greensboro, NC, USA

G.A. Uzochukwu et al. (eds), *Proceedings of the 2007 National Conference on Environmental Science and Technology*, DOI 10.1007/978-0-387-88483-7_34,

ethanol production means higher demand for corn, higher corn prices and, in consequence, higher corn acreage (McNew and Griffith, 2005). The increased corn acreage may in turn lead to higher GHG emissions if the corn production results in higher GHG emissions relative to the land uses it replaces.

The expansion of corn production may come from either the acreage transformed from grasslands, pasture, abandoned lands, or forests, or by increasing the incidence of corn in the rotations on the land that is already in row crop production. A few recent economic studies of the GHG mitigation potential of biofuels account for the former way of increasing corn acreage (McCarl et al., 2005; Secchi and Babcock, 2007), but the latter, corn production intensification effect, has not been adequately addressed. Yet the higher incidence of corn in traditional Midwestern crop rotations, such as a shift from the two-year corn-soybean (CS) rotation towards three-year corn-corn-soybean rotation and continuous corn (CC) usually leads to higher rates of nitrogen fertilizer application and lower corn yields. More corn in rotations also offers additional challenges for conservation tillage (Werblow, 2007) because of the need for higher tillage intensity to counter higher pest pressure and to facilitate preparation of soil for planting after the corn residue, which is heavier than the soybean residue. The greater tillage intensity could negatively affect soil and water resources due to the increased likelihood of nutrient and soil losses. The overall impact on CO_2 emissions is not, however, obvious. On one hand, tillage that is more intensive leads to higher CO_2 emissions from agricultural soils (see, e.g., West and Marland, 2002). On another hand, replacing soybean residue with the corn one may result in higher rates of soil carbon sequestration as the former is more fragile and is usually less in quantity on a mass per unit of area basis. Thus, a comprehensive study of the GHG mitigation implications of corn production intensification requires simultaneous consideration of both rotation and tillage effects. To address this need, our study quantifies the effect of the growing corn prices on crop rotations, tillage intensity, and CO_2 emissions for a major U.S. crop production region.

Methods

Study region and data. The Boone River watershed in Iowa is a natural choice for studying the effects of the biofuels industry growth as Iowa is currently the largest corn-producing state in the U.S. and has the highest number of ethanol plants. Land use and farming practice data were collected in a survey conducted by the Center for Agricultural and Rural Development at Iowa State University within the Boone River USDA/CSREES-funded project. The information on the current and previous crops grown and the tillage practices used in the year 2005 was collected on 8,275 fields corresponding to the USDA/NRCS Common Land Units (CLUs) (URL: http://datagateway.nrcs.usda.gov). The survey data have been supplemented by the GIS-based soil layer Soil Survey Geographic Database (SSURGO) (Soil Survey Staff, 2006) and the Corn Suitability Rating (CSR). The CSR is an index from 0 to

100 that measures land's productivity in corn production. CSR data were obtained from Iowa Cooperative Soil Survey (2003).

In 2005, the majority of the 592,456 acre (239,758 ha) watershed, over 90%, was under row crops, and our analysis covers all of the watershed's cropland. The size of the surveyed CLUs ranges from 0.01 to 551.12 acres (0.004 to 223.03 ha) with the mean of 64.43 acres (26.07 ha). Nearly 94% of the surveyed cropland is in a corn-soybean rotation, mulch tillage is used on the 93% of the acres, and corn was grown on 54% of the area in 2005. The land is highly productive for corn production: the average CSR in the sample is 77.0 with a standard deviation of 9.3.

Economic Model. We treat the year 2005 as the baseline, and compare farmers' rotation and tillage choices under alternative corn prices. In agreement with the common practice (see, e.g., Katsvairo and Cox, 2000; McCarl et al., 2005), we assume that profit maximization is the driving force in farmers' behavior and for simplicity abstract from the impact that the choices have on variance of production. For computational purposes, we assume that land is homogenous within each field (CLU).

The field-by-field farmers' choices are determined by comparing two-year cumulative expected net returns. Given the historical farming practices in the area, in any field and year, we consider two mutually exclusive crops: corn or soybeans, and three mutually exclusive tillage systems: conventional, mulch, or no-till. Conventional tillage usually involves the use of plowing and is defined as any tillage practice that leaves less than 30% residue cover after planting. Mulch tillage involves the use of tillage tools such as chisels, field cultivators, disks, sweeps, or blades. No-till leaves the soil undisturbed from harvest to planting except for nutrient injection and planter passes (West and Marland, 2002). The choice of the crops for the two years is assumed to completely define the rotation: the rotation is CC if corn is chosen both years and CS if corn is chosen one year and soybeans for the other. Continuous soybean is usually not practiced in our study area, due to the significantly increased likelihood of crop failure (because of pest infestations and disease), and thus is not considered in this analysis.

The expected net returns are the difference between expected revenue and the expected costs of production. The expected revenue is the product of crop price and expected yield. We consider corn prices ranging from the historically average \$2 to \$5 per bushel (\$78.74 to \$196.84 per t). Because soybean prices usually move in unison with corn prices (FAPRI, 2007), we assume a constant difference of \$3.5 per bushel between the two prices.

The expected yield is the maximum potential yield adjusted for tillage and rotation effect. Following Secchi and Babcock (2007), the maximum potential yields are approximated as $2.25*\text{CSR}$ bu/acre ($0.14 * \text{CSR}$ t/ha) for corn and $0.67*\text{CSR}$ bu/acre for soybeans ($0.05 * \text{CSR}$ t/ha). Both the crop previously grown on the field and the tillage affect yields, but the exact magnitude of the effect depends on a multitude of factors, including geographic location, soils, and climatic factors. In accordance with the results of agronomic studies conducted under the conditions similar to our study area (Katsvairo and Cox, 2000), we assume that the expected yields are equal to the maximum potential yields for corn after soybeans and soybeans after corn, under both conventional and mulch till. No-till corn after soybeans and no-till soybeans after corn reduce the expected yield by 5 and 20% from the maximum potential yield, respectively. Soybeans provide a proven soil nitrogen benefit to corn grown in the following year. Therefore, growing corn after corn leads to a yield reduction across all tillage categories: the expected yield is assumed to be 95, 90, and 80% of the maximum potential yield for conventional, mulch, and no-till, respectively.

Assumptions about representative machinery operations, fertilizer applications, and costs of production are based on Duffy and Smith (2007). Levels of fertilization are assumed independent of tillage, but are affected by previous crop to partially counter the corn-after-corn yield reduction effect. We follow Duffy and Smith (2007) who report fertilizer rates for average Iowa farms, and assume that Nitrogen (N), Phosphate (P), and Potash (K) are applied at the rates of 190, 55, and 45 lb per acre (213, 62, and 50 kg/ha) for corn after corn, and at the rates of 135, 60, and 50 lb per acre (151, 67, and 56 kg/ha) for corn after soybeans, respectively. For soybeans after corn, P and K are applied at the rates of 40 and 75 lb per acre (45 and 84 kg/ha), respectively. We are well aware that fertilizer applications significantly affect yields, but an explicit modeling of varying fertilizer levels is beyond the scope of the study and is left for future analyses.

For each corn price level under consideration, the outcome of the economic model is the prediction of the rotation and tillage choices associated with the each year in rotation, for each field (CLU) of the study area. The economic model predictions are subsequently used to assess the changes in CO_2 emissions, both from agricultural soils and from fossil-fuel consumption attributable to the use of farm machinery and fertilizer.

Soil Carbon. Average annual changes in soil organic carbon attributable to changes in rotations and tillage are estimated at each CLU using the Environmental Policy Integrated Climate (EPIC) model (Izzauralde et al., 2006), which simulates field-scale sized areas characterized by homogenous soil, landscape, weather, crop rotation, and management system parameters. The model operates on a continuous basis using a daily time step and can perform long-term simulations of hundreds of years. The required soil information to run EPIC came from the SSURGO data. The specific assumptions on farming operations and fertilizer levels used in the economic model were carried over to the EPIC model to ensure consistency. Two 30-year simulations were compared for each of the corn prices considered at each of the CLUs: first assuming the continuation of the 2005 baseline, and the second assuming the continuation of the rotation and tillage system chosen under the particular price scenario. We measure the change in soil carbon content attributable to the scenario as the annual average of the difference between the total soil carbon under the scenario and that under the baseline at the end of the simulation period.

Carbon Emissions from Farm Operations. Following West and Marland (2002), we also estimate the changes in CO_2 emissions from the fuel use for tillage operations and the emissions attributable to manufacturing, packaging and storing fertilizers. It should be noted that these are approximate estimates only as wide ranges of energy use have been found for both tillage and fertilizers. In this study, we use the mean CO_2 emission estimates reported in Lal (2004).

Results and Discussion

We find a replacement of the CS rotation with CC and a tightly corresponding replacement of the mulch with conventional tillage at corn prices above \$3.6 per bushel (\$141.73 per ton), with an almost complete (98.6%) switch to continuous corn and conventional tillage at \$5 per bushel (\$196.84 per t) (Table 1).

The watershed-average change in soil organic content, corresponding to the change from CS with mulch till to CC with conventional till, was estimated at –0.0353 tons per acre (–79.06 kg per ha). This compares favorably with the U.S. average data reported in West and Marland (2002). Interestingly, about 49% of the study area would actually sequester carbon under these conditions, probably because the interplay between the tillage effect that tends to lower carbon content and the effect of the replacement of soybean residue with corn residue that may lead to soil carbon accumulation. Further investigation of this finding is needed.

If the entire study area were to change from the baseline to continuous corn with conventional tillage, an average annual emission of CO_2 is estimated to occur at the rate of 18,808 tons (17,062 t) per year due to CO_2 release from agricultural soils, with an additional emission of 45,549 tons (41,322 t) per year due to the changes in farm operations. Although the tillage intensity and fertilization level change contributions are to be taken cautiously because of the average estimates as described above, these findings suggest that the CO_2 emissions from farm operations can contribute a significant portion to the overall GHG effect of the changes in cropping patterns and need to be taken into account in the overall assessment of the GHG effects.

Table 1 The impact of increasing corn prices on the amount of land planted to continuous corn, and associated tillage, soil carbon, and CO_2 emission impacts

Corn price, \$ per bushel	Land in continuous corn, percent of watershed total	Land under conventional tillage, percent of watershed total	Decline in soil carbon content, tons per year from the baseline	Increase in CO_2 emissions attributable to changes in tillage operations and, tons per year from the baseline
3.5	0.0	0.0	0.00	0.00
3.7	24.7	24.7	3,182	11,239.22
3.9	74.5	74.5	15,034	33,926.95
4.1	92.2	92.2	17,811	41,995.30
4.3	93.1	93.1	17,874	42,401.28
4.5	96.9	96.9	18,399	44,154.08

Conclusions

By considering the choice between corn-soybean rotation and continuous corn, and taking into account the potential yield reduction effects of alternative tillage systems, we found strong evidence for potential increases of corn in rotations and tillage intensity in the Boone River watershed in central Iowa, U.S., in response to the growth in corn prices. These changes are estimated to lead to sizable increases in CO_2 emissions, both from soil and from farm operations. The study points to the importance of GHG emission life-cycle accounting in estimating the GHG fossil fuel substitution effects of corn ethanol production, especially if GHG offset markets develop or if society chooses to reward biofuel carbon recycling via some incentive payments (McCarl et al., 2005).

Several caveats and limiting assumptions point in the direction for future research. Because of the lack of reliable data, we did not account for pesticide use in the farm operations component of CO_2 accounting. Pesticides have been shown to be CO_2 emission intensive (West and Marland, 2002) and as increase in tillage intensity may require fewer pesticides, accounting for this component would improve the accuracy of the overall assessment. The results are also conditional per the ability of these models to capture the changes modeled. As EPIC and similar models continue to evolve and are calibrated to newer field trials, the precision of the CO_2 emission assessment can be improved.

Acknowledgments Part of this research has been supported by the USDA. The findings do not reflect the views of the USDA or its staff. The authors would like to thank Todd Campbell for computational assistance and Chad Hart for valuable discussions.

References

Duffy, M., and D. Smith. 2007. "Estimated Costs of Crop Production in Iowa – 2007." File A1-20, Iowa State University Extension, Ames, Iowa. URL: http://www.extension.iastate.edu/agdm/crops/pdf/a1-20.pdf.

FAPRI Staff, 2007. "FAPRI 2007 U.S. and World Agricultural Outlook." FAPRI Publications 07-fsr1, Food and Agricultural Policy Research Institute (FAPRI) at Iowa State University. URL: http://www.fapri.iastate.edu/Outlook2007/

Iowa Cooperative Soil Survey. 2003. Iowa Soil Properties and Interpretation Database. URL: http://icss.agron.iastate.edu/.

Izzauralde, R.C., J.R. Williams, W.B. McGill, N.J. Rosenberg, and M.C. Quiroga Jakas. 2006. "Simulating Soil C Dynamics with EPIC: Model Description and Testing against Long-Term Data." Ecological Modeling 192:362–384.

Katsvairo, T., and W.J. Cox. 2000. "Economics of Cropping Systems Featuring Different Rotations, Tillage, and Management." Agronomy Journal 92 (3):485–493.

Lal, R. 2004. "Carbon Emissions from Farm Operations." Environment International. 30:981–990.

McCarl, B.A., D. Gillig, H.-C. Lee, M. El-Halwagi, X. Qin, and G.C. Cornforth. 2005. "Potential for biofuel-based Greenhouse Gas emission Mitigation: Rationale and Potential." In J. Outlaw, K.J. Collins, and J.A. Duffield (Eds.), Agriculture as a Producer and Consumer of Energy, pp. 300–316.

McNew, K., and D. Griffith. 2005. "Measuring the impact of ethanol plants on local grain prices." Review of Agricultural Economics. 27(2):164–180.

Secchi, S., and B.A. Babcock. 2007. "Impact of High Corn Prices on Environmental Quality: A Case of Iowa and the Conservation Reserve Program." CARD Working Paper 07-WP 447. Center for Agricultural and Rural Development, Iowa State University. May. URL: http://www.card.iastate.edu/publications/DBS/PDFFiles/07wp447.pdf.

Sibbing, J. 2007. "Setting America on the Right Biofuels Path." Journal of Soil and Water Conservation. 62(3):48A.

Soil Survey Staff, Natural Resources Conservation Service, U.S. department of Agriculture. 2006. Soil Survey Geographic (SSURGO) Database for Iowa. URL: http://soildatamart.nrcs.usda.gov.

Werblow, S. 2007. "More Corn: Is Conservation Tillage at Risk?" Partners: Quarterly Publication of the Conservation Technology Information Center. April, 25(2).

West, T.O., and G. Marland. 2002. "A Synthesis of Carbon Sequestration, Carbon Emissions, and Net Carbon Flux in Agriculture: Comparing Tillage Practices in the United States." Agriculture, Ecosystems and Environment. 91:217–232.

Comparative Effects of Sea Level Rise Versus Hurricane Event on Coastal Erosion

John B. Williams

Abstract Rising sea level from global warming raises concerns about coastal erosion. While gradual erosion from sea level rise may go almost unnoticed, combined effects of sea level rise and hurricanes can be devastating. For this reason, historical shoreline changes provide planners valuable tools to forecast and lessen impacts. Sakonnet Point, RI, USA, suffered the dramatic 'one-two punch' of a major hurricane and rising sea level and offers such a case history. This peninsular headland is the most seaward extension of RI near the Massachusetts border. The seaward extension of Sakonnet Point was historically a 300 m long peninsula of glacial till with large dunes. My study utilized a nearly 100-year photographic time-series and on-site observations to evaluate this dune system's disappearance. Since 1938, this peninsula was altered from a system of nearly fifteen-foot (4.6 m) high dunes to nearly complete submergence at high tide. Sea level rise along this New England coast was 0.25–0.3 m over the last century. Superimposed upon sea level rise was the catastrophic 1938 hurricane surge. Our results show that major storm damage to barrier dunes sets the stage for sea level rise to produce rapid beach disappearance. With accelerated sea level rise from global warming, Sakonnet Point serves as a valuable tool for coastal planners.

Introduction

Global warming has generated warning flags ranging from climate shifts and biodiversity changes to altered economies. For coastal populations, the most immediate concerns are related to shoreline damages connected to rising sea level and higher-magnitude hurricanes. Although coastal geologists have urged caution about ocean-side construction for decades, these concerns are magnified by the implications

J.B. Williams (✉)
Department of Biological & Physical Sciences, South Carolina State University, Orangeburg, SC, USA

G.A. Uzochukwu et al. (eds), *Proceedings of the 2007 National Conference on Environmental Science and Technology*, DOI 10.1007/978-0-387-88483-7_35,

of continued sea level rise (Bush, et al., 1996). Higher magnitude hurricanes with elevated storm surges have also been predicted to occur more frequently. Warmer ocean currents have the potential to 'fuel' stronger cyclones (Webster, et al., 2005). Foreseeing these future, destructive impacts on persons and property, government agencies are attempting preventative regulations and alternative management plans concerning coastal land-use and shoreline stabilization, e.g. Titus (1998),

Coastal planners must continue to realize that the topography of today will not be that of tomorrow. Planning only for the 'big one' can miss the chronic, but less dramatic (in the short-run) effects of rising sea level. Government entities must look beyond traditional '100 year storm' approaches and consider the more complex geomorphological effects of continual sea level rise. Sakonnet Point, Rhode Island has suffered the dramatic one-two punch effect of mega-storm surge and rising sea level and offers a valuable case-history. The objective of this study was to determine the relative effects of acute hurricane storm surge and chronic sea level rise on coastal erosion. Conclusions would be developed qualifying their interactive effects over long time spans.

Materials and Methods

Sakonnet Point is a peninsular headland and the most seaward extension of Rhode Island near the Massachusetts border. Its geological structure is of granite outcrops with abundant glacial till and morainel sand deposition. The historical seaward extension of Sakonnet Point was a *ca.* 300 m long peninsula of glacial till covered with large sand dunes. This prominent dune system marked the westward border of the Cape Cod glacial dune system and was a mapped coastal feature.

My study utilized a nearly 100-year time series record of ground level and aerial photography combined with site-specific observations to investigate this dune system's near disappearance. During this time span, this coastal feature has been altered from a system of nearly fifteen-foot (4.6 m) high sand dunes to almost complete submergence at high tide. Historical sea level rise along this southern New England coast has been about 0.25–0.3 m over the last 100 years. This rate is less than sea level rise along the Southeastern USA coast due to the opposing force of glacial rebound.

Superimposed upon these gradual erosional forces of sea level rise was a catastrophic hurricane event. Sakonnet Point was nearly at ground zero for maximum storm-surge from the 1938 category 3–4 hurricane and our photographic time series documents this major erosion event. Our time-series analysis was developed from aerial photographs from the 1930's to 2006, using mostly U.S. Geological Survey photos. Ground-level photographs were obtained for 1910–2006 and included assistance from the Little Compton Historical Society. Eyewitness accounts of the 1938 Hurricane effects upon Sakonnet Point were provided by D'Entremont (2006). Our direct field observations spanned the period from 1965 to 2006 and included substrate sampling and photography.

Results and Discussion

The 1938 Hurricane was the most powerful storm to strike New England and has been ranked by NOAA-NWS as one of the most powerful hurricanes to ever strike the U.S. Sakonnet Point was located in the forward right quadrant as the hurricane slammed into the coast with a forward velocity of 50–70 mph (80–113 km/h) and wind gusts reaching recorded extremes at Blue Hill, Massachusetts of 180 miles per hour (290 km/h) (Vallee and Dion, 1998). Its storm surge impact was magnified by its incredible forward velocity as it raced northward through the Atlantic at up to 70 mph (113 km/h). The surge also coincided with a maximum Autumn high tide. The keeper of the Sakonnet Point Lighthouse survived the storm surge wave and estimated its height at over 50 ft (15.3 m) (D'Entremont, 2006).

The well-developed dune system at Sakonnet Point prior to this hurricane is documented by the photograph from *ca.* 1910 (Fig. 1). Dune height along the mid-axis of the point was estimated to be fifteen-feet (4.6 m) high. The prominent granite headland at the right center of Fig. 1 serves as an important reference point in comparing the post-hurricane photo of 1938 (Fig. 2). One can easily see the magnitude of erosion produced by the surge (Fig. 2). The small building to the right of the rock formation (Fig. 1) was also destroyed. All of the dune system was removed except for a few remnant stands of grass along the point midline.

During the next 35 years, the outline of Sakonnet Point remained similar to its pre-1938 Hurricane dimensions, with some accretion of new dunes. However, beginning in the mid-1970's this remaining central dune line began to experience over-wash events of increasing frequency. Sea level rise had allowed prevailing wind-driven waves to reach a critical angle of approach; enhancing their eroding force. Gradually the aerial outline of Sakonnet Point eroded from its 'finger-like

Fig. 1 Sakonnet Point, RI *ca.* 1910. Prominent dune system extended to the tip of the point and was nearly 15 ft (4.6 m) high along the midline

Fig. 2 Sakonnet Point, RI,1938, after the hurricane storm surge. The barren rocky headland still provides a reference, but the point was stripped clean of its former dune system. Some remnant grass stands are visible near the former base of the eroded dunes. Courtesy of Little Compton Historical Society Col.

shape' of the previous century to a narrow recurved sand-spit. Sakonnet Point elevation had become so reduced by the gradual effects of sea level rise, 0.25–0.3 m over the last 100 years, that it was now almost completely covered at high tide. Figure 3 documented the extent of sea level rise effects by 2006.

Fig. 3 Sakonnet Point tip in 2006 showing extreme effects of gradual sea level rise. Photo was taken at low tide and with an average three foot tidal amplitude, this area is nearly all submerged at high tide. View is from the top of the rocky headland shown in Figs. 1 and 2 looking to the southwest

Conclusions

Our results show that major storm damage to barrier dune systems can set the stage for sea level rise to produce rapid disappearance of coastal beach systems. Even normal wave patterns could now impact the beach due to an elevated sea level. Without the soil retention of a dune-building plant community, rising sea level will erode beaches and barrier islands at an accelerated rate even with average wave activity and low-risk storms. This sets up a 'combination punch' that can produce a 'knockout blow'; removing a beach or barrier island more rapidly than predicted. Coastal planners should heed the warnings of Sakonnet Point and not take for granted time lines based upon predicted sea level rise alone as the controlling factor for beach erosion. With expected acceleration of sea level rise from global warming, this disappearance of Sakonnet Point serves as a valuable tool for coastal communities.

Acknowledgments The assistance of Chuck Cressy, Jonathan Williams, David Williams, and Steve DelDeo in conducting field studies of the Sakonnet Point shoreline over the last forty years was greatly appreciated. Aerial and ground-level photography help was also provided by: Jeremy D'Entremont, John Shelton, the Geospatial Extension Program of the University of Rhode Island, and the Little Compton Historical Society. Much thanks also goes to my mother, Jeanne F. Williams, who personally experienced the 1938 Hurricane and also directed me to historical records of the storm effects on Rhode Island and Sakonnet Point.

References

Bush, D.M., O.H. Pilkey, and W.J. Neal. 1996. *Living by the Rules of the Sea*. Duke University Press, Durham, NC 200p.

D'Entremont, J. 2006. *The Lighthouses of Rhode Island*. Commonwealth Editions. 240p.

Titus, J.G. 1998. *Rising Seas*. Maryland Law Review 57: 1279–1399.

Vallee, D.R. and M.R Dion. 1998. *Southern New England tropical storms and hurricanes, a 98-year summary, summary 1909-1997*.Taunton, Mass, National Weather Service.

Webster,P.J., G.J. Holland, J.A. Curry, and H.-R. Chang. 2005. *Changes in Tropical Cyclone Number, Duration, and Intensity in a Warming Environment*. Science 309: 1844–1846.

Global Warming in Asheville, North Carolina

George Ford, William McDaniel, and Aaron Ball

Abstract As predicted by Svante Arrhenius in 1896, global warming is taking place as evidenced by documented rises in average sea level of about 1.7 mm/year during the 20th century. There have been naturally occurring cycles of global warming and cooling throughout the history of the world. Much has been written about the catastrophe that global warming would present to humankind, but the effects of elevated atmospheric carbon dioxide concentrations upon ambient annual mean temperatures at the local level in western North Carolina are not easily recognized at this time. Observation of annual mean temperatures in western NC did not immediately indicate a detectable temperature increase over the period of analysis. However, annual weather data for Asheville, North Carolina from 1965 until 2006 indicated an upward trend in annual mean surface temperatures of about 1.3°F (0.72°C) while global atmospheric carbon dioxide concentrations have risen about 62 ppm. This paper will present an examination of regional ambient annual surface temperature trends in western North Carolina relative to global atmospheric carbon dioxide concentrations. The focus will be on analysis of data to determine cyclical patterns.

Introduction

Global warming due to human activity was first predicted by Svante Arrhenius in 1896 (NASA, 2007). The primary gases in the atmosphere, nitrogen and oxygen, will not reflect solar radiation, but many gases such as carbon dioxide, methane, nitrous oxide and halogens will trap infrared radiation emitted by the Earth's surface into the atmosphere causing global warming and climatic change. Currently carbon dioxide has the most significant radiative forcing index (Hofman, 2007) and will be the primary greenhouse gas considered in this discussion. Atmospheric con-

---

G. Ford (✉)
Western Carolina University, Cullowhee, NC, USA

G.A. Uzochukwu et al. (eds), *Proceedings of the 2007 National Conference on Environmental Science and Technology*, DOI 10.1007/978-0-387-88483-7_36,

centrations of carbon dioxide have been measured accurately since the late 1970s (Tans, 2007). Historical levels of atmospheric carbon dioxide and climatic change over the last 400,000 years have been estimated from Antarctica ice core records. Antarctic temperatures have varied from about −18°F to 7.2°F relative to present day levels (Thorpe, 2005). In addition, various methods have been used to estimate climatic change over the last one thousand years, and "the climate appears to have been noticeably lacking in significant variations" (Thorpe, 2005, p.4).

Global records have shown about 0.9°F to 1.44°F increase in mean surface temperature of the earth over the last one hundred years (NASA, 2007; Thorpe, 2005). Over the same period, atmospheric carbon dioxide concentrations have risen from about 290 parts per million (ppm) to about 383 ppm today (Tans, 2007). The basic theory behind global warming is that as greenhouse gas concentrations (primarily carbon dioxide) in the atmosphere increase, global temperatures will increase due to the entrapment of infrared radiation. Kaufman, et al. have written about the link between global surface temperature and atmospheric greenhouse gas concentrations (2006), but the global mean temperature increases are not easily observed by the population at large. An increase in the mean global temperature is amplified at the poles. In the Arctic, winter temperatures have risen 9°F (Appenzeller, 2007) causing ice sheets to melt with the potential of increasing sea level worldwide. In addition, this rise in temperature could cause desalination of the Arctic oceans, thereby interrupting currents that affect weather and climate. Greenhouse gases have long half-lives, and emitted gases could be around for more than a century.

Getting reliable future climatic predictions based upon atmospheric carbon dioxide concentrations is difficult because there are so many secondary processes, which are not understood completely (NASA, 2007). The effect of a darkened surface exposed by a melting glacier is one example of a secondary process. The effect upon global climate of desalination of arctic oceans due to melting arctic ice is another example. The difficulty of predicting and observing changing climatic conditions based upon atmospheric carbon dioxide concentrations and a need to explain global warming in layman's terms are primary subjects of this discussion. The world population will need to be convinced that global warming is an issue that will affect them individually to voluntarily make efforts to reduce greenhouse gas emissions. An examination of data for a specific locality pertaining to global warming may start the needed dialogue.

Methodology

Daily average dry bulb surface temperatures were obtained for Asheville, North Carolina for the period of January 1, 1965 to December 31, 2006 from the local airport weather station. The data was tabulated, and the annual mean temperatures and variances for each year were computed in an Excel spreadsheet. To simplify analysis, the additional data from leap year days were eliminated from the database

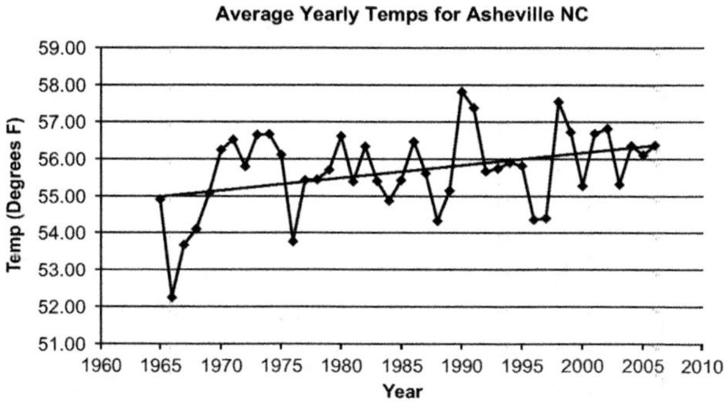

Fig. 1 Average yearly temperatures for Asheville, North Carolina

after checking the effects upon the computed means and variances. The effects of eliminating this data were minimal. These annual mean temperatures were graphed by year, and observations were made to determine trends. The method of least squares was used to create a trend line within the Excel spreadsheet.

A correlation analysis was also done to determine the correlation between mean surface temperature and global atmospheric carbon dioxide concentrations. The Pearson's Correlation Coefficient was determined for the data sets.

Results

Initial observation of the graph of annual mean temperatures did not immediately indicate a detectable temperature increase over the period of analysis. After additional research to determine the order of magnitude of the increase to be observed and reexamination of the data, a trend line indicated a temperature rise over the forty-one year period of about 1.4°F. The graph is shown in Fig. 1.

Discussion

The trend line for the period of analysis indicates an upward trend in surface temperature of 1.4°F, supporting a global warming hypothesis. But, the period under consideration is too short to make a conclusive decision about whether global warming is actually occurring in western North Carolina. If two data points are dropped from the data set, the trend line changes significantly, reflecting only a 0.9°F trend. The ability to change the analysis so drastically by making such minor changes to the database causes reconsideration of the validity of any conclusions drawn from the analysis.

In addition, if annual mean surface temperatures in Asheville are compared with global atmospheric carbon dioxide concentrations, little correlation is observed. A

Pearson Correlation coefficient of 0.36 was computed suggesting a low correlation between the variables.

Conclusions

The issue of elevated levels of atmospheric carbon dioxide is real. Melting glaciers and rising sea levels are projected for the 21st century of 8–19 inches (Meehl, et al., 2007); however, the impact in western North Carolina appears to be less severe. The predicted future 2°F or 3°F of average temperature rise do not seem significant. While cyclical patterns were identified, the data only represents a relatively narrow window along the local climatic timeline. Care must be taken when observing data due to outliers or exceptionally high or low average annual temperatures.

Additional research should be done to determine direct effects of global warming upon the populations of specific areas to point out the significance that the problem of global warming will or will not present to individuals.

References

Appenzeller, T. (2007). The big thaw. *National Geographic Magazine.* retrieved March 2, 2009 from Academic Search Premier Database.

Hofmann, D.J., 2007. The NOAA annual greenhouse gas index. *NOAA Earth System Research Laboratory*. Boulder, Colorado

Kaufmann, R.K., H. Kauppi, and J.H.Stock, 2006. Emissions, concentrations, & temperature: A time series analysis. *Climatic Change* 77, 249–278.

Meehl, G.A., et al. 2007. Global climate projections. *Climate Change 2007: The Physical Science Basis. Contributing Working GroupI to the Fourth Assessment Report of the Intergovernmental Panel on Climate Change.* Cambridge University Press, Cambridge, UK

NASA facts on-line. 1998. *NASA Goddard Space Flight Center.* Retrieved June 14, 2007 from *http://www.gsfc.nasa.gov/gsfc/service/gallery/fact_sheets/earthsci/eos/global_warming.pdf*

Tans, P. 2007. Trends in atmospheric carbon dioxide-Mauna Loa. NOAA/ESR. Retrieved June 14, 2007 from www.cmdl.noaa.gov/gmd/ccgg/trends.

Thorpe, A.J. 2005. *Climate Change Prediction: A Challenging Scientific Problem.* Institute of Physics, London, UK.

Part VI
Innovative Environmental Technology and Sensors

Tuckaseegee Watershed Observatory: A Collaborative Environmental Research Tool

Brian P. Howell, Roger Clapp, Jeff Marston, Chris Donaldson, and Brent McCoy

Abstract Current environmental regulations and scientific inquiry rely on single point catch samples or in-situ measurements to certify compliance or acquire basic data. Such methods do not lend themselves well for dynamic environmental water quality events of short duration. In addition, such measurements do not allow for the separation of such events as turbidity changes, nutrient flow, and physical parameter changes. This paper describes the progress on a system to address these issues.

Recent advances in both sensor technologies and communication methods allow for the deployment of high station count sensing networks of both the wired and wireless types. A joint team of Western Carolina University (WCU) and the Watershed Authority of the Tuckaseegee River (WATR) is working to create a comprehensive network for water quality measurements in the Tuckaseegee River basin. Initially targeted toward turbidity measurements for mountainside development monitoring, the system has been expanded in scope to include microclimate monitoring of the surrounding watershed, and multiple parametric measurement. When completed, the system will include multiple water monitoring stations on the Tuckaseegee and its source streams as well as microclimate sensors in various areas of the watershed. The system has expanded beyond basic compliance monitoring to include scientific information gathering for hydraulics, ecosystem modeling, and to that end is including outreach to the community and schools in the area for interaction and educational involvement.

The technologies used include multiple network topologies of sensor elements, multiple communications formats, sensor pallets, and data logging systems. This paper describes the overall design of the system, and addresses the primary technical and community issues which have had to be addressed to implement the system.

Brain P. Howell (✉)
Western Carolina University, Cullowhee, NC, USA

G.A. Uzochukwu et al. (eds), *Proceedings of the 2007 National Conference on Environmental Science and Technology*, DOI 10.1007/978-0-387-88483-7_37,

Introduction

Environmental observatories have emerged as a desirable method to allow multiple researchers access to remote areas for prolonged periods of observation. In addition, these observatories provide finer resolution for environmental features in both the spatial and time scales than previous field measurements/catch sampling methods are able to achieve. The Tuckaseegee Watershed Observatory (TWO) has emerged to achieve scientific, public awareness, compliance monitoring, and community education goals.

Objective. The objective of the system is two integrate water quality and climate data from multiple sources into an easily accessible, community involved, tool. The technology of the system has evolved from two individual projects in Western North Carolina. The first project has been the instrumentation to monitor sediment levels in the Tuckaseegee River. Excess sediment transport results in four major problems for the area. First, the tourism industry in Jackson, Macon, and Swain counties, rely on the perception of clear running rivers. Secondly, the watershed and sediment load drains into Fontana Lake. Next, industry which relies on river water must stop and use other sources for process water if the river water is not clean enough. Lastly, much of the Tuckaseegee River is rated for Trout fishing and must meet maximum allowable turbidity levels to keep healthy trout populations. There are several possible sources for sediment, both man made and natural. One detailed objective of the system is not only to detect turbidity events but to provide enough data to separate man made from natural causes.

The second project which has provided technology for the project has been a series of student projects at WCU. These initially focused on using Wireless Sensor Networks (WSN) to map microclimate data and to overlay that data with biological surveys to map the correlations of species population with climate data. It also became clear that a combined, general microclimate observatory could also answer question regarding plant speciation variation at various points on the mountains.

Sensory Data Fields. The two systems have somewhat different sensory pallets. Based on earlier work of Eads and Lewis (Eads and Lewis, 2002), a protocol for sampling has been established. Based on the earlier work, it became clear that the system would have at minimum the following specs:

Water Monitor Data fields:
Turbidity: wiped optical sensor, 0–1000NTU
Temperature: thermistor range, -20F-100F
Pressure: vented in-situ, 0–10 m depth calibrated
Conductivity: analog, 0–20000 usiemens
Sample/wiper interval, variable 1 min to 24 h.

Microclimate Data Fields:
Air Temperature: thermistor, range, -20F-100F
Light: Full Sun to Dark

Soil Temperature: -20-100F
Soil Moisture: 0–100% Content

Based on these sensory parameters, the main issues regarding the collection, storage, and communications of the information was next evaluated.

General Observatory Design Evolution

Given the two basic goals, measurement of water quality over the watershed and measurement of adjacent microclimate, the sensors above were chosen. With that selection, the next step was to identify meaningful locations primarily for the water monitoring systems. Even though the different aspects of the observatory can be presented individually, there exists connection which both slowed the design of the system but provided opportunities to reduce cost and increase the educational value of the system as well as the functionality.

Initial Site Selection. A map of the watershed area reveals possible sites. The Tuckaseegee is fed by several tributaries. Included in these are Scotts Creek, Caney Fork, and Savannah Creek. Scotts Creek in particular is unique in that its water is used for process cooling water by local business and flows through the middle of the town of Sylva, North Carolina. In addition, in Bryson City, there is a United States, Geological Survey (USGS) gage station. This station provides a wealth of historical water level data and was chosen as the area for site 2.

Savannah Creek meanders along US 441 for several miles before turning East then North into the Tuckaseegee near Webster, North Carolina. It was decided that for Units 3 and 4, one site in the upper reaches of the creek and another site close to the confluence with the Tuckaseegee River would be chosen.

Educational/Community Awareness. Early in this process, it was a stated goal of WATR to create signage to report in real time the turbidity of the water. One of the major problems with setting up a project of this magnitude is the development of trust within the community. From a practical standpoint, it means finding individuals and sites where equipment will not be randomly vandalized or worse yet sabotaged intentionally.

With the support of local business, including Jackson Paper in Sylva, two sites in town were selected. For the other sites, WATR spent two years developing relationships with local land owners, Jackson County officials, and school districts in the area. As a result of those meetings, specific locations were made available in the above areas.

Energy Sources. Because of accessibility to sites, energy management becomes a primary issue for the networking of the components. Also, the nature of the sensors determines to some degree the voltage and current requirements. Once again, at the sites in Sylva, there is access to 120 V AC power. At almost every other site and for all the microclimate locations, this is not an option, so small energy sources were evaluated. For version one of both the microclimate and the water monitors, solar power was selected to run the system. One aspect unique to this type of system is how the energy is directly related to the information density available and the reliability of the network. In other words, selections for methods for sensing impact the life of the instrument power, as well as the size of the battery required and energy generated, and the nature in which information can be communicated. In this case, the initial water monitor system required a 12 V source and approximately 150 ma during full operation. This gave a demand of 3.6AH/ day. The battery requirements appear modest, given an average solar operation of 6 hours/day, required a charging system of 7–10 watts to restore the battery each day. While not prohibitive, this increased the cost of the system and produced a requirement for a larger solar panel.

Several strategies were studied to reduce the energy demands. First, it was decided to independently power the sensor system from the processor so that in normal operation the system would leave most of the system turned off until the sample cycle. It was found that with as little as 20 s warm up, raw voltages from the sensors would stabilize to within 1 least significant bit (LSB) for the 10 bit A/D system over a span of input from 0 to 5 V. This dropped current demand to <20 mA under normal operation.

The other major power load was the requirement to use wireless as the communication source. Every transmitted packet would require approximately 50 mA of current during operation. Therefore, the total power demand could only be held closer to the 20 mA level by requiring the sensor and communications to sleep. Given that level, a solar source of <5 W can be utilized to maintain adequate charge along with a battery pack of <2AH also helping keep the monitor size small.

For the microclimate systems, the technology was based on "motes" created by Crossbow Corporation (Crossbow, 2004). These systems are native 3.3 V and are already designed to "sleep" unless sensing or communicating. These systems were already capable of field life of up to 9 months on 2 "AA" batteries. Converting to small solar panels (1/2 W) provides adequate power to keep the system on. The comparative hardware is shown in Fig. 1.

Communications/ Networking. The topology, access to infrastructure, and node spacing become the dominant features for any observatory system. Fundamentally, the systems have two different communications formats. The water monitor is inherently a point to point system albeit with long range between points, up to 20 miles line of sight. The motes are low power systems with range of only 100 m. This produces an immediate problem in connection.

The selection of the sites also produces networking problems. The sites in Sylva and Bryson City as well as at the schools have both land line and WiFi (802.11b-g) communications. The Savannah Creek sites have no local Internet access. In addition, all water based sites could not have extensive external cabling due to the potential for fouling from debris.

In addition to all this, WSN systems tend to concentrate information to the sink of the data, i.e., the gateway to the Internet (Mainwairing et al., 2002). That implies that the units closer to the gateway will have higher energy requirements because of more frequent transmission.

For the case of the water monitors, a point to point wireless gateway was selected. For the microclimate sensors, a local gateway in the microclimate sensor field was

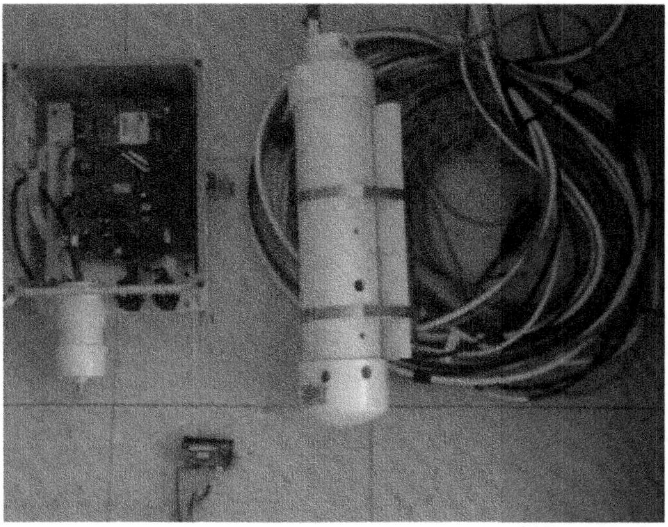

Fig. 1 Sample water monitor (*top*) and microclimate (*bottom*) hardware

chosen which will be connected to the commercial infrastructure by one of three ways: (1) Cellular Phone, (2) Point-Point Long range radio, (3) WiFi Connection. Students are establishing these as standard modules which can be mixed and matched according to future need. Each microclimate gateway links to these by standard Rs-232 port or serial voltage connection.

Data Storage. Because of the low bandwidths in the sensor fields and the requirement for data redundancy, data is being stored at two layers in the network. For individual sample cycles, the data is being stored as raw voltage values on the local sensor. Over 6400 samples with date time stamp can be stored locally. Servers are being established in the area to periodically interrogate the network and archive the historical data. These use SQL based database systems for storage and query management.

Data Mining. This feature is the ability to actively query the data base and the network itself in real time. This feature will be developed during 2009 as it requires modification of the routing algorithms in the individual sensors. One prototype is the TinyDB system in which a query from the Internet is split into the respective network parts to gather and compute the query results in the network itself. This would allow real time data mining of multiple data fields.

Initial Deployment Issues

Water Monitor. The water monitor portion was initially tested by students in 2006. Basic functionality was demonstrated, but housing issues on the sensors required a redesign of hardware and software during 2006–2007. Also at that time, the requirement was added for basic wireless communications. In May 2007, an initial deployment of the revised system was performed. It was at this time that an initial wireless 900 MHz modem was added. Also, additional memory was added to allow for local processing of raw sensor readings to engineering units for transmission to the database. The revised system has been water tested, and is awaiting completion of the Sylva signage for deployment in Scotts Creek next month.

Microclimate Sensors. The sensors were initially tested for brief periods in the 2006–2007 academic year with a field of 4 units. It was determined that the solar charging system was inadequate for long term deployment, and has been upgraded. Life testing with loads was performed in August, 2007. The adjacent Sylva nodes will be deployed approximately one month after the Sylva signage is deployed.

Future Plans

Upon completion of the Sylva signage and establishment of the database server in Sylva, sites 2, 3, 4 will be deployed by January, 2008. Microclimate sensors will be added to the nearby area. In addition, the point to point connections will be replaced with "mesh" type wireless connections to improve system reliability. Lastly, upon completion of discussions with the various school districts and additional local businesses, additional sites will be established to be used both as part of the observatory and for local classroom use. Individual researchers will be able to access both the serve database as well as bore into the network for real time data by January 2009. Enhancements to the network will be coming from three sources. WCU Kimmel School, Environmental Science Departments, and WATR as well as the local school districts will participate.

By reducing the energy load of the individual systems, alternate energy sources will be evaluated. Specifically, a micro-hydroelectric source has been identified for the water meters for a design topic of the WCU ET students.

Summary

A need was established to monitor basic water quality and microclimate parameters in the Tuckaseegee watershed area of Western North Carolina. By recognizing the resources available from the educational and commercial sectors, the opportunity was seen to create a multi purpose, modular regional observatory which would be a valuable tool for education, science, pollution monitoring, and public awareness. Such a system is now being deployed.

References

Crossbow Technology, Inc. 2004. *MPR/ MIB User's Manual.* Document 7430-0021-06 Crossbow Technology, Inc. San Jose, Ca. August 2004.

Eads, R.E. and J. Lewis 2002. "Continuous Turbidity Monitoring in streams of Northwestern California." *Turbidity and Other Sediment Surrogates Workshop.* Proc., Reno, Nevada, USA, May 2002.

Mainwairing, A., J. Polastre, R. Szewczyk, D. Culler, and J. Anderson 2002. "Wireless Sensor Networks for Habitat Monitoring." *Proc. 1st ACM International Workshop on Wireless Sensor Networks and Applications*, Atlanta, Ga. USA 2002.

Encapsulation of Potassium Permanganate Oxidant in Polymers

Stephanie Luster-Teasley, David Price, and Dereje Worku

Abstract The present work explores the ability to encapsulate chemical oxidants for environmental remediation. Several formulations were tested to create pellets with various release rates for the potassium permanganate. Potassium permanganate is a crystalline chemical oxidant that is currently dissolved in water for use in drinking water treatment and in the remediation of chlorinated solvents. Limitations for liquid permanganate remediation include safety hazards, the need to deliver the solution subsurface to contaminated areas, the need to repeatable inject potassium permanganate solutions to complete contaminant degradation, and plugging of subsurface soils with manganese oxide precipitates. Prototype pellet structures were produced. The pellets were capable of slowing releasing potassium permanganate over a 43-day period of testing. This technology provides new methods for potassium permanganate remediation including creation of subsurface permeable reactive barriers, wells packed with oxidation pellets, or reactors filled with pellets for pump and treat systems. This paper will summarize the development of this technology, release rate studies, and preliminary remediation studies for trichloroethylene using one preliminary prototype.

Background

Controlled release methods to encapsulate and slowly deliver pharmaceuticals or agricultural chemicals have proven to be highly effective (Akelah, 1996; Al-Zahrani, 2000; Ding and Liu, 1998; Segal, 1991; Uhrich, 1999). The ability to encapsulate chemicals and bacteria to produce a slow-release method for environmental systems is an emerging area in environmental engineering. Several researchers have developed controlled release systems for soil and water remediation using clay, waxes,

S. Luster-Teasley (✉)
North Carolina A&T State University, Greensboro, NC, USA

G.A. Uzochukwu et al. (eds), *Proceedings of the 2007 National Conference on Environmental Science and Technology*, DOI 10.1007/978-0-387-88483-7_38,

and gels. Previous studies have developed slow release oxygen for bioremediation, phosphate buffer encapsulation for pH control during denitrification of groundwater and sediment, alginate gel for Fenton photochemical oxidation, and encapsulation of bacterial cells (Brewster et al., 1995; Kang et al., 2004; Ross et al., 2005).

Highly reactive oxidants encapsulated and stabilized for remediation provides a technique for chemical oxidation treatments of water and soil. Currently, chemical oxidation methods use short-term liquids and gases which require repeated dosing of the oxidant to remediate contaminants. The ability to produce oxidant encapsulated structures provides a method for chemical oxidation treatment. Applications for this technology include formation of permeable reactive barriers through which groundwater plumes can travel for subsurface water and soil remediation. Pump and treat systems using pellet filled reactors additionally provide an option for remediation.

In the present work, several encapsulants were investigated for their potential to produce $KMnO_4$ encapsulated prototypes for environmental remediation. Prototypes of the encapsulated oxidants were produced to release $KMnO_4$ for extended periods of time. The delivery of solid oxidants from a encapsulated oxidant has several advantages. (1) It stabilizes the solid oxidants for emplacement in the subsurface to form permeable reactive barriers, packed in a series of wells where contaminated water will flow past the pellets for remediation, or packed in a reactor for a pump and treat system. (2) It reduces the need for maintenance associated with the gaseous and liquid oxidants. (3) It reduces the dangers associated with handling the oxidant by workers. (4) It extends the release of an oxidant without the need to re-inject the gas or liquid solutions.

The purpose of this paper is to summarize the laboratory research conducted to produce encapsulated $KMnO_4$ and examining the feasibility of this technology. Prototypes were investigated using batch studies with aqueous media. The release rates and duration of the $KMnO_4$ release were monitored for various controlled-release (CR) prototypes. A preliminary study was conducted using one of the prototypes for trichloroethylene (TCE) degradation.

Experimental Materials and Methods

Certified ACS $KMnO_4$ (Fisher Chemical, Fair Lawn, NJ) was selected as the oxidant to encapsulate. Six CR prototypes were tested during this exploratory research. The CR prototypes structures were capable of controlled-release of the oxidant. For this project, identification numbers will be assigned to the CR prototypes. These CR prototypes were designed to release the oxidant in the presence of water.

The stability of encapsulated $KMnO_4$ was evaluated by producing the CR prototypes and mounting them to glass slides. Digital photos were taken at 40x and 100x each month. The CR prototypes were stored in the dark at room temperature for 6 months. $KMnO_4$ release from the CR prototypes were studied using two methods. In the first method, bottles containing 50 ml of DI water and 0.5 g of CR prototypes pellets were stored for 3–4 days. After 3–4 days, the $KMnO_4$ concentration was measured using a Spectronic spectrometer at 525 nm wavelength. Following the $KMnO_4$ concentration reading, the pellets were rinsed with DI water three times and 50 ml of fresh DI water was added to the bottles with the same pellets. This process was then repeated by storing the reaction bottles for another 3–4 days and a $KMnO_4$ measurement was performed. A continuous study of $KMnO_4$ was performed by adding CR prototypes were added to 50 ml of DI water in reaction bottles. Measurements for $KMnO_4$ were conducted every 3–4 days. Following the measurements, the CR prototypes were returned the bottles, capped, and stored for another 3–4 days before the next measurement.

Results and Discussion

Encapsulant Selection

Several groups have investigated the potential for encapsulating $KMnO_4$ in waxes. Our goal was to investigate the use of an encapsulant that would be easy to handle and incorporate $KMnO_4$ into pellet shaped structures. Previous studies using wax argues the affordability, ease in pellet formation, and its hydrophobicity in water. However, there are drawbacks to using waxes which include a low melting temperature (47–64°C) and long-term stability of $KMnO_4$ encapsulated in the soft, waxy structures. Therefore our goal was to find an easy to use encapsulant capable of overcoming the limitations seen in waxy structures used to encapsulate $KMnO_4$. Criteria for encapsulant selection include the compatibility with the selected oxidant, the ability to form solid pellet structures and the ability of the encapsulant to release the oxidant.

Stability of $KMnO_4$

Prototype development was based on three criteria: (1) encapsulant compatibility with the oxidant, (2) the ability for diffusion of the oxidant and (3) the ability to form pellet structures feasible for use as fill material for a permeable reactive barrier or in a reactor system. $KMnO_4$ was tested for reactivity with the three encapsulants. One concern is to ensure that the oxidant is not reactive with the encapsulants. If the $KMnO_4$ is reactive with an encapsulant,, visible evidence would be a brown discoloration caused by MnO_2 from $KMnO_4$ degradation or visible degradation of the encapsulant. For this test, $KMnO_4$ was encapsulated and mounted on a glass slides. Photos at 40x and 100x were taken of the pellets using a digital microscope to observe any reactions or discoloration of the shell. Little to no reactivity was seen in the encapsulants used to produce the CRBP prototypes. Figure 1 shows that the $KMnO_4$ appears to be stable and non-reactive with $KMnO_4$ crystals in the target encapsulant material selected for PAB prototypes. Figure 1a is a photo of the

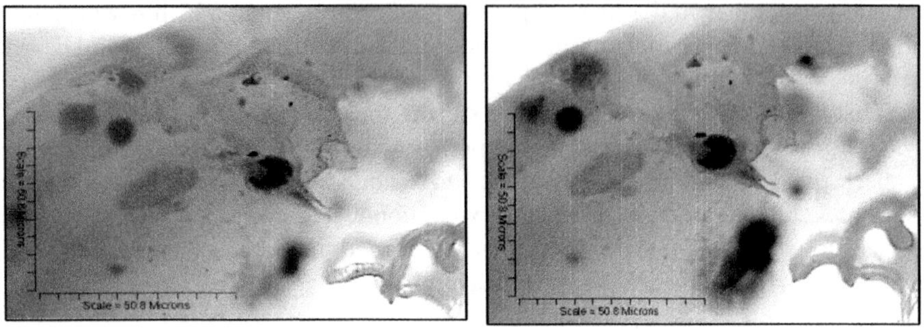

(a) Encapsulant with oxidant at start (b). Encapsulant with oxidant seven months later

Fig. 1 Photo of encapsulant with $KMnO_4$: (**a**) Month Zero and (**b**) Month Seven

$KMnO_4$ in the encapsulant photographed on May 16, 2006 and Fig. 1b shows the same sample on December 12, 2006.

Release of $KMnO_4$

Biphasic $KMnO_4$ release was observed in our $KMnO_4$ release studies. This is consistent with the observations seen in previous studies (Kang et al., 2004; Ross et al., 2005). A rapid release of $KMnO_4$ occurred during the initial 1–3 days followed by a controlled and slower release of $KMnO_4$ for an extended period of time. The photos in Fig. 2 show the pellets and the diffusing oxidant (Invention disclosure filed).

We were able to design controlled release of the oxidant based on the encapsulant. Figure 3 shows the $KMnO_4$ release exhibited in studies where fresh DI water was replaced every 3–4 days. Biphasic release of the oxidant was observed followed by a steady-state release of the oxidant occurring for 40–46 days. Figure 4 shows the continuous study for $KMnO_4$ over a 46 day period where $KMnO_4$ concentration was measured until visible MnO_2 precipitate was observed in the reaction bottles.

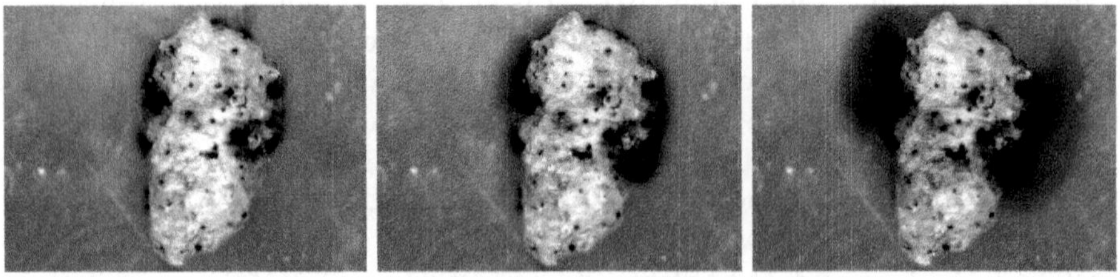

Fig. 2 Time series photos of pellet with oxidant diffusing from shell

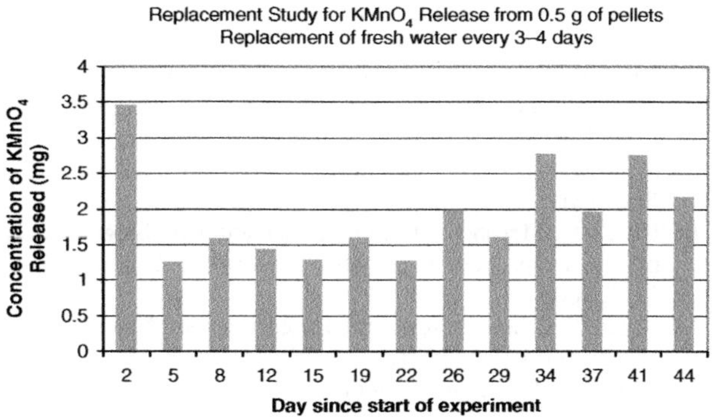

Fig. 3 Biphasic and extended release of 0.1 g of $KMnO_4$ from CR prototype

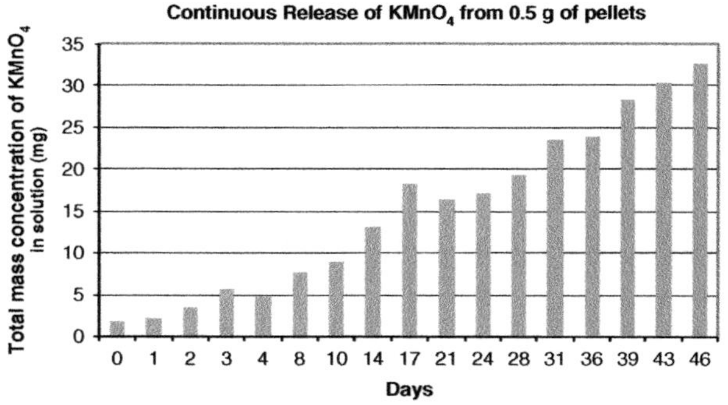

Fig. 4 Continuous release study of $KMnO_4$ from CR prototype pellets for 46 days

Conclusions

This research demonstrated the ability to encapsulate oxidants to produce controlled release structures. These structures have the potential to be used for remediation of various forms of contaminants using permeable reactive barriers, augmented soils containing pellets, and pump and treat systems.

Acknowledgments This work was funding through NSF Small Grants for Exploratory Research Grant #0640035. We would also like to acknowledge fund from NSF Louis Stokes Alliance for Minorities Program (NC LSAMP) which provided stipends for the undergraduate students participating in this research.

References

A. Akelah, "Novel utilizations of conventional agrochemicals by controlled release formulations," *Materials Science and Engineering*, vol. C, pp. 83–98, 1996.

S.M. Al-Zahrani, "Utilization of Polyethylene and Paraffin Waxes as Controlled Delivery Systems for Different Fertilizers," *Industrial Engineering Chemistry Research*, vol. 39, pp. 367–371, 2000.

M.A.G.G.O. Brewster, J.D. Redman, K.A. Sander, "Observed migration of a controlled DNAPL release by geophysical methods.," *Groundwater*, vol. 33, pp. 977–987, 1995.

J. Ding and G. Liu, "Water-Soluble Hollow Nanospheres as Potential Drug Carriers," *Journal of Physics and Chemistry B*, vol. 102, pp. 6107–6113, 1998.

N. Kang, I. Hua, and P.S. Rao, "Production and Characterization of Encapsulated Potassium Permanganate for Sustained Release as an In-Situ Oxidant," *Industrial Engineering Chemistry Research*, vol. 43, pp. 5187–5193, 2004.

C. Ross, L.C. Murdoch, D.L. Freedman, and R.L. Siegrist, "Characteristics of potassium permanganate encapsulated in polymer," *Journal of Environmental Engineering*, vol. 131, pp. 1203–1211, 2005.

M. Segal, "Patches, Pumps, and Timed Release: New Ways to Deliver Drugs," *FDA Consumer*, vol. 25, pp. 13–15, 1991.

K. Uhrich, "Polymeric Systems for Controlled Drug Release," *Chemical Reviews*, vol. 99, pp. 3181–3198, 1999.

C. Elegans Chemotaxis and Reproduction Following Environmental Exposure

Mulumebet Worku and Carresse Gerald

Abstract The free-living nematode *Caenorhabditis elegans* (C. elegans) is a well established biological model. Chemotaxis is an important behavior in enabling it to locate food sources such as E.coli. This study evaluates the chemotaxis of *C.elegans* to food following exposure to Moxidectin (Cydectin), Black Seed Oil, non-fat milk, or colostrum. A ring of bacteria (food) on nematode growth agar medium served as attractive signals to encourage *C. elegans* to move. Nematodes were placed in the center of the agar plate in Moxidectin, Black seed oil, colostrum or non-fat milk. Over the three day life cycle the reproduction (number of worms), movement (yes or no) and chemotaxis (number migrating to the ring of bacteria through the chemical) was recorded. Exposure to different chemicals differentially affected the total number of worms, movement and chemotaxis. Moxidectin and black seed oil inhibited migration and reproduction at the concentrations tested. Both colostrum and non-fat milk increased migration and reproduction. *Caenorhabditis elegans* is a sensitive multi-cellular bio and chemosensor. The bacterial ring assay is a very useful in vitro approach for evaluation of the effect of chemical exposure on chemotactic behavior and reproduction. Thus, this model can be used to better understand the molecular basis of the response to environmental exposure and for identifying molecular targets or biochemical pathways involved.

Introduction

Caenorhabditis elegans (C. elegans) are free-living, soil-dwelling nematodes that have been extensively studied. They are models for studies in genetics and developmental biology (Brenner, 1974; Riddle et al., 1997). Useful attributes include the

---

M. Worku (✉)
North Carolina A&T State University, Greensboro, NC, USA

G.A. Uzochukwu et al. (eds), *Proceedings of the 2007 National Conference on Environmental Science and Technology*, DOI 10.1007/978-0-387-88483-7_39,

small size (1 mm in length), relatively short life cycle, simple well characterized cell lineage, ease in culturing in vitro and the availability of the sequenced genome (Boyd et al., 1999).

Chemotaxis is an important behavior in enabling *C. elegans* to locate food sources such as E. coli. Being soil dwelling nematodes that live at an air-water interface they are exposed to chemicals in water and air. These nematodes have evolved chemosensory properties used for survival through modification of their behavior in response to chemical cues (Bargmann and Mori, 1997). Numerous volatile and nonvolatile chemicals have been screened to identify chemical attractants and repellents for *C. elegans* (Bargmann and Mori, 1997). These nematodes are able to distinguish among chemicals with differing structure. It has been suggested that water-soluble chemicals are used mainly for short-range chemotaxis to bacteria, and volatile chemicals for longer-range chemotaxis to distant food sources (Bargmann and Mori, 1997).

Studies support the use of *C. elegans* as a representative of free living bacterivore nematodes (Boyd and Williams, 2003) in toxicity screening. Thus, *C. elegans* has been used as a multicellular biosensor of biological toxicity for determining the effects of alkylphenols and organotin compounds (Tominaga et al., 2002). Further, *C. elegans* is increasingly being used as a toxicity testing organism in aquatic media and soils (ASTM, 2002; Graves et al., 2005; Boyd and Williams, 2003; Davies and McIntire, 2004). This is exemplified by the ASTM Standard Guide for Soil toxicity (ASTM, 2002).

Parasitic nematodes pose the single most significant threat to livestock world wide due to rising anthelmintic resistance. This resistance is defined as a heritable reduction in the sensitivity of a parasite population to the action of a drug. The reduction is expressed as the decrease of the frequency of individual parasites affected by exposure to the drug, compared to the frequency observed in the same population upon initial or prior to exposure (Condor and Campbell, 1995). The utility of *C. elegans* as a tool to study parasite gene function is important for efforts to address problems in animal and human health (Gilleard, 2004). For example, *Caenorhabditis elegans* receptors that regulate locomotor behaviour are functional orthologues of receptors in *Haemonchus contortus*, a strongylid nematode of small ruminants and cattle (Cook et al., 2006).

Model organism such as *C. elegans* are needed to study the mechanisms of resistance to anthelmintic classes and for new drug discovery. This can be accomplished by understanding the mode of action of anthelminthic compounds in disrupting the nematodes ability to locate food sources such as E. coli. Chemotaxis to an attractive odorant in vitro can be used to understand the effect of chemical exposure and for identifying molecular targets or bio chemical pathways mediating response. Davies and McIntire (2004) described a simple bacterial ring assay for the isolation and behavioral analysis of mutants of *C. elegans* that are resistant to the depressive effects of ethanol on locomotion.

Objectives: This study evaluates the chemotaxis of C. elegans to food following exposure to Moxidectin(Cydectin), Black Seed Oil(Niagra Sativa), non-fat milk or colostrum using the bacterial ring assay. Over the three day life cycle the reproduction (number of worms), movement (yes or no) and chemotaxis (number migrating to the ring of bacteria through the chemical) was recorded.

Materials and Methods

Caenorhabditis elegans: *C. elegans* N2strain plate culture was purchased from Carolina Biological Supply Co (CBSC), Burlington NC.

E. Coli Preparation: K-12 *E. coli* obtained from CBSC was incubated overnight at 37°C in nutrient broth. Two hundred μl of the *E. coli* culture was applied and spread onto Nematode Growth Agar (NGA) plates using sterile techniques and placed in an incubator for 24 h at 37°C for inoculation.

Plating C. elegans: A 1 cm^3 block of medium from the *C. elegans* plate culture (CBSC) containing live *C. elegans*, was cut with a scalpel and placed on an inoculated NGA plate, nematodes containing surface facing down. The Nematodes were left to reproduce and mature for ~4 days at room temperature before treatment.

Bacterial Ring Assay: The bacterial ring assay was conducted as described by Davies and McIntire (2004) with slight modifications. Briefly a ring of bacteria (food) on NGM agar medium served as attractive signals to encourage *C. elegans* to move. The center of the agar plate was treated with 0.016 mg/ml (molarity = 25 μM) Moxidectin (Quest®Gel, Cydectin)(Wyeth, NJ)/sterile water solution, 25 ul Black Seed Oil (Black Sunnah), 10 ul Colosturm (Symbiotics) or 10 ul Non-fat carnation instant milk (Nestle) were used for the treatment. Control plates did not have any chemicals on them. Nematodes were placed in the center of the agar plate. Worms were maintained at room temperature.

Observations: Over the three day life cycle the reproduction (number of worms), movement (yes or no) and chemotaxis (Number migrating to the ring of bacteria through the chemicals) was observed under a stereoscope (CBSC) and recorded. Nematodes that had reached the food were picked individually to new seeded plates and allowed to recover and reproduce. Averages of three experiments were compared.

Results and Discussion

Caenorhabditis elegans is a sensitive multi-cellular bio and chemo-sensor. The utility of the *C. elegans* bacterial ring assay described by Davies and McIntire 2001 to determine differential chemotactic responses was ascertained using four chemical modulators of the nematode life cycle (See Table 1 and Fig. 1).

A 3 % decrease in chemotaxis was observed with Moxidectin treatment, an 88% decrease with blackseed oil treatment a 72% and 70% increase with colostrum and non-fat milk respectively (Table 1). Reproduction was observed on all plates in the absence of treatment. The numbers of worms increased from day 1 to day 3 (See Table 2). However treatment resulted in differential effects that reflect reported

Table 1 Effect of selected chemicals on *C.elegans* migration on day 3 of the life cycle: Numbers of nematodes

	Moxidectin (25 uM)	Black Seed Oil (25 ul)	Colostrum (10 ul)	Non-fat milk (10 ul)
Control	355	86	73	194
Treated	204	10	264	649
Number of repeats	3	1	1	2

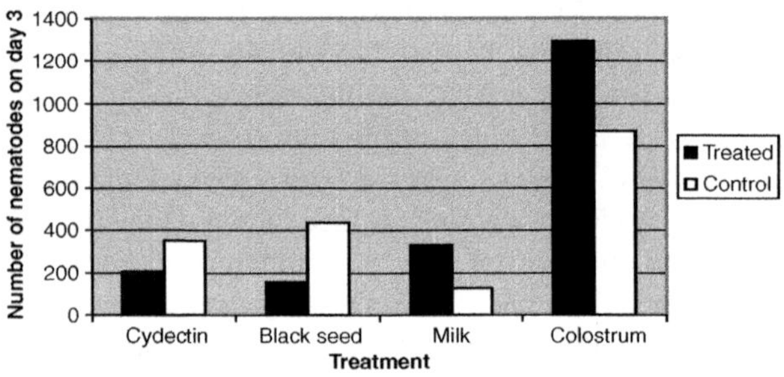

Fig. 1 Differential effects of chemical treatment on *C.elegans* reproduction. Values represent the total numbers of nematodes counted in treated plates – total numbers of nematodes counted in respective control plates

effects of each chemical. Moxidectin and Black Seed oil resulted in reduced numbers of nematodes when compared to their respective controls. Further, exposure to Colostrum and Non-fat instant milk increased the numbers of nematodes (See Table 2).

In this study water soluble Moxidectin (Cydectin, Quest Gel) was used. This Macrocyclic lactone represents the main treatment option for treating nematodes in livestock and is also used to control external parasites. Macrocyclic lactones are applied as broad-spectrum anthelmintics and ectoparasiticides in horses, cattle, sheep, pigs. Macrocyclic lactones are taken up by many nematodes via the cuticula and by oral ingestion. In blood-sucking parasites oral absorption is more important anthelmintics (Wolstneholme and Rogers, 2005). The macrocyclic lactones are the biggest selling and arguably most effective anthelmintics currently available.

Table 2 Effect of Chemicals on total numbers of nematodes (Migrated + non migrated)

	Moxidectin (25 uM)	Black Seed Oil (25 ul)	Colostrum (10 ul)	Non-fat milk (10 ul)
Day 1 Treated	38	9	130	204
Day 3 Treated	204	158	331	1297
Day 1 control	16	9	130	178
Day 3 control	355	436	97	873

According to the manufacturer (Fort Doge, NJ), QUEST 2% Equine Oral Gel used in this study, acts by interfering with chloride channel-mediated neurotransmission in the parasite. This results in paralysis and elimination of the parasite. Moxidectin (Cydectin) treatment resulted in a 40 % reduction in total number of worms following first time exposure. Migration of worms to a food source was reduced by 10% on first time exposure to 25 micromoles of Moxidectin. Exposure to 50 micromoles of Moxidectin killed all nematodes (results not shown).

When nematodes were exposed to 25 ul of Black Seed oil there was an 88 % reduction in total numbers of worms following first time exposure. The total number and Migration of worms to a food source was reduced (see Tables 1 and 2) on first time exposure when compared to nematodes maintained in PBS. Black Seed Oil (Sweet Sunnah Black Seed Liberty, NY 12754) was used in this study. Black Seed (Nigella sativa) has a high concentration of essential fatty acids associated with health benefits. Nigella sativa seeds were demonstrated to contain 36–38% fixed oils, proteins, alkaloids, saponin and 0.4–2.5% essential oil (Ali and Blunden, 2003). Much of the biological activity of the seeds has been shown to be due to thymoquinone, the major component of the essential oil, but which is also present in the fixed oil. The analysis of the volatile oil (essential oil) content in the black seed reveals chemical compounds that have been proven to exert anti-inflammatory action.

Milk protein-derived bioactive peptides are claimed to be health enhancing components that can be used to reduce the risk of disease or to enhance a certain physiological function (Davidson, 1999). *C. elegans* is maintained in the laboratory either on solid or in liquid growth medium. Milk has been found to be an excellent component of axenic (liquid media) for growth. These nematodes can be grown on agar plates using *E. coli* as a food source, or in liquid cultures with or without bacteria (Hope, 1999; Link et al., 2000). Milk has been established as an ingredient in C.elegans growth media. Use of a chemically defined medium (C. elegans Maintenance Medium (CeMM)) now allows standardization and systematic manipulation of the nutrients that animals receive including milk and its components (Clegg et al., 2002).

Colostrum is a form of milk produced by the mammary gland in the first few days after giving birth. It is a dietary supplement and has been linked to having curative powers (Davidson, 1999). In this study, colostrum (Symbiotics from Naturade, Inc., CA) that was derived from the first two milkings and had a guaranteed minimum compositon of 1.5% Lactoferrin (14 mg), 25% Immunoglobulins (240 mg IF), and 3% Proline-Rich Polypeptides (29 mg PRP), was used. Colostrum treatment resulted in increased migration to E. coli when control plates were compared to treated plates (See Table 1). An earlier response (day 1) was observed in colostrum treated nematodes. Higher numbers of migrated nematodes were observed with colostrum treatment (Fig. 1). Colostrum may provide nutrients for nematode growth. Further this C. elegans model may be used to study the effect of nutrients in Colostrum in disease related and other genes.

Non-fat instant dry milk (Nestle Wilkes-Barre, PA) is evaporated milk, it is fresh milk which is carefully heated to remove about half the water. As a result of the evaporation of water, the milk becomes richer and creamier with an increased amount of protein and calcium. Both the control and the treated plates showed chemotaxis, however, more nematodes migrated to the E. coli in the presence of milk. In addition, reproduction was affected as a greater number of nematodes grew in the presence of the milk. The addition of milk improves growth of C. elegans in axenic liquid culture in the absence of E.coli as a food source. In the present study it augmented the chemotaxis and reproduction on agar plates. Thus this may allow for the development of rapid chemotaxis assays to observe factors influencing chemotaxis. Nonfat Carnation Instant milk is a good medium for growth and does not inhibit the chemotaxis of C. elegans.

Summary: The differential effects of the chemicals used is presented in Fig. 1. The bacterial ring assay is a very useful in vitro approach for evaluation of the effect of chemical exposure on chemotactic behavior and reproduction. Thus, this model can be used to better understand the molecular basis of the response to environmental exposure and for identifying molecular targets or biochemical pathways involved. The versatility and simplicity of this approach can be used to evaluate the genetic basis for the effect of these and other chemicals in conjunction with mechanisms affecting development of adaptation.

References

Ali, B.H. and G. Blunden. 2003. "Pharmacological and toxicological properties of Nigella sativa", Phytotherapy Research 17, 299–305.

American Society for Testing and Materials. 2002. "Standard guide for conducting laboratory soil toxicity tests with the nematode Caenorhabditis elegans". In: Annual book of ASTM standards. Philadelphia, PA, 11.05, 1606–1616.

Bargmann C.I. and I. Mori. 1997. "Chemotaxis and thermotaxis". In: D.L. Riddle, T. Blumenthal, B.J. Meyer, and J.R. Priess (eds.), C. elegans II, Cold Spring Harbor Laboratory Press, Cold Spring Harbor, NY, 717–737.

Boyd, W. A., G.L. Anderson, D.B. Dusenbery, and P.L. Williams. 1999. "Computer tracking method for assessing behavioral changes in the nematode Caenorhabditis elegans" In: F.T. Price, K.V. Brix, N.K. Lane (eds.), Environmental Toxicology and Risk Assessment: Recent Achievements in Environmental Fate and Transport, vol. 9, ASTM STP 1381, American Society for Testing and Materials, West Conshohocken, PA, 2000, pp. 225–238.

Boyd, W.A. and P.L. Williams. 2003. "Comparison of the sensitivity of three nematode species to copper and their utility in aquatic and soil toxicity tests": Environmental Toxicology Chemistry 22:2768–2774.

Brenner, S.J. 1974. 'The genetics of *Caenorhabditis elegans*". Genetics, 77, 71–94.

Clegg, E.D., H.F. LaPenotiere, D.Y., French, and M. Szilagyi. 2002. "Use of CeHR axenic medium for exposure and gene expression studies". Reproductive Hazards Laboratory, U.S. Army Center for Environmental Health Research, Fort Detrick, MD. https://drum.umd.edu/dspace/bitstream/1903/2204/1/CeHR_medium.htm.

Condor G.A. and W.C. Campbell. 1995. "Chemotherapy of nematode infections of veterinary importance, with special reference to drug resistance". Advances in Parasitology 35:1–84.

Cook, A., N. Aptel, V. Portillo, E. Siney, R. Sihota, L. Holden-Dye, and A. Wolstenholme. 2006. "Caenorhabditis elegans ivermectin receptors regulate locomotor behaviour and are functional orthologues of Haemonchus contortus receptors". Molecular and Biochemical Parasitology, 147(1):118–125.

Graves, A.L., W.A. Boyd, and P.L. Williams. 2005. "Using transgenic *Caenorhabditis elegans* in soil toxicity testing". Archives of Environmental Contamination Toxicology 48(4):490–4.

Davidson A. 1999. Oxford Companion to Food, "Beestings", p. 69.

Davies, A.G., and S.L. McIntire. 2004. "Using C. elegans to screen for targets of ethanol and behavior-altering drugs". Biol Proced Online 6:113–119.

Gilleard, J.S. 2004. "The use of Caenorhabditis elegans in parasitic nematode Research". Parasitology. 128:S49–S70 Cambridge University Press Copyrigh. Cambridge University Press.

Graves, A.L., W.A. Boyd, and P.L. Williams. 2005. "Using Transgenic *Caenorhabditis elegans* in Soil Toxicity Testing". Archives of Environmental Contamination and Toxicology. Archives of Environmental Contamination Toxicology 48:490–494.

Link, E.M., Hardiman, G., Sluder, A.E., Johnson, C.D., and Liu, L.X. 2000. "Therapeutic target discovery using Caenorhabditis elegans. Pharmacogenomics" The total numbers of nematodes counted on the third day of the life cycle is presented in Table 2, 1:203–217.

Hope, I.A. 1999. "*C. elegans*". Oxford University Press, Oxford.

Riddle, D.L., T. Blumenthal, B.J. Meyer, and J.R. Priess, (eds.). 1997. "C. Elegans II." Cold Spring Harbor Laboratory Press, Plainview, New York

Tominaga, N., M. Tomoeda, S. Kohra, Y. Takao, M. Nagae, K. Ueda, Y. Ishibashi, T. Kai, and K. Arizono. 2002. "A convenient sublethal assay of alkylphenol and organotin compounds using the nematode Caenorhabditis elegans." Journal of Health Science 48:555–559.35.

Wolstneholme, A.J. and A.T. Rogers. 2005. "Glutamate-gated chloride channels and the mode of action of the avermectin/milbemycin anthelmintics". Parasitology. 131 Suppl:S85–95.

Storing Weather Data and Dissemination Via the Web

Kawana O. Fuller and Austin Frazier

Abstract As part of a much larger project, significant progress has been made toward making localized weather data available to the research community and even to the general public through the Internet. A weather station, installed on the roof of McNair Hall on the campus of North Carolina A&T State University, captures weather data and a C++ program accesses the data and stores it in a MySQL database. As of now, the data is accessed using PHP and made available in various forms. The data is being published on the Web in fixed intervals throughout each day. With KML, weather statistics are also available on Google Earth. There, users have access to the weather data through an icon at the location of the sensor, which, when clicked, displays a link to our weather page. Users also have the option of viewing calculated statistics or current weather data for the past 24 h, 7 days, weeks, or even months. Though significant, this work is only a small first step toward using grid computing and multiagent systems for real-time sensor control and data management in large sensor networks.

Introduction

The project reported here involved initial implementation work in the Distributed Architectures research area of the NOAA Interdisciplinary Scientific Environmental Technology (ISET) Cooperative Science Center. ISET collaborates with NOAA's Earth System Research Laboratory (ESRL) and will help fulfill NOAA's mission to provide the data needed to address climate and weather related concerns.

The goal was a simple yet reliable source of live weather data, storage of the data in a data base, and Web-access to the live data as well as the stored data and aggregate data. Since state-of-the-art coordination and data sharing across geographically

K.O. Fuller (✉)
Department of Computer Science, North Carolina A&T State University,
Greensboro, NC, 27411, USA

G.A. Uzochukwu et al. (eds), *Proceedings of the 2007 National Conference on Environmental Science and Technology*, DOI 10.1007/978-0-387-88483-7_40,

distributed nodes involves the Web and markup languages, another goal was to gain experience with a markup language handling geospatial information. This effort is constrained to use affordable commercial sensors and computing platforms that use free standard software components.

Materials and Methods

The sensor used for this project is a Davis Instruments VantagePro2 Weather Station with wireless connection to a console, as shown in Fig. 1. The console in turn is connected to a USB port of either a computer running Windows or a computer running Linux. Both operating systems were used to ensure flexibility in future development. Note that only one computer is connected to the console at a time. The data provided by the weather station via the console includes temperature, humidity, dewpoint, wind direction and speed, barometric pressure, rain rate and totals, wind chill, temperature-humidity-wind (THW) index, and heat index. On the Windows machine, we installed the WAMP software bundle, which includes specifically for Windows applications the Apache HTTP Web server, the MySQL database management system, and support for the PHP server-side scripting language. (The acronym "WAMP" is formed from the first letters of "Windows", "Apache", "MySQL", and "PHP".) On the Linux machine, we separately installed the Apache server, MySQL, and PHP, giving a configuration known as LAMP (WAMP with Linux—hence the "L"—instead of Windows).

The weather station comes with Windows software that, at adjustable intervals, downloads data from the console, writes the data to an archive file, and updates an HTML document generated from an editable template file. To capture data from the console in a form we can store in a MySQL table, we had to install the VantagePro2

Fig. 1 The configuration consisting of the weather station, console, and Internet-accessible PC

DLL 2.3, downloaded from the Davis Web site, and write a C++ program using it. For the Linux machine, we downloaded wview, which is a collection of Linux daemons interfacing with the weather station to retrieve archive records and current conditions; wview was customized to interface also with MySQL.

The weather station was attached to the roof of McNair Hall on the campus of North Carolina A&T State University, and the console was placed in a lab six floors below the roof, just beside the PCs to which it is connected. No loss of data has ever been observed on the wireless connection. Since the folder with the VantagePro2-generated HTML document is not under the server root, an alias for the folder (as well as a Directory element) was added to the configuration file for the Apache server. We developed a C++ program that accesses data from the console at regular intervals (currently five minutes) and stores it in a MySQL table. On the Linux machine, this is largely achieved by configuring wview.

PHP scripts were written to generate replies to requests from Web browsers with data retrieved from the database table. The PHP scripts also produce aggregate values, such as averages and trends, from the stored data, although some processing is done with MySQL's procedural language. Finally, some scripts respond to requests to update tables that store data derived from the table with the raw data.

Since the work reported here is a modest beginning for a large project that will make extensive use of markup languages for the Web, some experience was sought in linking our recorded sensor data with a server supplying documents in some markup language other than the ubiquitous HTML. The Keyhole Markup Language (KML) is an XML-based language for managing the display of three-dimensional geospatial data. In particular, KML is the language used for documents that are requested and rendered (with a great deal of imagery supplied by Google) by Google Earth (available for free download as a browser plug-in). In addition, KML shares much of its syntactic structure with GML, which is defined by an Open Geospatial Consortium (OGC) standard. We thus wrote a KML file specifying an icon at the location of our weather station. This file includes a description element containing a CDATA section with HTML code. The HTML includes an anchor element with a hyperlink to the HTML document generated by the Davis-supplied software. Clicking the icon in Google Earth opens a box with a rendering of the HTML in the KML file. Clicking the link in this box causes the browser to render the generated HTML document. The box could be made to contain links to any or all of the PHP scripts we developed at the cost of cluttering its layout. Our KML file resides in a folder that is a child of the server root. For the KML file to be transported over the Web and recognized by Google Earth, we had to add to the Apache configuration file a line mapping the .kml filename extension to the MIME type application/vnd.google-earth.kml.

Results and Discussion

All the weather data is being successfully stored in the database table for the raw data. The PHP scripts that provide various raw and aggregate data values have excellent response time. Space restrictions allow us to present only a few of the renderings. Figure 2 shows the webpage generated by wview, which contains, in addition to the raw data values, several images generated from this data. It also contains links to an almanac; weather trends for the last 24 h, 7 days, 28 days, and 365 days as well as a Doppler radar image of the Greensboro area. Figure 3 shows the view of McNair Hall in Google Earth. The pin icon has been clicked, opening the box that displays a brief description of the location and the link to the generated HTML document.

The Distributed Architectures research area of the ISET center is part of the Information Technology Applications thrust area. The other research areas in this thrust area are Data Fusion and Data Mining. The real-time data supplied by our weather station and available to our PCs can support initial experimentation with data fusion, and the data stored in our database tables is extensive enough for testing out certain data mining techniques.

Our weather station could be incorporated into a sensor network, which is a computer-accessible network of spatially distributed devices using sensors to monitor conditions at different locations. Our work with a Web server, database tables, and KML moves this project in the direction of a sensor web, which is a web-accessible sensor network with archived sensor data that can be discovered and

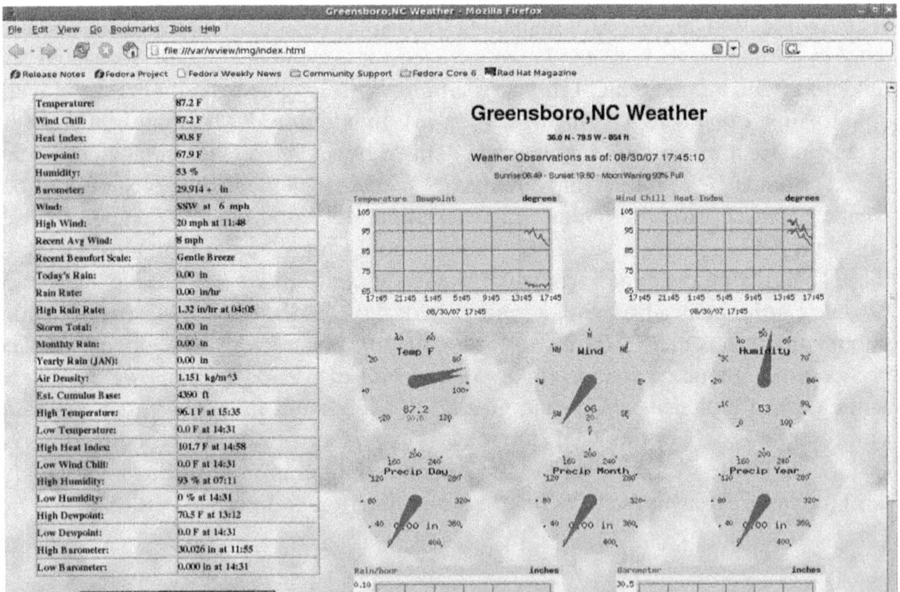

Fig. 2 Webpage generated by wview

Fig. 3 Google earth with McNair Hall icon selected

accessed using standard protocols. The OGC Sensor Web Enablement (SWE) initiative is building a framework of open standards for sensor webs (Botts et al., 2006).

One way this project could evolve is in the direction of a data broker. The MetBroker data broker (Ninomiya et al., 2007), for example, seamlessly integrates legacy weather databases with sensor networks. MetBroker provides client applications with details of available data, receives requests from them, queries remote databases, and returns results to the client applications. To incorporate data from a Field Server (which includes multiple sensors coupled with a Web server) with data from legacy databases, the Field Server data is written to an archive, from where it is regularly copied to a database that is part of the Field Server and also is MetBbroker accessible.

The Distributed Architectures research area, however, will eventually turn to grid computing and multiagent systems. Grids provide an infrastructure for dynamic and geographically distributed virtual organizations (VOs). They aim to enable "resource sharing and coordinated problem solving in dynamic, multi-institutional VOs" (Foster and Kesseleman, 2004). An agent "is an encapsulated computer system that is situated in some environment, and that is capable of flexible, autonomous action in that environment in order to meet its design objectives" (Wooldridge, 1997). Interest is generally not in single agents but in multiagent systems, where coordination and collaboration are key. Foster, Jennings, and Kesselman (Foster et al., 2004), have noted a convergence between the grid community (which historically has concentrated on "brawn") and the agents community (which has concentrated on the "brain"). Of course, the first step in either camp is to have reliable data available, and progress is greatly facilitated if we can use affordable hardware and free standard software.

Conclusions

This project succeeded in providing a reliable source of live weather data, which is provided by a Davis Instruments VantagePro2 Weather Station. There is a wireless connection from the weather station to a console, which is connected to a nearby PC running either Windows or Linux. Besides the software supplied by Davis, we installed the WAMP software bundle on the Windows machine and a LAMP bundle on the Linux machine; in both cases, this involved the Apache Web server, the MySQL database management system, and the PHP server-side scripting language. We wrote C++ code to copy weather data from the console to a database table. Our PHP scripts provide Web-access to the live data as well as the stored data and aggregate data, some of which is stored in additional tables. All aspects of this project proved to be reliable and robust, and performance has not been an issue. The total equipment cost for this project was a little over $1000, all related to the weather station. All the software used beyond that which came with the weather station can be freely downloaded and is the most commonly used software in its class.

We wrote a document in the geospatial markup language KML, which is used for documents requested by Google Earth, and put the file under the server root to make it Web-accessible. The KML file sspecifies an icon at the location of our weather station and associates a hyperlink with the text for the icon; when the link is clicked, one of our data pages is rendered by the browser. This has given us experience with a markup language handling geospatial information.

This project is an initial step in the NOAA ISET Distributed Architectures research area. The real-time data we supply can support initial experimentation with data fusion, and the data in our database tables can be used to test certain data mining techniques. In the future, we shall turn to grid computing and multiagent systems, where our weather station will be just one not very remarkable resource among many.

Acknowledgments Dr. Albert Esterline and Dr. Yaohang Li of the Dept. of Computer Science at North Carolina A&T State University supervised this project. We also thank the NOAA ISET Cooperative Science Center for its support. Dr. Matt Heaver and Mr. Ed Knuth of the University of Alaska Southeast shared their experience with weather station configuration; without their help, this project would have taken significantly longer. Mr. Donald Rustchmann and Mr. Paul Bajari of the Dept. of Computer Science at North Carolina A&T assisted greatly with installation.

References

Botts, M., G. Percivall, C. Reed, and J. Davidson. 2006. OGC Sensor Web Enablement: Overview and High Level Architecture. Open Geospatial Consortium Inc. White Paper, OGC 06-050r2 Version: 2.0.

Foster, I. and C. Kesselman, C. (Eds.). 2004. The Grid: Blueprint for a New Computing Infrastructure, 2nd ed. Morgan Kaufmann, San Francisco, CA.

Foster, I., N.R. Jennings, and C. Kesselman. 2004. "Brain Meets Brawn: Why Grid and Agents Need Each Other." Proceedings of the 3rd Int. Conf. on Autonomous Agents and Multi-Agent Systems (AAMAS 2004), New York, pp. 8–15.

Ninomiya, S., T. Kiura, A. Yamakawa, T. Fukatsu, K. Tanaka, H. Meng, and M. Hirafuji. 2007. "Seamless Integration of Sensor Network and Legacy Weather Databases by MetBroker." International Symposium on Applications and the Internet Workshops, 2007 (SAINT Workshops 2007), pp. 68–68.

Wooldridge, M. 1997. "Agent-based Software Engineering." IEEE Proc. on Software Engineering. 144:26–37.

Rapid and Simple Method for the Encapsulation of *Lactobacillus reuteri* in the Production of Lactic Acid

Salam A. Ibrahim, Somphavanh Phetsomphou, Chyer Kim, Abolghasem Shahbazi, Chung W. Seo, Amer AbuGhazaleh, and Moussa M.E. Salem

Abstract Lactic acid (LA) is the most widely used multifunctional organic acid and approximately 50% of it is produced by biotechnological process. Immobilizing microbial cells not only improves cell retention but also protects cells from harsh environmental conditions during LA production. The objectives of this research were therefore to develop a simple method to encapsulate (immobilize) *Lactobacillus reuteri* for the continuous production of LA and to evaluate LA production in three different types of media using free and immobilized *L. reuteri.*Five strains of *L. reuteri* grown in *lactobacillus*MRS broth at 37°C for 24 h were washed in peptone water and suspended in 2% sodium alginate solution. Encapsulation of cells was performed by dripping the mixture of sodium alginate and culture into ice-cold (2°C) 0.4 M calcium chloride solution using a separator funnel. The beads were then subjected to each of 500 ml MRS, modified MRS and whey-based broth and then incubated at 37°C for 12 h. Samples were withdrawn at 2-hour intervals during incubation period and analyzed for LA as represented in pH. Results show that the developed method is a rapid and simple microbial encapsulation procedure for the continuous production of LA. The efficacy of LA production as measured in pH was not significantly different in all tested media. At the end of fermentation process, pH of whey medium containing conventional (free) and encapsulated cells reached to 4.20 and 3.85, respectively. This indicates that higher amount of acid is yielded with encapsulated cells than free cells. In addition, immobilized cell strain MM2-3 produced the highest pH (3.5) while free cell strain SD2112 produced the lowest pH (4.05). Hence, results from this study suggest that we were able to develop a simple and rapid method for the encapsulation of *L. reuteri*.

Keywords Rapid method · Encapsulation · *Lactobacillus* reuteri

---

S.A. Ibrahim (✉)
North Carolina A&T State University, Greensboro, NC, USA

G.A. Uzochukwu et al. (eds), *Proceedings of the 2007 National Conference on Environmental Science and Technology*, DOI 10.1007/978-0-387-88483-7_41,

Introduction

LA is widely used in the food, pharmaceutical and chemical industries (Newman and McBurney, 2004). Several species of lactic acid bacteria (LAB) have been commercially used for LA production through the fermentation process of sugars, typically glucose (Wee et al., 2006). One of the expanding industrial applications of LA is polymerization of LA to polylactic acid (PLA), which is used to manufacture a variety of products, including biodegradable plastics and textile fiber (Wee et al., 2006). These applications have increased interest in developing more efficient production for LA.

Several technologies have been used for the efficient production of LA. One of those methods is a conventional batch fermentation process, which results in low rates of cell growth and productivity due to its inhibitory acid accumulation. Therefore, immobilization of viable LAB has attracted considerable interest for a decade. Successful microencapslation depends on the right selection of the wall-material and encapsulation technique for a specific core material (Scannell et al., 2000). Many properties of the microencapsulated system such as the retention of viable bacteria cells or the protection of core materials are related to the porosity and the integrity of the microcapsules (Scannell et al., 2000). Chronopoulous et al. (2002) and Champagne et al. (1987) also observed that cell growth in core coated with gelatine after immobilization yielded high cell concentrations per gram of encapsulated support system and immobilization of LAB by entrapment within a gel bead greatly improved LA production.

Calcium alginate due to its low cost has been commonly used to immobilize LAB for the production of LA (Elezi et al., 2003; Kyung et al., 2003). The most common method used to encapsulate LA cells into core-coated capsules is through syringe needle injection using airless spray gun (Kyung et al., 2003; Scannell et al., 2000). In this method, bacteria cells are pumped through a needle to generate droplets which are then collected in ice cold calcium chloride solution. However, this method may produce beads that are less uniform in diameter size and shape.

A great amount of whey is produced as a byproduct worldwide in the manufacture of cheese or casein and its disposal has been an issue of environmental pollution. In general, the manufacture of one ton cheese or casein produces 8 or 25 tons of whey, respectively. The biochemical oxygen demand of whey is approximately 35,000–45,000 g/L and therefore, 100 L of whey has a polluting strength equivalent to the sewage produced by 45 people (Swaisgood, 1982). However, whey contains a wide range of biologically active proteins (enzymes, vitamin-binding proteins, metal-binding proteins, immunoglobulin, etc.), lactose, minerals and water-soluble vitamins (Björck and Hopkin, 1991), which may serve as a fermentable medium for the production of LA.

The objectives of this research were (1) to develop a simple and rapid method to uniformly encapsulate *L. reuteri* cells for the production of LA, (2) compare whey based broth with two different types of MRS broth as fermentation medium for the production of LA, and (3) evaluate production of LA using free and immobilized cells.

Materials and Methods

Bacterial Strains and Their Preparation

Five strains of *L. reuteri* (CF2F, DSM20016, MM2-3, MM7 and SD2112) obtained from Biogaia Biologics, Inc. (Raleigh, NC) were used in the study. Each strain was cultured individually in 10 ml of whey broth and incubated at 37°C for 24 h.

Media

Whey, MRS and modified MRS (mMRS) broth were used for this study. Whey broth was consisted of 34% premium deproteinized whey (60 g/L, Davisco, Foods International Inc., Le Sueur, MN), yeast extract (2.5 g/L, Sigma-Aldrich Co., St. Louis, MO), L-cysteine hydrochloride anhydrous (0.5 g/L, Sigma-Aldrich), manganese sulphate (0.03 g/L, Thermo Fisher Scientific, Fair Lawn, NJ), and tween 20 (1 ml/L, Thermo Fisher Scientific). Modified MRS (mMRS) was prepared with MRS broth (55 g/L, DifcoTM Lactobacilli MRS broth, Becton Dickson and Co., Sparks, MD), manganese sulphate (0.05 g/L), tween 20 (5 ml/L), and L-cysteine chloride hydrochloride anhydrous (0.5 g/L). MRS broth was prepared with MRS broth (55 g/L) and L-cysteine chloride hydrochloride anhydrous (0.5 g/L). The broth was then autoclaved at 121°C for 15 min.

Immobilization Procedure

Overall procedure of cell immobilization is shown in Fig. 1. A 72-hour culture of each strain with cell density (A_{610})of 1.2 containing 1.2×10^6CFU/ml was centrifuged (Sorvall RC5B *plus*, Thermo Scientific, Asheville, NC) for 10 min at 4°C ($5{,}000 \times g$). The pellets were washed with sterile 1% peptone water and mixed with 2% (w/v) sodium alginate (Sigma-Aldrich Co.). The homogenous mixture of the bacterial cell suspended in alginate solution was then extruded drop by drop into 0.4 M calcium chloride solution (Thermo Fisher Scientific) at pH 7.1 using a 500 ml separatory funnel (ID: 1 cm, Kimax, Kansas City, MO). Beads were slowly stirred on a magnetic stirrer (Thermo Fisher Scientific) at a speed setting of < 3 rpm. After completion, the beads were rinsed with sterile 1% peptone water and were allowed to be further hardened for 30 min at 4°C.

Bead Uniformity

Surface area and volume measurements of individual beads were based on water displacement method in which 17 beads were dropped into 3 ml water in a graduated cylinder, in which water rise was observed and recorded for calculation. Bead

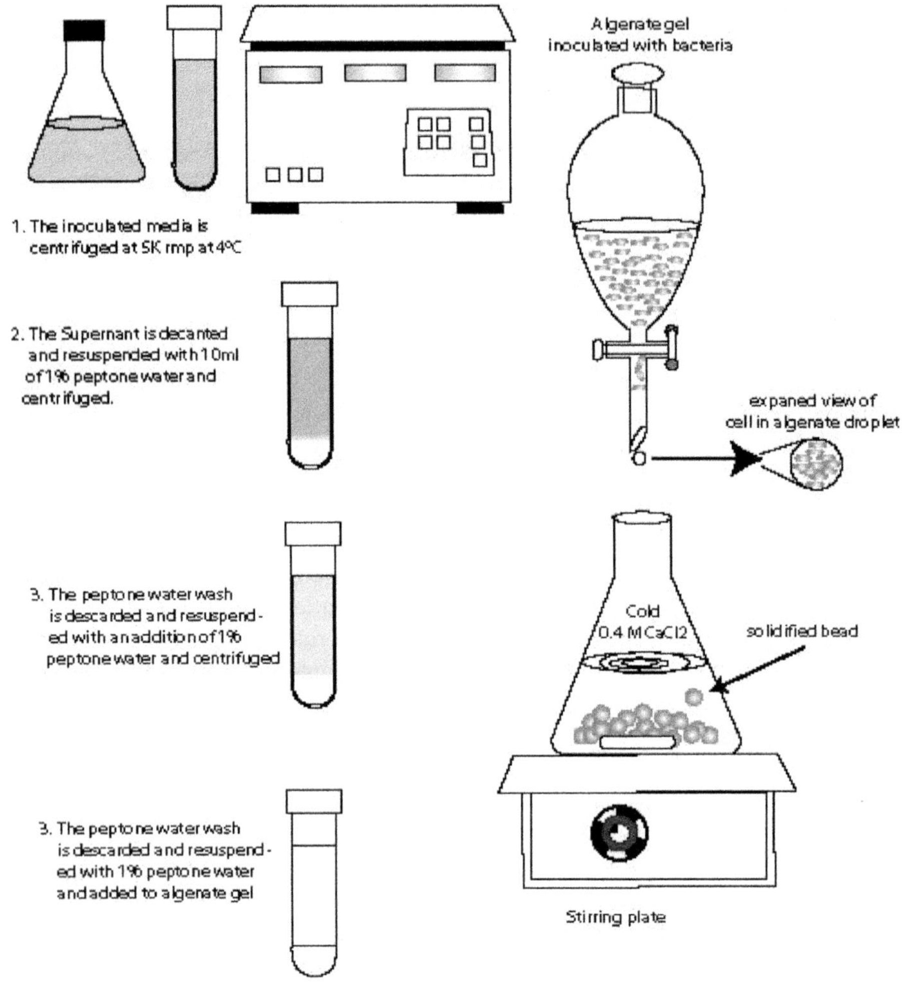

Fig. 1 Cell immobilization procedure

integrity was determined based on maximum force (N) needed to rupture the outer surface of individual bead using a Texture Analyzer (TA-XT2, Texture Technologies Co., Scarsdale, NY). The study was carried out at room temperature (24±2°C).

Comparison of Batch Free and Immobilized Cell Fermentation Systems

2.5 ml of either immobilized beads or free cells of the five individual strains of *L. reuteri* was inoculated into 500 ml of whey, MRS and mMRS broth, respectively, and incubated 12 h at 37°C for fermentation. The production of LA was then determined as monitored by changes in pH of both fermentation systems. The pH of samples was analyzed every 2 h during the 12-hour fermentation process using a pH meter (Accumet AR 60, Thermo Fisher Scientific).

Statistical Analysis

All experiments were replicated three times. Data were analyzed by the general linear model procedure of the Statistical Analysis System procedures. Comparison of means was performed using Duncan's multiple range test.

Results and Discussion

The bead uniformity analysis indicated that beads manufactured by the developed method were uniform in size and shape with average diameter of 3.53±0.10 mm, surface area of 39.13±0.10 mm^2 and volume of 1,012.57±0.10 mm^3. The results also illustrated that a maximum force of 30.10 Nand time of 3.53 s were required to rupture the outer surface of the gel coating of the beads.

Changes in the pH of whey, MRS, and mMRS broth used for the growth of immobilized *L. reuteri*strains during 12-hour fermentation are shown in Fig. 2. Batch immobilized cell (bead) study on the production of LA as represented in pH changes revealed that there was no significant difference of LA production among whey, MRS and mMRS medium during 12-hour fermentation process (Fig. 2). In other words, whey may be utilized as a cheaper alternative for LA production to commercial MRS and mMRS. Several scientists (Lui et al., 2005; IDF, 1991) indicate that whey contains a wide range of enzymes, proteins, and lactose and therefore may be a suitable nitrogen and carbon source for the growth of LAB while the addition of yeast extract and manganese (Mn^{2+}) improves the production of LA.

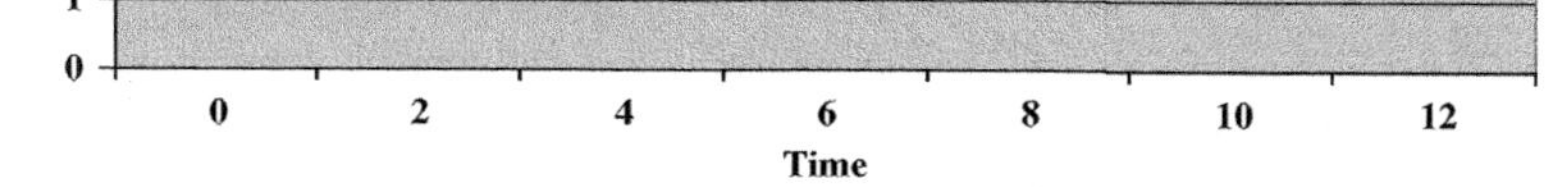

Fig. 2 Changes in pH during 12 h fermentation of immobilized *L. reuteri* (DSM20016) using three different medium (whey, MRS and mMRS broth) at 37°C

Table 1 Effect of cell immobilization by *L. reuteri* strains on the pH of whey broth during 8 h fermentation at 37°C*

Strain	Cell type	pH due to fermentation time (h) 2	4	8	12
CF2F	Free	5.20	4.40	4.40	4.40
	Immobilized	4.00	4.00	4.00	3.98
DSM20016	Free	4.95	4.30	4.30	4.30
	Immobilized	4.20	4.10	4.00	3.98
MM2-3	Free	4.78	4.18	4.10	4.00
	Immobilized	3.82	3.50	3.50	3.50
MM7	Free	5.10	4.30	4.20	4.20
	Immobilized	3.70	3.70	3.70	3.70
SD2112	Free	5.20	4.30	4.20	4.10
	Immobilized	4.20	4.10	4.10	4.10

*Initial pH of whey broth was 7.00.

Effect of cell immobilization by five strains of *L. reuteri* on the pH of whey broth during 12-hour fermentation at 37°C is shown in Table 1. When whey broth was inoculated with each of five strains of *L reuteri* and subjected to 12-hour fermentation process, on average, lower pH (~3.85) values were obtained from the media inoculated with immobilized cells than those (~4.20) of free cells. These results are in accordance with the findings of Elezi et al. (2003) that LA production using immobilized cell system, particularly calcium alginates, improved yields in comparison to free cell systems. Moreover, Table 1 indicates that pH reading obtained by fermentation of both free and immobilized MM2-3 was the lowest and therefore the strain might be more acid-tolerant and could be used more effectively for the production of LA than other strains used in this study.

Conclusion

The combination of *L. reuteri* (strain MM2-3) and whey based medium makes batch bead fermentation a promising and economical approach for high production of LA. Our results show that it is possible to develop a simple and rapid method for the encapsulation of *L. reuteri*.The addition of tween 20 to whey medium helped substrate diffusions and therefore increased LA productivity. Overall, based on the promising results found from this study on the application of cell immobilization and whey utilization to the production of LA, further research on alternative supplements promoting the growth of LAB is suggested.

Acknowledgments This research was supported by funds allocated to North Carolina Agricultural and Technical State University through the Cooperative State Research, Education and Extension Service of the United States Department of Agriculture, Project No. NC.X-173-5-02-170-1. The authors also would like to express gratitude to Dr. C.S. Turner, Associate Dean for Research for her support and Dr. K.J. Gruber for his suggestions on this manuscript.

References

Björck, L. and E. Hopkin. 1991. Significance of the indigenous antimicrobial agents of milk to the dairy industry. Int. Dairy Fed. 264, 2–19.

Champagne, P.C., and B.C. Cote. 1987. Cream fermentation by immobilized lactic bacteria. Biotechnol. Lett. 9, 329–332.

Chronopoulous, G., A. Bekatorou, E. Bezirtzoglou, A. Kaliafas, A.A. Koutinas, R. Marchant, and I.M. Banat. 2002. Lactic acid fermentation by *Lactobacillus casei* in free cell form and immobilized on gluten pellets. Biotechnol. Lett. 24, 1235–1236.

Elezi O., Y. Kourkoutas, A.A. Koutinas, M. Kanellaki, E. Bezirtzoglous, Y.A. Barnett, and P. Nigam P. 2003. Food additive lactic acid production by immobilized cells of *Lactobacillus brevis* on delignified cellulosic material. J. Agric. Food Chem. 51, 5285–5289.

Kyung, W.L., C.B. Seung, H.C. Woon, and W.K. Hyun. 2003. Structural observation of microencapsulated *Lactobacillus acidophilus*by optical and scanning electron microscopy. Food Sci. Biotechnol. 12, 14.

Lui, X., Y.K. Chung, S.T. Yang, and A.E. Yousef. 2005. Continuous nisin production in laboratoery media and whey permeate by immobilized *Lactococcus lactis*. Process Biochem. 13–24.

Newman, K.D. and M.W. McBurney. 2004. Poly (D, L lactic-*co*-glycolic acid) microspheres as biodegradable microcarriers for pluripotent stem cells. Biomaterials. 25, 5763–5771.

Scannell, A.G.M., C. Hill, R.P. Ross, S. Marx, W. Hartmeier, and E.K. Arendt. 2000. Continuous production of lacticin 3147 and nisin using cells immobilized in calcium alginate. J. Appl. Microbiol. 89, 574–576.

Swaisgood, H.E. 1982. Chemistry of milk proteins. In: P.F. Fox (Eds.). Developments in dairy chemistry.1, 1–60. Applied Science Publishers, New York. 1–60.

Wee, Y.J., J.N. Kim, J.S. Yun, D.H. Park, and H.W. Ryu. 2006. Pilot-scale lactic acid production via batch culturing of *Lactobacillus* sp. RKY2 using corn steep liquor as a nitrogen source. Food Technol. Biotechnol. 44, 293–298.

Part VII
Pollution Prevention/Solvents and Processes

Analysis of Partition Coefficients and Ternary Liquid-Liquid Equilibrium Data for Highly Non-Ideal Systems

Wahid Islam and Vinayak Kabadi

Abstract The objective of this research was to study the change in partition coefficient with concentration of solute in a liquid system consisting of two relatively immiscible solvents. This would require a consistency between data for infinite dilution partition coefficients and ternary LLE data at finite concentrations of the solute. Moreover an attempt would be made to model the infinite dilution and finite concentration behavior of these systems by a single thermodynamic model. During the course of our work, we found 35 ternary systems in the literature for which the infinite dilution partition coefficients and the ternary LLE data could not be predicted accurately by the currently available liquid state models.

Ternary liquid-liquid equilibrium data for hexane+1-butanol+water, carbon tetrachloride +propanoic acid+water, and water+propanoic acid+hexane were measured from very low concentrations of the solute to finite concentrations at 25°C. The measured partition coefficients showed good agreement with the literature values. UNIQUAC and modified UNIQUAC activity coefficient models were used to represent the data. The results showed large errors, exhibiting the inability of these models to correlate the data. Attempts are being made to apply some modified versions of quasi-chemical activity coefficient models to these systems.

Literature Studies

The Partition Coefficient (K_{sw}) of a solute is defined as the ratio of the concentration of the solute in water saturated solvent (organic) phase to the concentration of the solute in solvent saturated water phase. In Chemical Thermodynamics, researchers usually refer to Distribution Coefficient (D_{sw}), which is the ratio of mole fraction

W. Islam (✉)
Department of Chemical Engineering, North Carolina A&T State University, Greensboro, NC, USA

G.A. Uzochukwu et al. (eds), *Proceedings of the 2007 National Conference on Environmental Science and Technology*, DOI 10.1007/978-0-387-88483-7_42,

of solute in organic phase to its mole fraction in aqueous phase. Hence the relation between partition coefficient and distribution co efficient is given by

$$K_{SW} = \frac{C_i^S}{C_i^w} = \frac{\overline{V}^W \gamma_i^W}{\overline{V}^S \gamma_i^S} = \frac{\overline{V}^W}{\overline{V}^S} D_{SW} \tag{1}$$

This study was aimed at all available ternary systems with water as the third component and at a temperature of 25°C. Extensive collections of binary, ternary and quaternary data are available in "Liquid – Liquid Data Collection", Chemistry Data Series, DECHEMA (Sorensen and Arlt, 1979–1980; Macedo and Rasmussen, 1987). Volume I of this series contains binary systems and volumes II, III and IV contain ternary and quaternary systems. The DECHEMA books also contain the common and specific UNIQUAC parameters for all the systems given in the book.

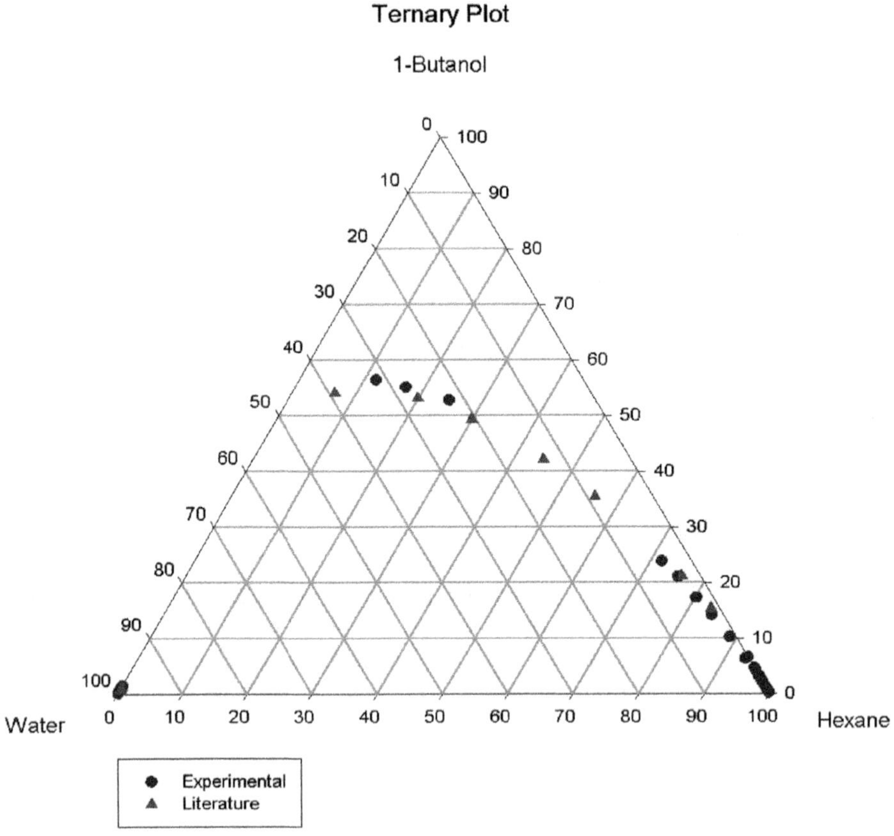

Fig. 1 Ternary diagram for Hexane-1-Butanol-Water

49 ternary systems were selected from these data books for study. For 18 of these systems, the infinite dilution distribution coefficient and ternary data at finite concentration could be related using common UNIQUAC parameters. These systems were classified as consistent systems (i.e a ternary system for which common UNIQUAC parameters given in literature are able to predict ternary data as well as partition coefficient at infinite dilution). The remaining 31 systems for which the infinite dilution distribution coefficient and the ternary data at finite concentration could not be related were classified as inconsistent systems. For these systems, common UNIQUAC parameters as given in literature are not able to predict infinite dilution properties and finite ternary data. They require the use of specific parameters that have been regressed for certain ranges of data. The wide disparity between the values indicates that UNIQUAC is unable to predict D_{sw} at infinite dilute concentrations. There are two types of UNIQUAC parameters like common and specific parameters. Common parameters for a binary system, AB, are UNIQUAC

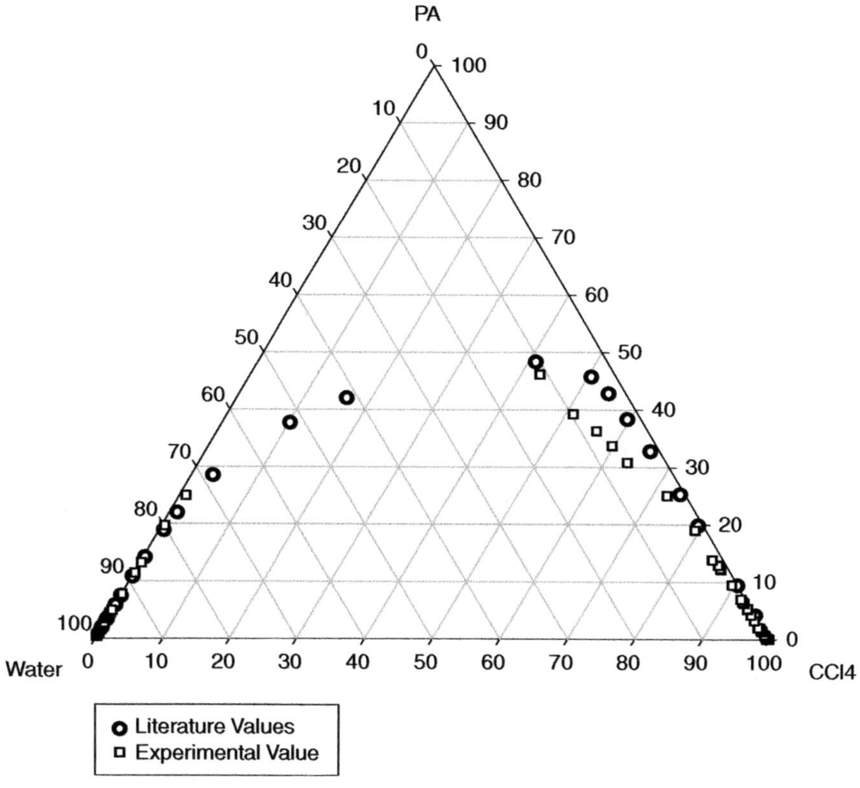

Fig. 2 Ternary diagram for Water-Propionic acid-CCL_4

parameters which may be used in LLE calculations for any system containing components A and B. On the otherhand specific parameters are UNIQUAC parameters fitted individually to each ternary system. These parameters work better than common parameters but the major disadvantage is that they apply to the particular system and cannot be extended to any other system.

Results and Discussion

Ternary diagrams for hexane-1-butanol-water, water-propionic acid-CCL_4, and water-propionic acid-hexane are shown in Figs. 1, 2 and 3 respectively. Data in these figures are from Sorensen and Arlt, 1995, Javvadi, 2000 and Rizvi, 2003. These three non-ideal systems were chosen for study in this work.

For all three of these systems the distribution coefficients available in the literature (Hansch and Leo, 1979) deviate by an order of magnitude from the corresponding values from infinite dilution activity coefficients calculated by the UNIQUAC model with the specific parameters of Sorensen and Arlt, 1979–1980 (see Table 1). This is because the UNIQUAC parameters are not predicting the concentration well. In order to understand the behavior of the system at very low concentration of solute

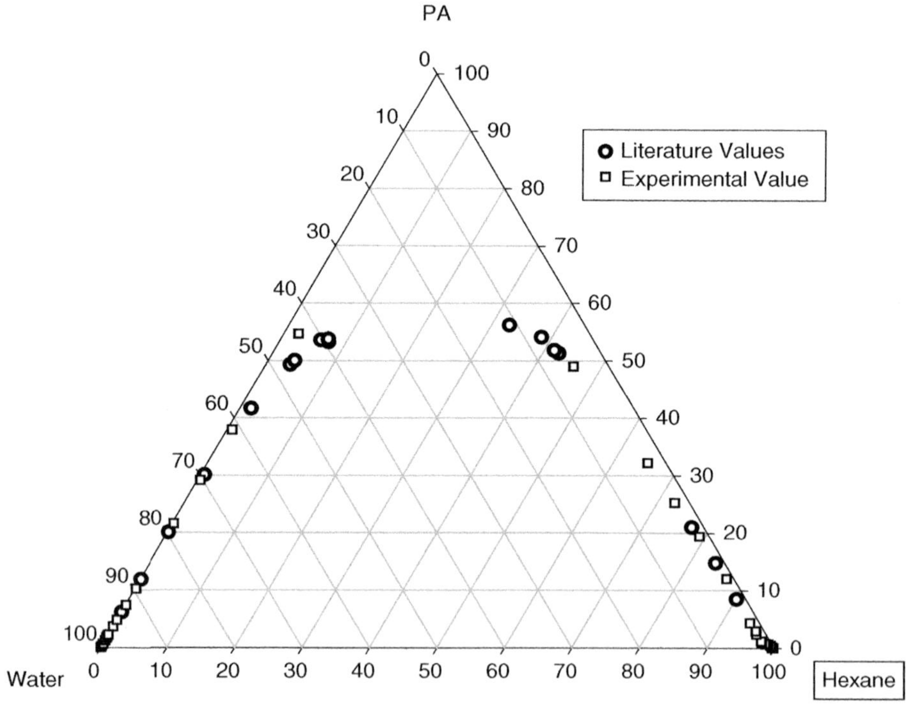

Fig. 3 Ternary diagram for water-propionic acid-hexane

Table 1 Distribution Coefficient activity coefficient values at infinite dilution from literature and calculated

	Dsw		γ^{∞}			
			Literature value (Tiegs et al., 1986; Gmehling et al., 1994)		Calculated value*	
System	Literature value (Hansch and Leo, 1979)	Calculated value*	$\gamma^{\infty,S}$	$\gamma^{\infty,W}$	$\gamma^{\infty,S}$	$\gamma^{\infty,W}$
Hexane-1-Butanol-Water	1.212	14.634	35.8	51.60	3.49	51.06
CCL_4 – Propanoic Acid – Water	0.068	1.1473		51.7	3.19	3.66
Hexane-Propanoic Acid – Water	0.02	1.651	2300	51.7	.2598	.3908

*Calculated by Uniquac Model by using specific parameter getting from Dechema (Sorensen and Arlt, 1979–1980)

in feed, we have shown in Fig. 4(a, b, c, d, e, f) for all three systems (He-PA-Water, Hexane-1-Butanol Water and CCl_4-PA-Water). Analyzing our data for CCl_4-PA-Water and Water-PA-Hexane we can clearly see in Fig. 4(b,f) both at infinite dilution the line appears to start at a different slope as compared to the literature. After a certain point, the literature values and the experimental values both has the same slope. Just before the slopes match (i.e, Calculated vs. Experimental), there is a bump either going up for the top phase and going down for the bottom phase. We can also note from Fig. 4(d) the behavior of the solute in the two phases (solvent and water rich phase) at very dilute concentrations in our experimental data compared with the results predicted by the UNIQUAC equations. For Hexane-1-Butanol-Water system, in the aqueous phase we note that the activity coefficient of solute in water at infinite dilution is predicted quite well by UNIQUAC in comparison with the literature value. However in the hexane phase, there is large order of variation between the values predicted by UNIQUAC and literature.

Conclusion

Our study indicates that Dechema values cannot be extrapolated to infinite dilution concentrations accurately. Consequently UNIQUAC parameters regressed with data outside the infinite dilution region are also not good in predicting infinite dilution properties. Hence more study is required to apply some modified models for these

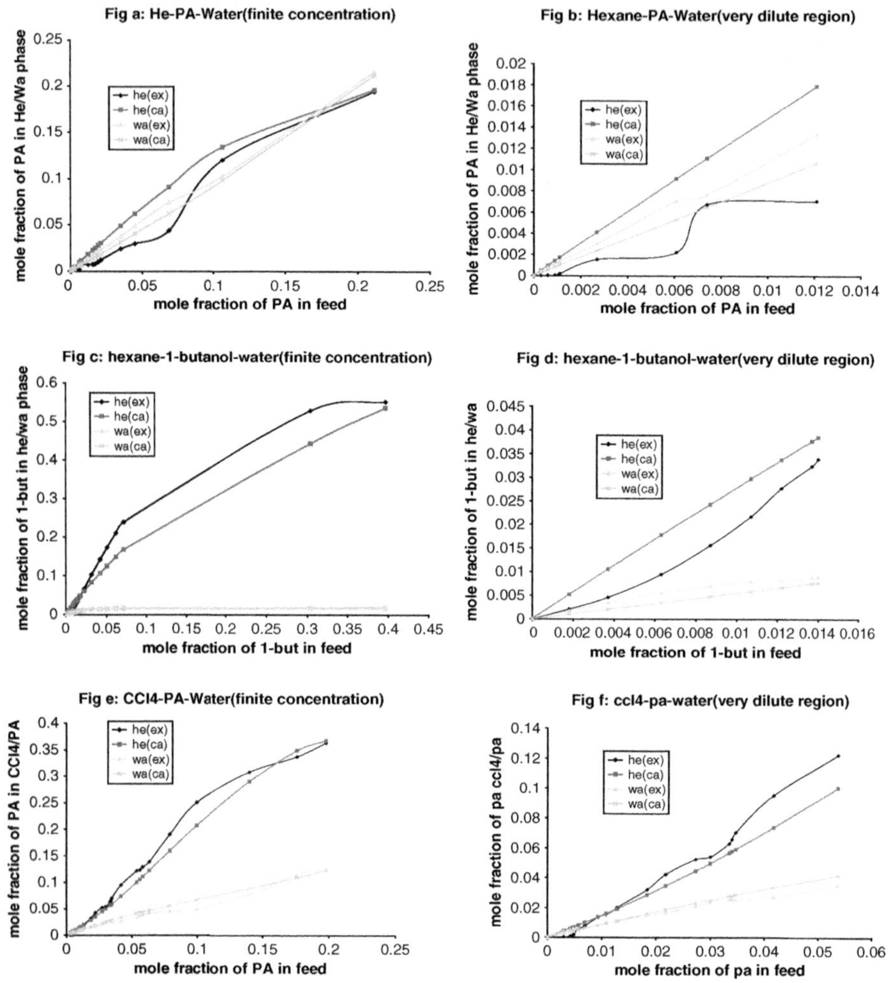

Fig. 4 (**a**, **b**, **c**, **d**, **e** and **f**): Experimental vs. calculated data for Hexane-PA-Water, Hexane-1-Butanol-Water and CCl_4-PA-Water at finite concentration and at very dilute region

highly non ideal systems to get consistency between finite concentration and at infinite dilution.

References

Gmehling, J., Menke, J., and Schiller, M. *Activity Coefficients at Infinite Dilution, DECHEMA, Chemistry Data Series*, **1994**, Vol. 9, part 4.

Hansch, C., and A. Leo, *Substituent Constants for Correlation analysis in Chemistry and Biology,* **1979**, Wiley – Interscience NY

Javvadi, A. Partitioning of Higher Alcohols in Alkane – Water at Dilute Concentrations, Masters thesis, **2000**, Department of Chemical Engineering, North Carolina A&T State University

Macedo, E.A., and P. Rasmussen, *Liquid – Liquid Equilibrium Data Collection,* DECHEMA, Chemistry Data Series, **1987**, 5, Part 4.

Rizvi, S. Z., Measurement of Liquid-Liquid Equilibrium and Infinite Dilution Partition Coefficients Data for Two Non-Ideal Systems, Masters Thesis, **2003**, Department of Chemical Engineering, North Carolina A&T State University.

Sorensen, J.M., and W. Art, *Liquid – Liquid Equilibrium Data Collection,* DECHEMA, Chemistry Data Series, **1979–1980**, 5, Part 1–3.

Tiegs, D., Gmehling, J., Medina, A., Soares, M., Bastos, J., Alessi, P., and I. Kikik, *Activity Coefficients at Infinite Dilution,* DECHEMA, Chemistry Data Series, **1986**, Vol. 9, part 1.

Application of Chitosan in Remediation of Dyes

Gbekeloluwa B. Oguntimein, Olatunde Animashaun, and Ivie Okpere

Abstract Chitosan is derived from Chitin, which is the exoskeleton of a crab or shrimp shell and some other crustaceans. This project explores this characteristic behavior of chitosan in the removal of dye from solution as a model for the potential application in textile industry wastewater treatment. Azure dye was used as the model dye used in the textile industry wastewater (TIW). The adsorptive characteristics of different chitosan particles sizes for Azure dye using 0.025 M acetate buffer at varying pH values were investigated. The relationship between q =(Co-Ce)/M the ratio of dye adsorbed to the mass of chitosan to the equilibrium concentration Ce was investigated for the different particle sizes.

The experimental data were fitted to three adsorption models – Linear, Freundlich and Langmuir. The data fitted the linear model with highest correlation coefficient, the highest slope values corresponded to the smallest particle size used during the experiment (particle size <0.25 mm diameter) at the lowest pH- 3.5. The data obtained was used in the design a fix-bed process, the mass of chitosan required for an EBCT (Empty Bed Contact Time) of 5 min was calculated to be 4000 kg and a volumetric wastewater treatment of 3.3×10^6 Liters.

Introduction

Water is our most precious natural resource and it is impossible to live without it. The potential beneficial uses of water are lost due to changes in its composition as a result of human activity. Water pollution is still a problem in the US thirty five year after the clean water act. EPA estimated that 40% of the waters in the US did not meet the water quality standards set for them and that over two thirds of the population lives within ten miles of impaired waters (U.S. EPA, 2006). Industrial wastewater pollution is still a major concern as a result of the potential carcinogenic nature of chemicals released (Barss, 1999).These industrial pollutants are synthetic

G.B. Oguntimein (✉)
Department of Civil Engineering, Morgan State University, 21251, Baltimore, Maryland, USA

G.A. Uzochukwu et al. (eds), *Proceedings of the 2007 National Conference on Environmental Science and Technology*, DOI 10.1007/978-0-387-88483-7_43,

organic compounds that can be easily absorbed by the surrounding environment. An industry of concern is the textile industry which releases toxic dyes in its wastewater. Between 10 and 15% of the total dye used in the dyeing process may be found in wastewater (Rodriguez et al., 1998). Azo dyes are the world's most commonly used dyes in the dye using industries. Azo dyes are characterized by the one or more azo groups (-N=N-) present in the dye molecules. They play an important part in almost every type of application and account for almost 60% of the total number of commercially manufactured dyes. Their high industrial usage, together with about 10% loss of their total production makes them an environmental pollution problem. Some azo dyes are manufactured from certain chemicals aromatic amines that are known or suspected carcinogenic. Once azo dyes have undergone azo bond cleavage they form aromatic amines, which are mostly toxic and carcinogenic. Their discharge to the environment may cause serious and long lasting damages. Azo dyes, after passing through the washing process in the textile industries are passed into the environment; it is estimated that between 10 and 15% of the total dye used in the dyeing process may be found in wastewater (Brown et al., 1981).

Maryland waters, are rich with crustacean life like crabs and lobsters which are a delicacy in Maryland restaurants and as such the crab processing is a major economic industry. The waste from this industry is the exoskeleton, the disposal of these shells has posed an environmental problem since they have to be disposed in landfills (Markstahler et al., 1993; Bi-State, 2005). The exoskeleton of crab is chitin which is the second most abundant polysaccharide in nature after cellulose. Chitin mainly consists of the amino sugar N-acetylglucosamine, which is partially deacetylated (Fig. 1). The mostly deacetylated form of chitin is called chitosan (Hirano, 1999; Peter, 1995). Chitosan is very similar to chitin, the difference is that chitosan has an amine group instead of an amide group which means that chitosan doesn't have any carbons double bonded to oxygen and chitin does (Figs. 1 and 2). Chitosan has primary amino groups that confer interesting pH-solubility behavior; at low pH, it is polycationic, water soluble, and unreactive while at neutral pH, it is insoluble in water, nucleophilic and reactive.

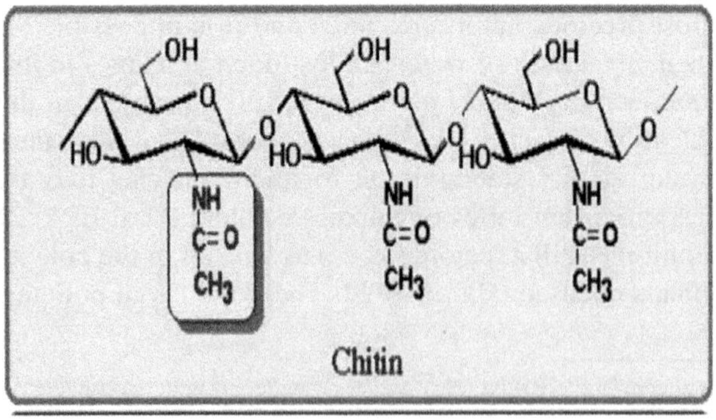

Fig. 1 Structure of Chitin

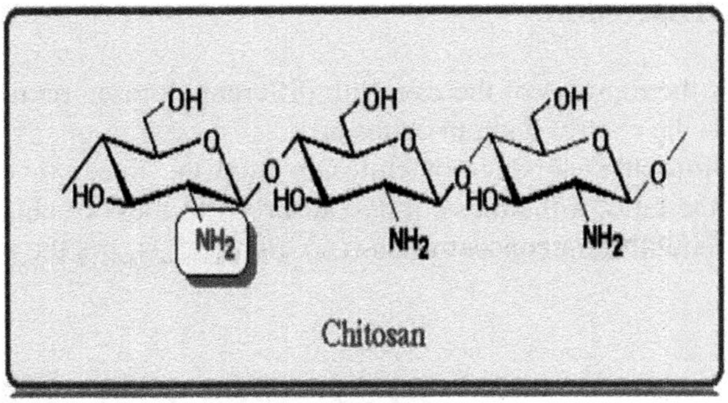

Fig. 2 Structure of Chitosan

Research has been performed to utilize the binding capacity of chitosan for the removal of metals from industrial wastewater. Since chitosan's amino group acts as a primary binding site for metals, this feature is being examined for the removal of soluble silver from industrial wastewater (Gerente et al., 2007), Arsenic in water (Chen and Chung, 2006) and for the remediation of coal mine waste water (CMW) (Lasko and Hurst, 1999). It is therefore the objective of this study to investigate the adsorption of dye in Textile Industry wastewater (TIW) using a model dye to investigate the adsorption characteristics of chitosan.

Materials and Methods

Chitosan and Azure Dye were purchased from Sigma Chemical Company St Louis MI. The solution mixture was centrifuged in an IEC Centrifuge and the absorbance at 600 nm of the supernatant was measured with Spectronic Gensys 5 spectrophotometer. Different particle seizes of chitosan were obtained by milling and sieving through various mesh sizes . Different pH values were investigated using 0.025 acetate buffer.

A flow chart of the procedure is shown in Fig. 3.

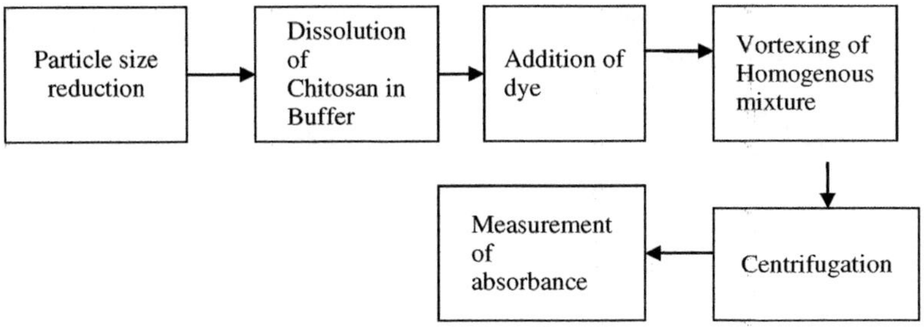

Fig. 3 Flow chart for experimental procedure

Results and Discusion

Figure 4 shows the removal of the dye with different chitosan particle sizes at pH 5.0 compared to the control without chitosan.

The adsorption characteristics of chitosan fitted the linear model when $q = (Co - C_e)/M$ the ratio of the mass of dye adsorbed to mass of chitosan was plotted against the equilibrium concentration (C_e). Figure 5. shows the plot of q versus

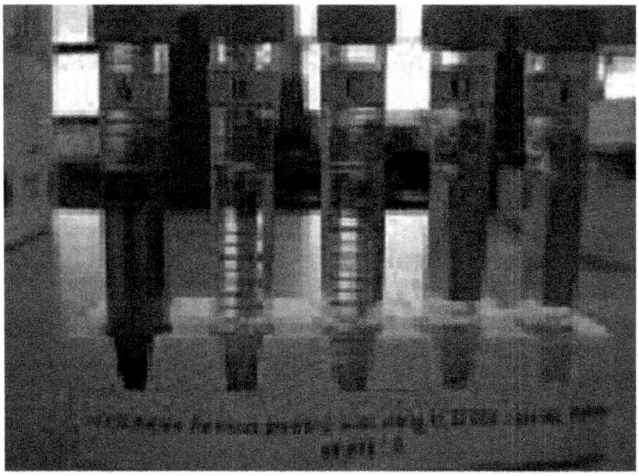

Fig. 4 Chitosan particle size 0.025 μm exposed to fixed concentration of azure dye at pH 5.0

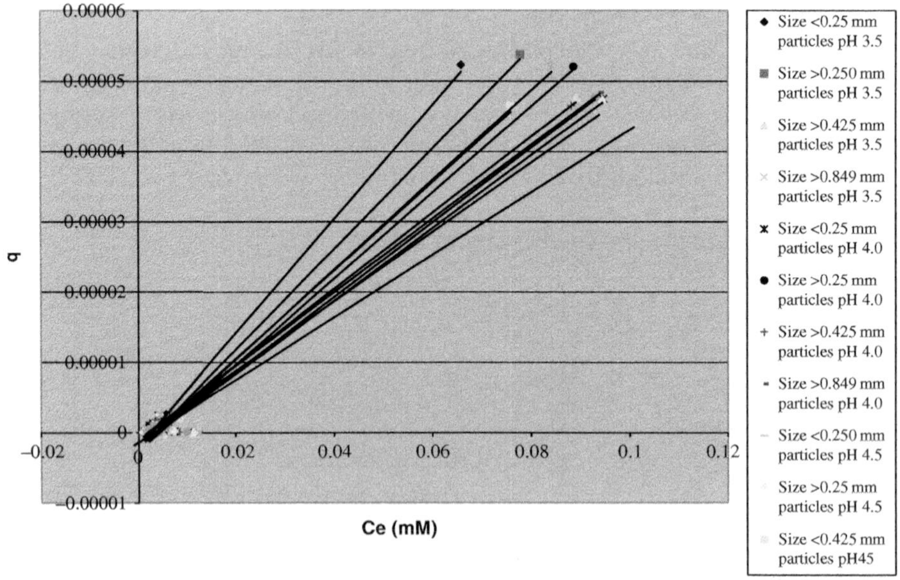

Fig. 5 Plot of q versus equilibrium concentration C_e

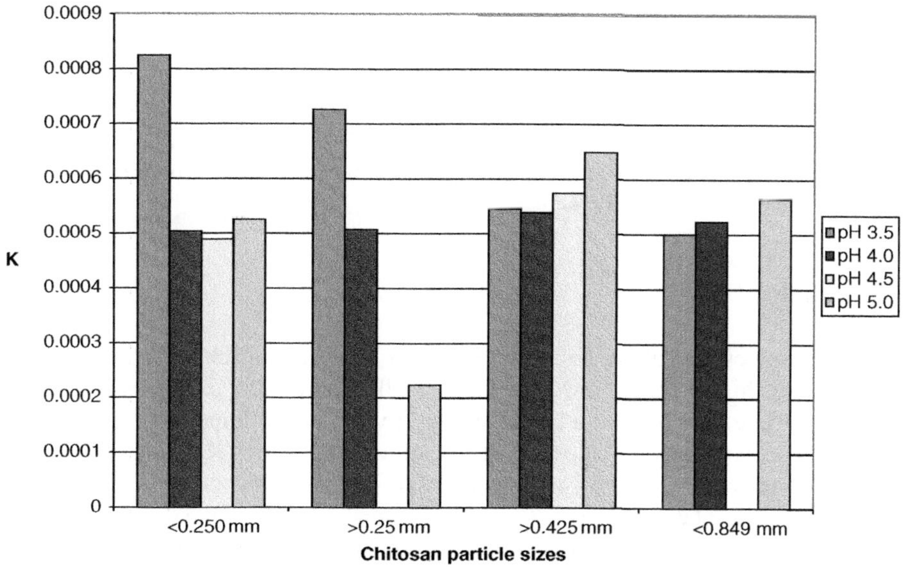

Fig. 6 Effect of Chitosan particle sizes on K values

C_e for different particle size and at various pH.. The plot of the slope K which corresponds to the adsorption coefficient for different particle seizes is shown in Fig. 6. The highest adsorption coefficient was obtained for particle size $< 0.025\ \mu m$ at pH 3.5.

Design:

For a batch operation using the linear model and the design Equation (1),

$$\frac{V}{M} = \frac{KC_e}{(C_o - C_e)} \qquad 1.$$

the amount of chitosan required to treat 1000 L/min of TIW containing 10 ppm of dye such that the final effluent is 0.05 ppm is 347,350 kg/day. For a continuous operation in a fixed bed column the chitosan usage rate (CUR) is calculate using Equation (2).

$$\frac{m}{Q} = \frac{C_0 - C_e}{KC_0} \qquad 2.$$

Where m is the mass of chitosan (g), Q is the volumetric flow rate (L/h), C_o is the initial dye concentration.

The specific throughput rate (STR) of Textile industry wastewater (TIW) is given by Equation (3).

$$\mathrm{STR} = \frac{Qt}{m_{chtiosan}} = \frac{V_b t}{EBCT^* m_{chitsosan}} \quad 3$$

Where EBCT is the Empty Bed Contact time in minutes, V_b is the volume of the fixed bed. The mass of chitosan required for an EBCT (Empty Bed Contact Time) of 5 min was calculated to be 4000 kg and a volumetric TIW treatment of 3.3×10^6 Liters.

Conclusion

From the experimental results, the histogram showed that chitosan's adsorptive capacity was highest at the lowest pH and smallest particle size. This is so because smaller particle sizes provide a larger surface area per unit volume for chitosan's particles to react with the dye. The highest adsorption coefficient (K) value of 0.000825 (L/mg) corresponded to the smallest chitosan particle size, <0.25 mm diameter, using 0.025 M acetate buffer at pH 3.5. With respect to operational design, the batch column for required a large amount of chitosan 347,350 kg/day for effective operation. This amount is not practical for use so the batch column was disregarded. For the fix-bed column, the CUR was calculated as 1.2 g/L for the highest K value. The required mass of chitosan for an EBCT of 5 min was 4000 kg for a volumetric the treatment of 3.3×10^6 L TIW.

References

Barss, K. Clean Water. New York: Chelsea House Publishers, (1999).

Brown, D.H., Hitz, H.R. and Schafer, L. (1981) The assessment of the possible inhibitory effect of dyestuffs on aerobic wastewater bacteria. Experiment with a screening test. Chemosphere 10:245–26.

Bi-State Blue Crab Technological Advisory Committee: (2005) http://www.chesbay.state.va.us/Publications/CBC-CRAB-05.pdf

Chen, C. and Chung, Y. 2006. Arsenic Removal Using a Biopolymer Chitosan Sorbent. Journal of Environmental Science & Health, Part A: Toxic/Hazardous Substances & Environmental Engineering 41(4):645–658.

Gerente, C., Lee, V.K.C., Cloirec, P. Le, McKay, G. 2007 Application of Chitosan for the Removal of Metals From Wastewaters by Adsorption—Mechanisms and Models Review. Critical Reviews in Environmental Science & Technology 37(1):41–127.

Hirano, S. 1999. Chitin and Chitosan as Novel Biotechnological Materials, Polymer International, 48:732–740.

Lasko, C.L., and Hurst, M.P. 1999. An Investigation into the Use of Chitosan for the Removal of Soluble Silver from Industrial Wastewater. Environmental Science & Technology 33(20): 3622–3626.

Peter, M.G. 1995. Journal of Medical Science Pure Applied Chemistry. A32 v4, 629–640.

Markstahler, E., Newell, T., Snyder, M. 1993. Commercial *food waste* from *restaurants* and grocery stores. Resource Recycling 12(2):56.

Rodriguez, E., Pickard, M.A., and Vazquez-Duhalt, R. 1998. Industrial Dye Decolorization by Laccase from Ligninolytic Fungi. Current Microbiology 38:27–32

U.S. EPA, 2006. Overview of Current Maximum Dilly Load-TMDL- Program and Regulations. http://www.epa.gov/owow/tmdl/overviewfs.html

An Apparatus for Density and VLE Measurements for Gas-Liquid Systems

Rashawn K. Washignton and Vinayak N. Kabadi

Abstract The Objective of this research project was to design and build and apparatus to measure density and vapor liquid equilibrium data for gas-liquid systems. This work was done in support of the National Science Foundation Science and Technology Center's (NSFSTC) goal of identifying and enabling sustainable process and products using CO_2-related technology. The apparatus consists of a syringe pump, a multi-port multi-position valve, a high pressure cell with multiple sampling ports and fitted with a magna-drive stirring system. A precision temperature oven, a pycnometer fitted with a 4-port 2-position valve and a GC with TCD and FID detectors and with on-line sampling capability. For the density measurements a required amount of liquid solvent and gas are delivered to the high pressure cell by the syringe pump, which also maintains the required pressure of the system. The cell and pycnometer are placed in the oven which maintains the desired temperature. The mixture is well-stirred and the composition is measured precisely by the on-line GC. The liquid is then passed through the trapped inside the pycnometer. The pycnometer is disconnected and weighed. The density is obtained from the mass of the liquid and the volume of the pycnometer. The apparatus can also be operated for VLE measurements. Static vapor-liquid equilibrium is achieved in the high pressure cell which is fitted with multiple dip-tubes for sample extraction. Upon equilibration the liquid and vapor samples are sent to GC for analysis. The density data for CO_2-ethyl lactate system have been measured in the temperature range 25–80°C and at pressures from atmospheric to 2000 psi for a few compositions of the liquid mixtures. Measurements are continuing to cover the entire composition range for the binary system.

R.K. Washignton (✉)
Chemical Engineering, North Carolina A&T State University, Greensboro, NC 27411, USA

G.A. Uzochukwu et al. (eds), *Proceedings of the 2007 National Conference on Environmental Science and Technology*, DOI 10.1007/978-0-387-88483-7_44,

Introduction

The use of organic solvents throughout the chemical, gas processing, petrochemical and pharmaceutical industry is a necessary step in different processes. However, the uses of organic solvents have various environmental implications. The Montreal Protocol identified the need to re-evaluate chemical processes with regard to their use of volatile organic compounds and the impact that these compounds have on the environment. Green solvents were developed to reduce and even eliminate the widespread use of toxic solvents, poisonous metals and other hazardous substances.

One such green solvent is ethyl lactate; it is derived from the processing of corn and an ester of lactic acid. The chemical properties of ethyl lactate make it a suitable replacement of organic solvent used in the coatings, adhesive, and chemical cleaning industry. The chemical properties of ethyl lactate include a high solvency power, high boiling point, low vapor pressure, low surface tension, and excellent cleaning potency. Although ethyl lactate is a biodegradable compound, it has replaced many hazardous organic solvents, and has been implemented extensively in the manufacture of semiconductor wafers.

Research and development in the use of supercritical fluids in various applications have significantly increased. This is especially true for the microelectronics and semiconductor industry. The use of supercritical CO_2, has gained popularity amongst coffee manufacturers looking to move away from some of the classic decaffeinatingused solvents of the past, to produce better tasting, and healthier decaffeinated coffee. Other applications of super critical CO_2 include dry cleaning. Dr. Joseph DeSimone, professor of chemical engineering department at North Carolina State University, developed a method of using super critical CO_2, instead of perchloroethylene (PCE or *Perc*) or water in the cleaning process. The focus on supercritical CO_2 is due to its relatively low cost, superior energy efficiency, ease of recovery of the solvent, low surface tension and transport properties.

Utilizing various forms of high pressure equipment, supercritical CO_2 can be easily implemented into semiconductor manufacturing, and removed from wafer-cleaning by-products due to the high vapor pressure at near-ambient conditions used in the normal manufacturing process of semiconductors. It is for this reason recycling of the solvent is an attractive prospect. During the photolithography stage, a pattern is created on the wafer by using polymer-based liquids in which photoactive molecules are dissolved in a carrier solvent called a photoresist. Applications of supercritical CO_2 at this stage include deposition, development and stripping of photoresist, drying to prevent pattern collapse, post-etch cleaning and restoration of dielectrics and chemical fluid deposition and etching of metals.

Objectives

The objectives of this apparatus were designed to compliment and expand research in discovering the thermodynamic properties of CO_2 containing systems. Initially, the apparatus was designed to do static vapor-liquid equilibrium (VLE) experiments, but it was then noticed with a few modifications, to the design, the apparatus could be used to also obtain high pressure density measurements of gas-liquids systems. The use of a static equilibrium experiment can produce or generate of more sound data than its Continuous Flow system counterpart (Kologani, 2002). The density of a binary system composing of one relative gas, and liquid, has been a challenge for most scientists, and there is a lack of data in this field, which sparked the effort to devise a method of obtaining this data. This apparatus has a temperature range of 25–80°C, and a pressure range of 20 psig – 2000 psig

Procedure

Density

To collect the high pressure density data, a pycnometer is placed along side of the pressure vessel, connected by the multi-port/position valve, and a 4 port valve. Instead of using the multi-port/position valve as a means of collection, it's used as a means of distribution. The pycnometer and pressure vessel is pressured by the syringe pump, and heated to the set temperature by the oven. The material is recirculated by flowing from the syringe pump, into the pressure vessel. The materials is then sent to the multi-port/position valve, were it's sent to 4 port valve on top of the pycnometer. At the 4 port valve, the fluid can be selected to either fill the pycnometer, or by pass it. When filled, the pycnometer is maintained at the same pressure and temperature as the pressure vessel. Once filled the fluid can then be discharged from the system, or recirculated back through the syringe pump, to complete the loop. Once the pycnometer is full, and has had enough time to be recirculate, the 4 port valve is switched to trap the material within the pycnometer, and the multi-port/position valve is switched to bypass the pycnometer section, which is still at the same temperature and pressure as the pressure vessel. The pycnometer is then taken out, weighed, and compared to that of the empty pycnometer, to get the weight of the fluid inside. This weight is divided by the volume of the pycnometer as function of pressure and temperature, to give the corresponding density.

Vapor Liquid Equilibrium

To collect the VLE data, equilibrium is attained when the pressure in the pressure vessel is stable and the composition of vapor and the liquid sample is consistent. The liquid samples are taken from the equilibrium cell by mean of dip tubes at predetermined heights, while the vapor sample was collected directly from one of the top ports on the pressure vessel. The liquid sample lines are connected to a multi-port/multi-position valve, then to a liquid sampling valve, while the vapor sample is sent directly to a vapor sampling valve, before entering the GC, to be analyzed, and data obtained.

Schematic

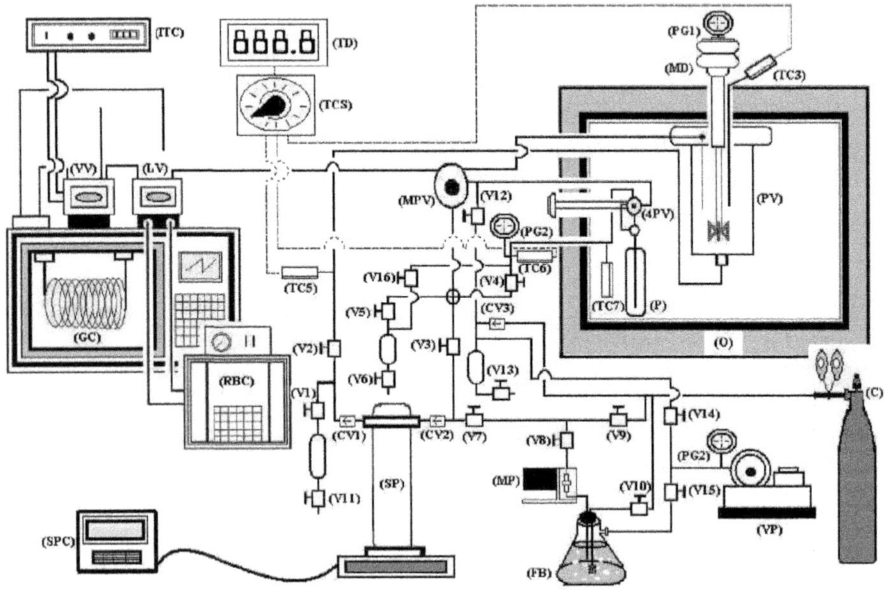

Legend

(GC) Gas Chromatograph; **(O)** Oven; **(SPC)** Syringe Pump Controller; **(SP)** Syringe Pump; **(FB)** Feed Bottle; **(RBC)** Refrigerated Bath; **(ITC)** Temperature Controller; **(4PV)** 4-Port Valve; **(TD)** Thermocouple Display; **(TCS)** Rotary Thermocouple Selector; **(V#)** Valves; **(MPV)** Multi-Port/Position Valve; **(VV)** Vapor Sampling Valve; **(TC#)** Thermocouples; **(LV)** Liquid Sampling Valve; **(MP)** Metering Pump; **(CV#)** Check Valves; **(PG#)** Pressure gauge; **(PV)** Pressure Vessel; **(MD)** MagneDrive

Results

Currently the calibration of the pycnometer, utilizing degasified distilled water, and the NIST database, (Webbook.nist.gov, 2007) has been plotted as a function of temperature vs. volume, while holding pressure constant in Fig. 1. The data for pure ethyl lactate has been obtained, utilizing the volumes obtained from the water calibration, and plotted in Fig. 2.

Conclusion

In conclusion, utilizing degasified distilled water to calibrate the pycnometer, for volume, and measuring the empty weight of the pycnometer, the density measurements for pure ethyl lactate has been obtained experimentally and compared

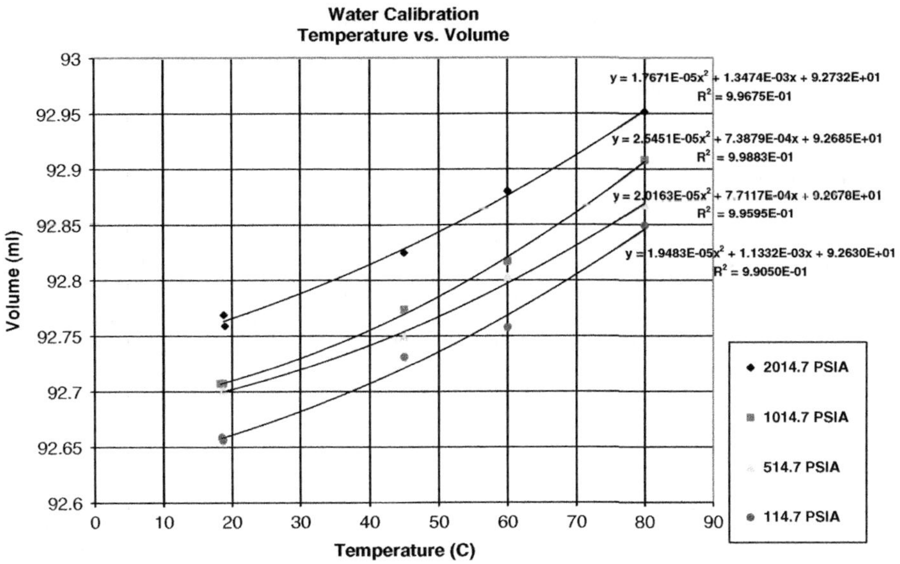

Fig. 1 Water calibrations for volume of pycnometer as function of temperature and pressure

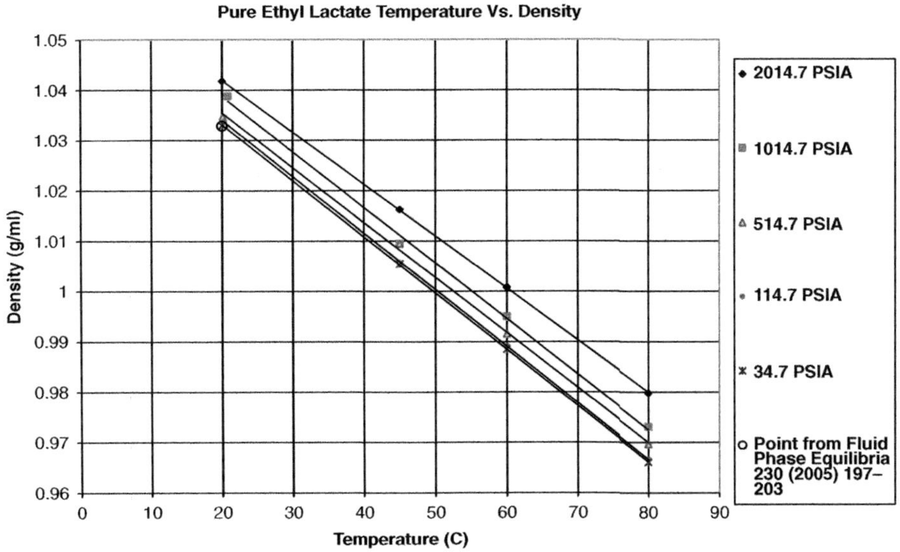

Fig. 2 Density data for pure ethyl lactate

successfully with available literature data (Pena-Tejedor et al., 2005). The data obtained experimentally matched with the one literature data obtained from The Fluid Phase Equilibria (2005, 230, 197–203). The trend is quite conclusive and substantial, given the supporting evidence; however, the data of the CO_2-Ethyl Lactate mixture must now be obtained, and placed in a similar fashion on the data table. Once the density data has been completed, the Static VLE experimental data can be sought after with the suggested modification to this apparatus.

References

Kolagani, R., A Flow Apparatus for VLE Measurements of Carbon Dioxide Containing Systems, M.S. Thesis, North Carolina Agricultural and Technical State University, 2002

Pena-Tejedor, S., Murga, R., Teresa Sanz, M., Beltran, S., Vapor-liquid equilibria and excess volumes of the binary systems ethanol + ethyl Lactate, isopropanol + isopropyl lactate and n-butanol + n-butyl lactate at 101.325 kPa, Fluid Phase Equilibria, 2005, 230, 197–203.

MSE Walls in Solid Waste Applications

Thomas M. Yanoschak

Abstract As expanding existing landfills and siting new landfills becomes increasingly difficult, landfill owners and operators are continually looking for new ways of increasing the effective disposal capacity at their existing facilities. The use of mechanically stabilized earth (MSE) walls is one potential means by which landfill operators can add additional waste disposal airspace with little or no additional land requirements. This paper introduces the concept of MSE walls by providing examples of how incorporation of the walls can increase disposal capacity of an existing landfill. MSE wall design is briefly addressed with discussions on the types of geogrid reinforcement, geogrid properties, factors affecting MSE wall stability, and the types of facing options that are available for MSE walls. Design examples are provided illustrating the use of MSE walls to expand the capacity of both lined and unlined landfills. An example showing an MSE wall incorporated into a lateral expansion is also provided.

Introduction to MSE Walls

Mechanically Stabilized Earth (MSE) technology allows the construction of near vertical earthen walls, therefore reducing the amount of land required for embankment construction. MSE walls are also generally less expensive to construct than traditional concrete or other types of retaining walls. For these same reasons, MSE walls have been used in landfill applications where additional air space was needed but the purchase of additional land was either impossible or prohibitively expensive.

The steeper slopes constructed with MSE walls are made possible by the use of soil reinforcement during embankment construction. Soil reinforcement serves a similar function as steel reinforcement serves in reinforced concrete structural members: to provide tensile strength in a material that inherently has very little

---

T.M. Yanoschak (✉)
HDR, Raleigh, North Carolina, USA

G.A. Uzochukwu et al. (eds), *Proceedings of the 2007 National Conference on Environmental Science and Technology*, DOI 10.1007/978-0-387-88483-7_45,

to no tensile strength. Whereas a typical unreinforced soil may be able to achieve maximum stable slopes of between 3:1 (horizontal:vertical) and 2:1 (18.4° through 26.6°), soil reinforcement allows construction of much steeper slopes of up to approximately 1:6 (80.5°).

Example Solid Waste Application

Figures 1 and 2 illustrate how incorporation of a MSE wall can increase the waste disposal capacity of an existing landfill without significantly increasing its footprint. Figure 1 shows a plan view and cross section of a hypothetical landfill that has reached capacity. The gross volume of the landfill is approximately 2,400,000 cubic yards (cy), (1,835,000 m^3).

Figure 2 shows a possible MSE wall configuration that could be used to gain additional airspace. In this design, the outer toe of the perimeter berm is kept constant, but soil reinforcement is used to increase the outer slope of the perimeter berm from 3:1 (18.4°) to 1:3 (71.6°). In this example, approximately 700,000 cy (535,000 m^3)

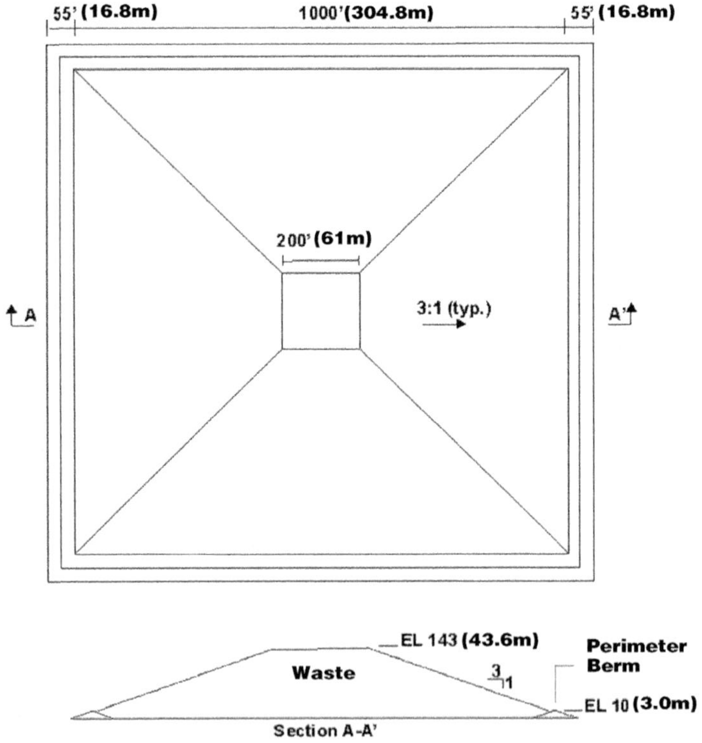

Fig. 1 Plan view and cross section of a hypothetical landfill at capacity

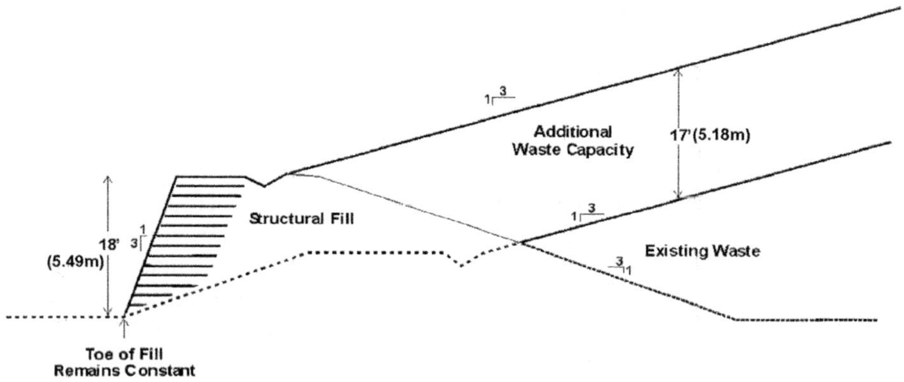

Fig. 2 Possible MSE wall configuration

of additional volume is gained. This is equivalent to a 29% gain in volume when compared to the original volume of the landfill illustrated in Figure 1.

MSE Wall Design

The following discussion introduces the reader to the various design components of MSE walls. It is meant as a general discussion. There are many in depth design references available to the reader, for example, Koerner (1997a), Tensar Corporation (1988), and Christopher (1990).

Geogrid Reinforcement. The most commonly used reinforcement today are geogrids. The definition of a geogrid according to Koerner (1997b) is "a geosynthetic material consisting of connected parallel sets of tensile ribs with apertures of sufficient size to allow strike-through of surrounding soil, stone, or other geotechnical material".

Types of Geogrids. Geogrids can be described as uniaxial or biaxial, depending on the orientation of the load carrying ribs, and as either flexible or rigid, depending on the method of manufacture.

Uniaxial geogrids are designed to carry tensile loads along one principal direction while biaxial geogrids are designed to carry tensile loads in two principal directions located 90° apart.

Rigid geogrids are manufactured from sheet plastic, usually high density polyethylene (HDPE) for uniaxial geogrids and polypropylene for biaxial geogrids. The geogrid is formed by punching holes in the sheet and cold stretching (post-tensioning) the sheet in the direction of strength. Uniaxial geogrids are therefore stretched in one direction while biaxial geogrids are stretched in two.

Flexible geogrids are made of high tenacity polyester yarns that are sprayed with a protective coating, such as latex or polyethylene, in order to make the grid easier to handle and to avoid ultraviolet and chemical degradation of the polyester yarns.

Geogrid-Soil Interaction. Stress transfer between geogrids and backfill soils can be evaluated in the laboratory using pullout tests. Generally granular backfill soils exhibit greater stress transfer and therefore are preferred for MSE wall construction. The cohesive component of shear stress, c, in clayey soils is often ignored in MSE wall design in order to include an additional factor of safety.

Other Geogrid Properties. Many other properties are reported by geogrid manufacturers in their product specifications. Although these properties are important in order to insure product conformance with specified values, they are generally not used directly in wall design unlike the strength and geogrid-soil interaction properties previously described. These properties may include rib/strand count, aperture size, open area, thickness, weight, carbon black content, flexural rigidity, junction strength, and junction efficiency.

MSE Wall Stability. A pullout force is exerted on the geogrid due to the steepened face of the MSE wall that may otherwise fail if the geogrid were not present. This pullout force is transferred to the geogrid by means of the frictional resistance between the geogrid and the soil on both the upper and lower surfaces of the longitudinal and transverse ribs and by means of bearing resistance against the front surface of the transverse ribs. To counteract the pullout force, resisting forces are developed within the geogrid in the same manner as the pullout forces except that they act in the opposite direction. For a stable condition to exist, resisting forces must exceed pullout forces.

Figure 3 conceptually illustrates the function of geogrid reinforcement within an MSE wall. For simplicity, a single layer of geogrid is assumed to exist within the wall. If the geogrid layer were not present, a failure plane would develop behind the wall due to the steep face of the wall.

The wedge of soil that would otherwise move outward is held in place by the geogrid. The length of geogrid located between the face of the wall and the failure plane, designated as "Pullout Length" or PL, is where the pullout tensile stress (TP) is developed within the geogrid. In order to keep the unstable wedge of soil from moving

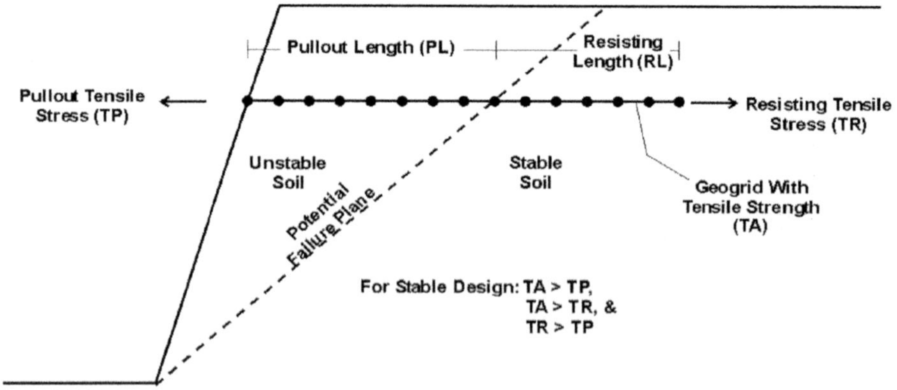

Fig. 3 Conceptual illustration of geogrid in MSE wall

outward, an equal or greater resisting tensile (TR) stress must be developed within the stable soil mass within the "Resisting Length" or RL of the geogrid. If TR < TP, then the geogrid embedded in the stable soil will pull out and the wall will fail.

Another important concept illustrated in Fig. 3 is that the pullout and resisting stresses developed within the geogrid must not exceed the tensile strength of the geogrid. If this occurs, the geogrid will either excessively deform or break and the wall will fail.

Facing Options. Several MSE wall facing options are available. The most expensive option is the use of precast concrete facing panels. Decorative concrete block facing units, timbers, and gabions may also be used as facing options. The most often used MSE wall facing option used in solid waste applications involves extending the biaxial geogrid layers beyond the edge of the wall and wrapping it over the layer of soil that is placed over the geogrid. The advantages of this facing option include its low cost when compared with the precast concrete panel and decorative block options and the ability of the face to be hydroseeded to support vegetative growth.

Design Examples

The following design examples have been encountered by the author at actual landfill sites and are presented to demonstrate the use of MSE walls to increase the capacity of existing landfills.

MSE Wall Over Lined Landfill. Figure 4 shows a situation where an existing lined landfill could not be laterally expanded due to a right of way and stormwater channel located outside of the toe of the existing perimeter berm. Hydraulic

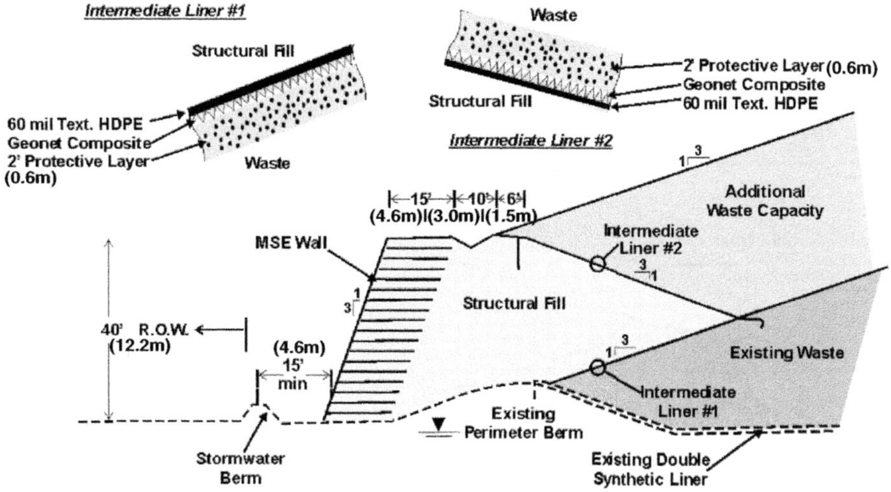

Fig. 4 MSE wall over lined landfill

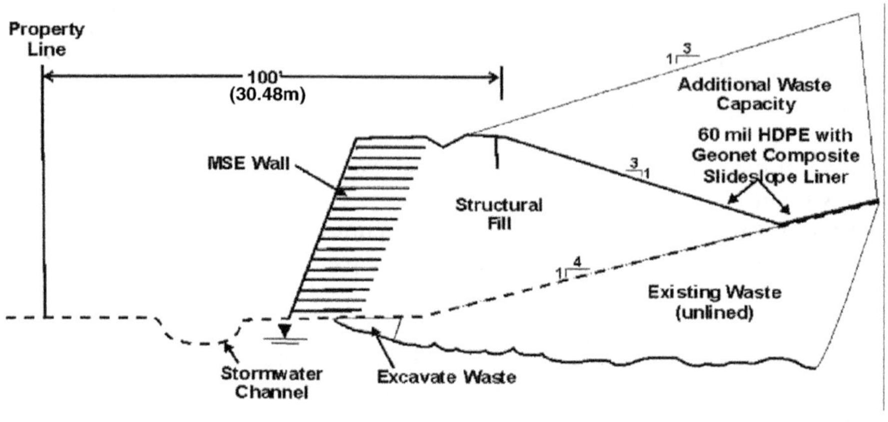

Fig. 5 MSE wall over unlined landfill

modeling indicated that the width of the stormwater channel could be reduced to 15 ft (4.6 m), thus allowing construction of an MSE wall within the remainder of the channel.

MSE Wall Over Unlined Landfill. Figure 5 contains an example where lateral expansion of an unlined landfill was not possible because of a regulatory buffer requirement that a distance of 100 ft (30.48 m) be kept between the property line and the edge of waste. Purchase of the neighboring property was not feasible so construction of an MSE wall was proposed. The unlined landfill was operated before the adoption of the buffer requirement and therefore solid waste existed beneath the proposed location of the MSE wall. The MSE wall design required excavation of waste from beneath the reinforced portion of the wall and replacement with structural fill.

MSE Wall for Lateral Expansion. The MSE wall design shown in Fig. 6 illustrates a situation where an MSE wall was proposed for a lateral expansion of an

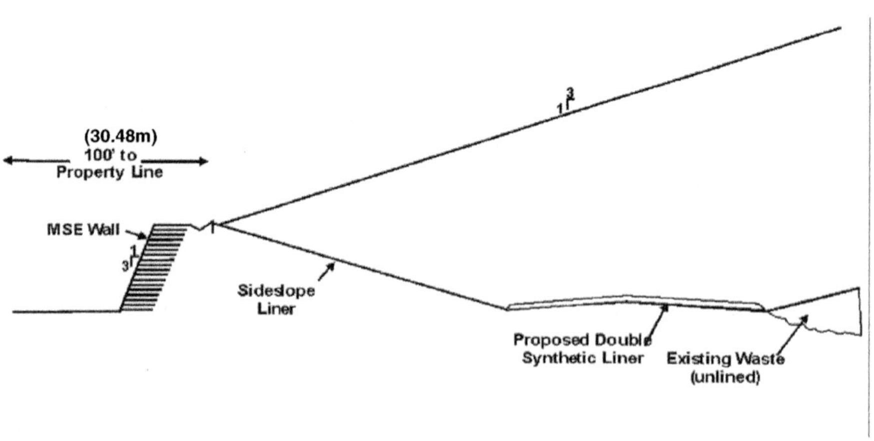

Fig. 6 MSE wall for lateral expansion

existing landfill. A wall was proposed because the extent of the lateral expansion was limited by the requirement that 100 ft (30.48 m) be kept between the property line and the edge of waste and because the adjacent property would not be available for purchase.

Conclusion

MSE walls provide the landfill owner or operator with an option for potentially expanding a landfill when the traditional lateral or vertical expansion is either not feasible or not cost effective. Successful incorporation of an MSE wall in a landfill design, however, requires careful evaluation of the many design variables as outlined in this paper. A thorough and accurate cost analysis should be performed during the conceptual design phase to ensure that an MSE wall will be cost effective.

References

Christopher, B.R. 1990. "FHWA Guideline for Reinforced Soil Design, Theory and Practice." Federal Highway Administration. Washington, D.C.

Koerner, R.M. 1997a. *Designing With Geosynthetics – Fourth Edition*. Prentice Hall. Upper Saddle River, New Jersey, pp. 355–365.

Koerner, R. M. 1997b. *Designing With Geosynthetics – Fourth Edition*. Prentice Hall. Upper Saddle River, New Jersey, p. 315.

The Tensar Corporation. 1988. "Guidelines for the Design of TENSAR® Geogrid Reinforced Soil Retaining Walls." Tensar Technical Note TTN:RW1.

Study of the Molecular Weight Dependence of Glass Transition Temperature for Amorphous Poly(L-Lactide) by Molecular Dynamics Simulation

Jian Zhang, Jizhong Yan, Leonard Uitenham, and Jianzhong Lou

Abstract Molecular dynamics simulation has been used to investigate the molecular weight dependence of glass transition temperature for amorphous poly(L-lactide). Amorphous PLLA systems were created using molecular modeling and NPT ensemble MD simulations were carried out using the modified OPLS-AA force field. The molecular weight dependence of glass transition temperature was studied and the good agreement between the simulation results and experiments was obtained.

Introduction

The disposal problem due to non-degradable petroleum-based plastics has raised the demand for biodegradable polymers. (Richard and Bhanu, 2002) Poly(lactide) (PLA) is a biodegradable aliphatic polyester derived from 100% renewable resources, such as corn and sugar beets. Moreover, it has unique physical properties that make it useful in diverse applications including paper coating, fibers, films, and packaging. (Ray et al., 2000) Currently, PLA is primarily used for medical applications such as drug delivery devices, absorbable sutures, and as a material for medical implants and other related applications. (Uhrich et al., 1999) PLA, functionalized PLA and PLA block copolymers can be prepared by direct condensation or by the ring-opening polymerization. (Wang et al., 2006) The Stereochemical microstructure of polymers can be controlled by the monomer composition in the feed or by the stereochemical preference of the initiating/catalytic system. (Majerska et al., 2004) Phase transition behavior, morphology, miscibility and other physical properties of PLA-based nano-scale structures are widely studied recently. (Sun et al., 2006) All these nano-scale research and applications make the fundamental understanding of the PLA-based material at the molecular level a necessity.

J. Lou (✉)
Department of Chemical Engineering, North Carolina A&T State University, Greensboro, North Carolina 2711, USA

G.A. Uzochukwu et al. (eds), *Proceedings of the 2007 National Conference on Environmental Science and Technology*, DOI 10.1007/978-0-387-88483-7_46,

Molecular dynamics simulation has been used extensively in the study of different aspects of polymer structures and properties. Recently, the resistance of PLA to hydrolysis based on the PLLA and PDLA blends and compatibility of PLLA and PVA blends were studied by molecular modeling simulations. (David and Yiqi, 2006; Sheetal and Tejraj, 2006) A new PLA force field developed by O'Brien et al using quantum mechanic calculations demonstrated quantitative improvement performance compared to existing models. (Christopher, 2005) Several approaches to determine the glass transition of polymers by computer simulations are reported. One common approach is to determine the kink in a graph of the specific volume versus temperature originating from the change of the thermal expansion coefficient at Tg when passing from a glassy to a rubbery system. Other approaches also use the increase of potential energy at Tg or the temperature dependence of the mean square displacement of polymer chains below and above Tg. (Karl et al., 2005)

In this paper, the usability of the new PLA force field developed by O'Brien et al (Christopher, 2005) was examined by glass transition temperature determination using molecular dynamics simulation. The fractal dimension was calculated. Furthermore, the molecular weight dependence of glass transition temperature, self diffusion coefficient and shear viscosity were studied. Our computational findings are supported by experiments and theories and provide insight into glass transition behavior of polymers.

Simulation Details

The system potential energy was calculated using the GROMACS implementation of the modified OPLS-AA force field (Equation 1). The force field parameters were obtained from the new PLA force field developed by O'Brien et al.

$$E_{PE} = \sum_{bonds} k_r (r - r_0)^2 + \sum_{angles} k_\theta (\theta - \theta_0)^2 + \sum_{impropers} k_\xi (\xi - \xi_0)^2 + \sum_{n=0}^{5} C_n (\cos(\Psi))^n + \sum_{i} \sum_{j>i} f_{ij} \left\{ \frac{q_i q_j e^2}{r_{ij}} + 4\varepsilon_{ij} \left[\left(\frac{\sigma_{ij}}{r_{ij}} \right)^{12} - \left(\frac{\sigma_{ij}}{r_{ij}} \right)^{6} \right] \right\}$$

The amorphous PLA box containing 32 PLA chains were generated using Materials Studio 4.0 and the resulting output coordinate files were modified to make them compatible with GROMACS. The resulting structures were energy-minimized using steepest descents method. This was done to remove strain in the polymer backbone and alleviate high energy close contacts. The resulting periodic systems (as shown in Fig.1) were the starting configurations for all MD simulations.

All MD simulations were done using the GROMACS 3.3 simulation package on a 40-node IBM xSeries Linux Cluster. For the determination of Tg for a certain molecular weight (polymerization degree N = 10, 20 or 30), different simulation boxes (32 PLA chains) are generated, minimized and the lowest energy

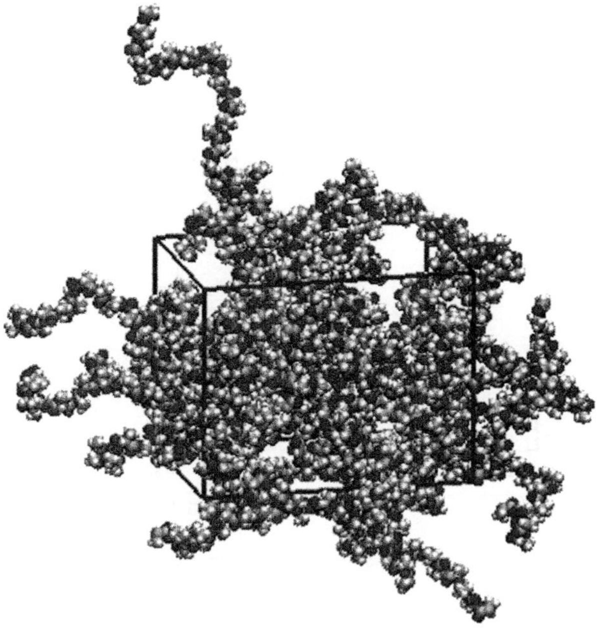

Fig. 1 Snapshot of PLA box starting configuration

configuration is chosen. This configuration is further relaxed for 10 ns under NPT conditions 550 K 1 bar to obtain a well relaxed start structure with the correct density. Afterwards, a cooling process is initiated by lowering the temperature stepwise by 50 K until a temperature 200 K is reached. At each temperature 2 ns NPT ensemble dynamics is carried out and the final configuration of this run is used as the starting structure for dynamics at the next (50 K lower) temperature.

Results and Discussion

The procedure to determine the glass transition temperature Tg is demonstrated in Figs. 2, 3, and 4 showing the computed specific volume v as a function of temperature T. The specific volume v was the reciprocal of the density of the box. The v values are well equilibrated for each temperature within the chosen simulation period (2 ns). Plotted values are mean values of v averaged over the last 1 ns of data sampling at each temperature taking a snapshot every 5 picoseconds. The increase of v with increasing T is less pronounced below Tg yielding to a kink in the V vs. T curve with the position of it determining Tg. A first rough estimate of the kink position is obtained by visual inspection of the data points. Then, these points are divided in two parts corresponding to the region below (glassy 200 K, 250 K, and 300 K) and above (rubbery 400 K and 450 K) of this estimate and each set of data is fitted to a straight line by linear regression. The final value of Tg results from

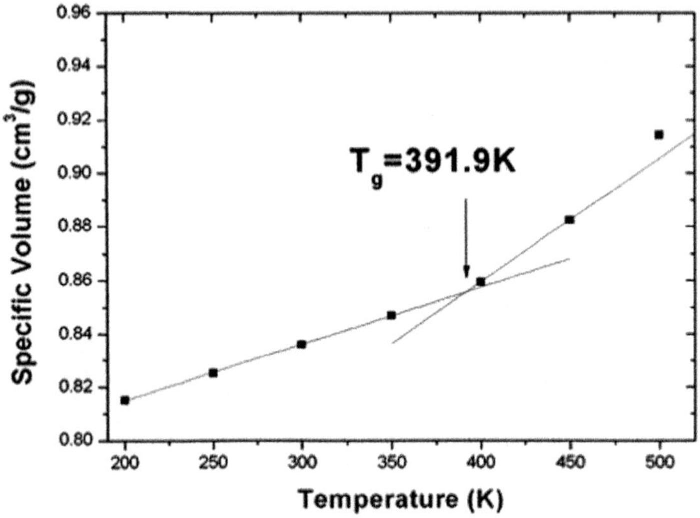

Fig. 2 The computed specific volume v as a function of temperature T for N = 30

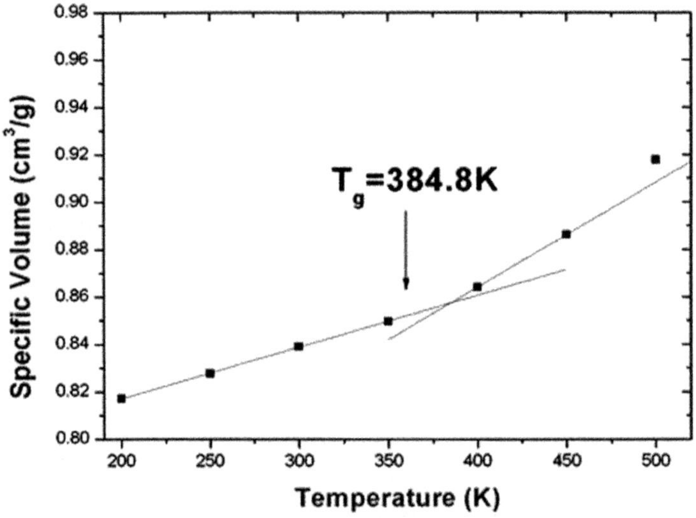

Fig. 3 The computed specific volume v as a function of temperature T for N = 20

the intersection of these two lines. In Fig. 5 the glass transition temperature vs. reciprocal degree of polymerization plots of three systems with different degree of polymerization N (N=10, N=20 and N=30) are given. In 1950, Thomas G. Fox, Jr. and Paul J. Flory found that the main effect of varying N is a shift of the glass transition temperature (Equation 2), where $T_g(\infty)$ is glass transition temperature for infinity molecular weight and C is a constant. The glass transition temperature data

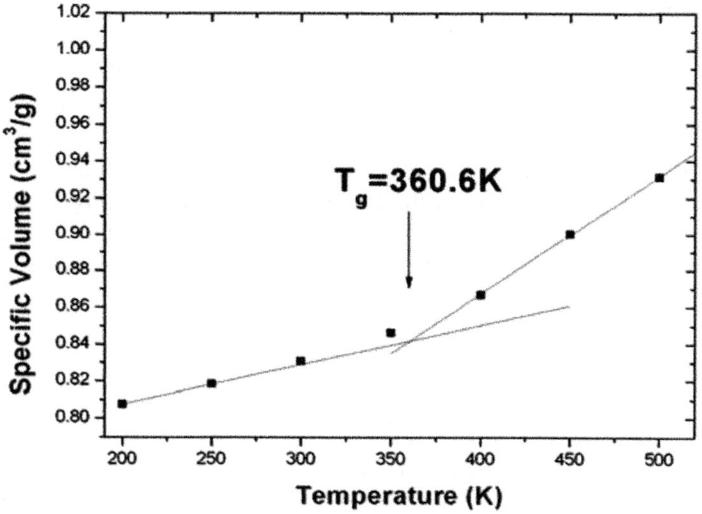

Fig. 4 The computed specific volume v as a function of temperature T for N = 10

were fitted to equation 2 by linear regression and R value of the regression is larger than 0.99.

$$T_g(N) = T_g(\infty) - \frac{C}{N} \tag{2}$$

Where, $T_g(\infty) = 408(K); C = 473$

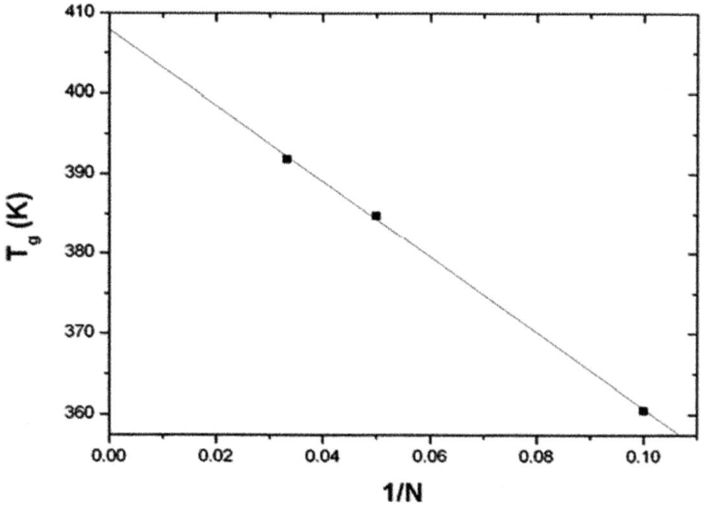

Fig. 5 The glass transition temperature vs. reciprocal N plots of three systems (N = 10, N = 20 and N = 30)

Conclusion

Molecular dynamics simulation has been used to investigate the molecular weight dependence of glass transition temperature for amorphous poly(L-lactide). The new PLA force field developed by O'Brien etc was feasible for the amorphous PLLA systems. The increase of v with increasing T is less pronounced below Tg yielding to a kink in the V vs. T curve with the position of it determining Tg. The molecular weight dependence of glass transition temperature was studied and the good agreement between the simulation results and experiments was obtained.

References

Christopher POB. 2005. Quantum and Molecular Modeling of Polylactide, PhD dissertation, Clemson University.

David K, Yiqi Y. 2006. Molecular modeling study of the resistance of PLA to hydrolysis based on the blending of PLLA and PDLA. *Polymer* 47:4845–4850.

Karl GW, Martin M, Andreas K, Gerhard Z. 2005. Glass transition temperature of a cationic polymethacrylate dependent on the plasticizer content – Simulation vs. experiment. *Chemical Physics Letters* 406:90–94.

Majerska K, Duda A. 2004. Stereocontrolled polymerization of racemic lactide with chiral initiator: Combining stereoelection and chiral ligand-exchange mechanism. *Journal of the American Chemical Society* 126(4):1026–1027.

Ray ED, Patrick RG, David EH. 2000. Polylactic acid technology. *Advanced Material* 12:1841

Richard AG, Bhanu K. 2002. Biodegradable polymers for the environment. *Science* 297:803.

Sheetal SJ, Tejraj MA. 2006. Molecular modeling simulations and thermodynamic approaches to investigate compatibility/incompatibility of poly(l-lactide) and poly(vinyl alcohol) blends. *Polymer* 47:8061–8071.

Sun L, Ginorio JE, Zhu L, Sics I, Rong L, Hsiao BS. 2006. Phase transitions and honeycomb morphology in an incompatible blend of enantiomeric polylactide block copolymers. *Macromolecules* 39(24):8203–8206.

Uhrich KE, Cannizzaro SM, Langer RS, Shakesheff KM. 1999. Polymeric systems for controlled drug release. *Chemical Reviews* 99(11):3181–3198.

Wang L, Jia X, Yuan Z. 2006. Synthesis and characterization of novel functionalized polylactides with pendent hydroxyl arms. *Polymer* 47(20):6978–6985.

Channel Catfish Estrogenicity and Sewer Overflows; Implications for Xenoestrogen Exposure

Conrad Daniel Volz, Frank Houghton, Nancy Sussman, Diana Lenzner, Devra Davis, Maryann Donovan, Talal El Hefnawy, and Patricia Eagon

Abstract Effluent from wastewater-treatment plants contains compounds that possess estrogenic activity. The southwestern Pennsylvania area has over 400 sewer overflows (SOs) that release untreated sewage. We sought to determine if the estrogenicity index (EI) of channel catfish from dense areas of SOs differed from catfish that are less impacted by SOs, using MCF-7 and BT-20 cell cultures. The MCF-7 human breast cancer line is estrogen receptor (ER) positive, while the BT-20 line is ER negative. The EI is based on the ratio of MCF-7 proliferation from application of fish extract to the response achieved from physiological levels of estradiol. Catfish caught near dense concentrations of SOs had significantly higher MCF-7 EIs than catfish from areas of less dense SOs, ($p=0.02$). The ER negative BT-20 cell line exhibited no proliferative response. We hypothesize that fish caught in concentrated areas of SOs have bioaccumulated more xenoestrogens than fish caught in less SO impacted areas. River water from SO contaminated areas is the primary source of drinking water for Allegheny County residents, potentially exposing large population groups to xenoestrogens. Our data suggest that evaluation of the estrogenicity of fish should be incorporated into risk assessment paradigms. Estrogen-screen evaluation of channel catfish is proposed as one model for further development.

Background

There is overwhelming evidence that effluents from sewage treatment plants exhibit strong human and animal in vitro and in vivo and in situ wildlife estrogenic activity (Aerni et al., 2004; Jobling et al., 2006; Liney et al., 2006; Petrovic et al., 2002). A wide variety of estrogenic agents is contained in municipal wastewater including

C.D. Volz (✉)
University of Pittsburgh Cancer Institute, Center for Environmental Oncology, Pittsburgh PA, USA; University of Pittsburgh, Graduate School of Public Health, Pittsburgh PA, USA; Scientific Director, Center for Healthy Environments and Communities, Pittsburgh PA, USA

G.A. Uzochukwu et al. (eds), *Proceedings of the 2007 National Conference on Environmental Science and Technology*, DOI 10.1007/978-0-387-88483-7_47,

pharmaceutical estrogens, phthalates, bisphenol-A, pesticides and detergent breakdown products like nonylphenol (Braga et al., 2005; Carballa et al., 2005, 2004; Cargouet et al., 2004). Many xenoestrogens can bioaccumulate in aquatic systems because they are highly lipid soluble (Hemond and Fechner-Levy, 2000).

The "Three Rivers" area of Pittsburgh has approximately 317, combined sewer overflows (CSOs) and sanitary sewer overflows (SSOs) – more than any other city in the United States, which release untreated municipal waste directly into receiving water during wet weather events (National Research Council, 2005). An estimated 16 billion gallons of sewage and stormwater are discharged yearly into receiving streams and main stem rivers in the Allegheny County Sanitary Authority (ALCOSAN) sewershed (Three Rivers Wet Weather, 2002).The main stem rivers affected by these CSOs and SSOs are the Allegheny and Monongahela Rivers and the river that they combine to form, the Ohio River, at Point State Park in downtown Pittsburgh. Rainfall amounts of as little as 1/10 inch can trigger opening of sewer outfall gates. Since CSOs and SSOs are released untreated, we hypothesized that they contain levels of pharmaceutical estrogens and xenoestrogens in excess of levels known to be released from wastewater treatment plants (WWTP) and further that channel catfish caught in areas of dense concentrations of CSOs and SSOs would display estrogenicity indexes that are higher than those caught in river areas that are less impacted by CSOs/SSOs.

Objectives

This study's objectives were:

1. To determine if the estrogenicity index of channel catfish fillet plus fat significantly varies by proximity to dense concentrations of CSOs and SSOs, as measured by the MCF-7 human breast cancer proliferative assay.
2. To determine if the channel catfish E-screen model can be used to help identify the sources of pharmaceutical estrogen and xenoestrogen pollution.
3. To involve semi-subsistence and recreational anglers in a Community Based Participatory Environmental Research Project.
4. To demonstrate to policymakers the known biological effects and potential environmental public health consequences of failure to update Allegheny County's and southwestern Pennsylvania's antiquated sewer systems.

Materials and Methods

We employed a community-based participatory environmental research approach (CBPR), working with community partners Venture Outdoors, Clean Water Action of Western Pennsylvania, fishing groups such as Bassmasters and individual anglers to understand common angling locations, fish taken and eaten by semi-subsistence and recreational fishers and to help determine sampling locations. All catfish were caught using rod and reel by local anglers as it has been demonstrated that scientists may catch smaller fish on average, which have bioaccumulated lower concentrations of contaminants, leading to potential underestimates of risk (Burger et al., 2006).

All CSOs and SSOs in the ALCOSAN service area were identified by latitude and longitude from a list provided in Appendix A and B of the US Department of Justice Consent Decree with ALCOSAN (USDOJ, 2007). CSO/SSO locations in the Kittanning area were identified from Pennsylvania Department of Environmental Protection records. Geospatial plots of CSOs and SSOs were made with Arc View 9.1. CSO/SSO densities were calculated for channel catfish sampling locations on the Allegheny, Monongahela and Ohio rivers based on the number of CSOs/SSOs within a 1 mile radius, within the watershed of each sampling point, which includes their feeder streams. The number of CSOs/SSOs at various distances upstream of each sampling point was also calculated.

The MCF-7 human breast cancer cell line is alpha-estrogen receptor positive and provides a well-established reporter cell line to screen for estrogen and estrogen mimicking chemicals (Blom et al., 1998). MCF-7 is known to respond to estrogen stimuli by increasing DNA-synthesis *in vitro*. MCF-7 cells are frequently used to study cell proliferation by estrogens and estrogen mimicking substances (Mellanen et al., 1996; Soto et al., 1992; White et al., 1994). Cell proliferation assays (CPA) were performed under the direction of Patricia Eagon PhD at the University of Pittsburgh, School of Medicine. Briefly, a one gram section of the fish fillet and fat (taken at maximum body depth, from backbone to stomach) was homogenized with a chloroform-methanol solution (80:20,v/v) to extract hormonally active substances. The chloroform methanol was evaporated with nitrogen, leaving a residue which was solubilized in ethanol-glycerol (3:1, v/v). A stock solution 1/100 fish extract was formed by combining 13.720 ml of a steroid free medium (95% RPI without phenol red and 5% dextran-charcoal treated serum) and 2.8 ml of 1/1 fish extract. MCF-7 and BT-20 cells were seeded at 5,000 cells per well. The experimental group wells were treated with the 1/100 dilution of fish extract, a 1 nM solution of estradiol was added to estradiol positive controls, and negative controls received only steroid-free medium. After incubation and further treatment the absorbance of each well was then measured at 490 nm with a Bio-Tek Synergy HT well plate reader. The 490 nm absorbance is directly proportional to the number of living cells in culture.

The presented estrogenicity index is based on the ratio of the MCF-7 cell proliferation achieved from application of fish extract to the mean proliferative response of the estradiol (E2) controls. The mean estrogenicity index of fishes from areas of dense- high CSO/SSO ($n = 10$) were compared with those from low CSO/SSO density areas ($n = 9$) using ANOVA with SPSS Version 12.0.

Results and Discussion

Figure 1, Arc GIS Map of Combined and Sanitary Sewer Overflows in the ALCOSAN Study Area, presents all CSO/SSO locations on the Monongahela, Allegheny and Ohio Rivers and on their feeder streams as well as catfish sampling locations. The map outlines all watersheds within Allegheny County, which includes

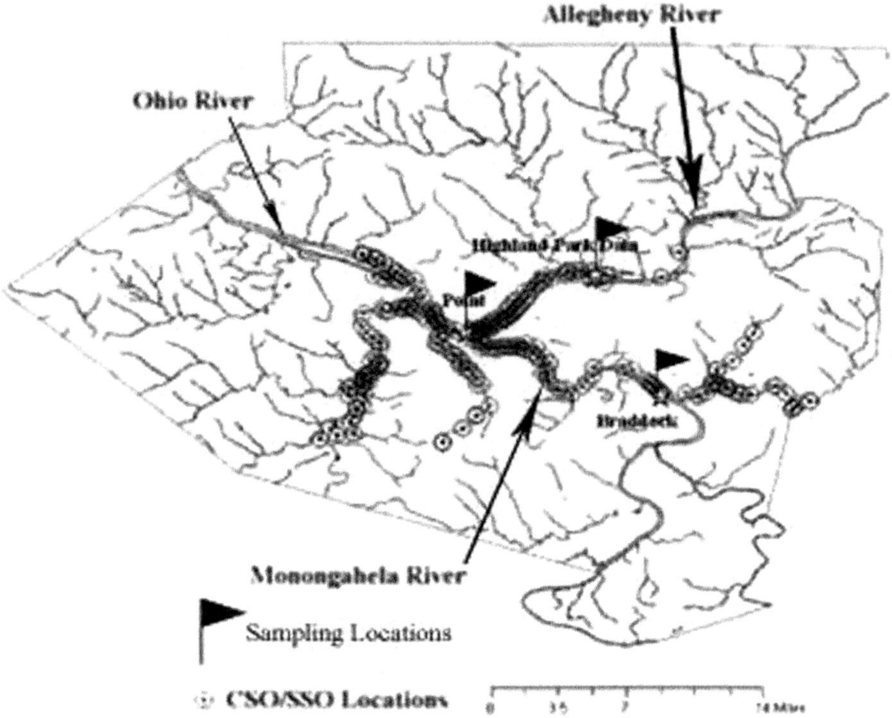

Fig. 1 Arc GIS map of combined and sanitary sewer overflows in the ALCOSAN study area

the City of Pittsburgh; 69% of the urbanized population (approximately 1,282,000 people) of the 10 county Southwestern Pennsylvania area lives in Allegheny County (Southwestern Pennsylvania Commission, 2007). Table 1 Sampling Locations, CSO/SSO Densities and Estrogenicity Data presents the locations of channel catfish catch, river of catch, number of CSOs/SSOs within a 1 mile radius (within watershed) of the sampling location, number of CSOs/SSOs within 1 mile upstream from the sampling location, number of samples and the mean, maximum, minimum and standard deviation of the estrogenicity index (EI) for channel catfish caught by location.

The 1-mile radius density of CSO/SSO outfalls was greatest at Point State Park (PSP) and at the Braddock Dam on the Monongahela River. CSO/SSO outfall density significantly decreased near the Highland Park Dam in Sharpsburg and even more so near Kittanning PA, both on the Allegheny River. The sampling locations at PSP and at the Braddock Dam on the Monongahela River each have a total of 53 CSO/SSO outfalls within a 1 mile radius of the sampling site, within its watershed. As you move upstream on the Allegheny River from PSP there are 72 CSO/SSO outfalls in the 7 miles between PSP and the Highland Park Dam. Upstream from PSP on the Monongahela River there are 62 CSO/SSO outfalls in the 11 miles between PSP and the Braddock Dam. CSO/SSO density falls near the Highland Park Dam

Table 1 Sampling locations, CSO/SSO densities and estrogenicity data

Sampling location	Point state park	Braddock dam	Highland park dam	Kittanning dam
River	Allegheny/ Monongahela Confluence, Ohio	Monongahela	Allegheny	Allegheny
CSOs/SSOs Radius of 1 Mile	53	53	17	3*
CSOs/SSOs 1 mile upstream	32	22	6	1
Number of Samples	4	6	3	6
Mean Estrogenicity Index, 1/100 Dilution	0.54	0.40	0.23	0.31
Min EI	0.31	0.17	0.13	0.20
Max EI	0.67	0.67	0.34	0.44
Std. Dev. EI	0.15	0.19	0.11	0.11

* within a radius of 3 miles.

sampling site as there are 17 CSO/SSO outfalls within a 1 mile radius of the site; of these only 4 CSOs and 2 SSOs are within the ALCOSAN system upstream from this site. There are only 3 identified CSO sites within 3 miles of the Kittanning sampling site, located approximately 40 miles upstream of PSP, on the Allegheny River.

ANOVA revealed that catfish caught at locations of dense concentrations of CSO outfalls (Point State Park and Braddock Dam) had significantly higher MCF-7 estrogenicity indexes than catfish from areas of less dense CSO effluents (Highland Park Dam and Kittanning Dam). This relationship was significant at $p = 0.02$. Catfish extract and estradiol application produced no proliferation of the BT-20 estrogen receptor negative cell line, suggesting that the fish extracts contain no non-estrogenic proliferation enhancers.

Conclusions

We hypothesize that extracts from fish caught in the densely concentrated areas of CSOs/SSOs are more estrogenic because they have bioaccumulated more pharmaceutical estrogens and xenoestrogens than fish caught in less CSO impacted areas. While the specific estrogenic compounds in the fish extracts are not yet identified; there is a growing literature indicating that a wide range of household and personal care products, pharmaceuticals, and lawn and garden pesticides with estrogenic potential are being emitted from waste water treatment plants and are contained in raw sewage emitted from CSOs and SSOs during wet weather events.

The environmental public health significance of this work is considerable: the source water for most of the population of the Greater Pittsburgh Metropolitan area is surface water from the Allegheny, Monongahela and Ohio Rivers. In fact the West View Water Authority source water intake pipes, serving over 200,000 people, in 31 municipalities, in portions of Allegheny, Beaver, and Butler Counties (West View Water Authority, 2007), are located just downstream from the ALCOSAN WWTP release point and have approximately 180 CSOs/SSOs within a 5 mile radius of these water intake pipes. There is increasing evidence that endocrine disrupting chemicals can affect human populations. Populations that consume fish from areas high in xenoestrogens are at higher risk for developing cancer, neurological impairment, and reproductive problems (Colborn and Soto, 1993). Testicular cancer and other male reproductive problems are increasing and may be attributed to endocrine disruptors in the environment as well (Davis et al., 2007; Toppari et al., 1996; Adami et al., 1994). Other studies show that over the past 50 years there has been a steady decline in human sperm count which could be attributed to estrogenic effects (Carlsen, 1992; Auger, 1995; Van Waeleghem, 1996). Additional studies have shown that endocrine disrupting chemicals are linked to developmental deficiencies and learning disabilities in children (EPA, 2001). Currently neither WWTPs nor water finishing plants are required to either sample for or treat source water for pharmaceutical estrogens and xenoestrogens

Naturalistic wildlife and holistic ecosystem studies provide valuable risk information concerning cumulative estrogenicity at the level of a system or population. An important aspect of the total risk to humans and the environment from ingestion of xenoestrogens is the profusion of substances with estrogenic activity that are being introduced into water through municipal (household) sources. The EPA must incorporate appropriate methodologies to evaluate total estrogenic risk from environmental mixtures of common xenoestrogens. E-screen evaluation of the channel catfish is proposed as one model for further discussion and development.

Acknowledgments The authors express their appreciation to the DSF Charitable Trust, the Heinz Endowments and Highmark Foundation, and University of Pittsburgh Cancer Institute (UPCI) for funding this research through the Center for Environmental Oncology of UPCI. CDV and DD are also funded through the Centers for Disease Control and Prevention's Environmental Public Health Tracking (EPHT) Program through a grant to the University of Pittsburgh, Academic Center of Excellence in EPHT. Many thanks go to Joanna Burger, PhD and Michael Gochfeld, MD, PhD for impressing upon CDV the importance of ecological receptors as biomonitors in exposure assessment studies.

References

Adami, B., Mohner, 1994.Testicular cancer in nine northern European countries. International Journal of Cancer 59: 33–38.

Aerni, H. R., Kobler, B., Rutishauser, B. V., Wettstein, F. E., Fischer, R., Giger, W., Hungerbuhler, A., Marazuela, M. D., Peter, A., Schonenberger, R., Vogeli, A. C., Suter, M. J. F., Eggen, R. I. L., 2004. Combined biological and chemical assessment of estrogenic activities in wastewater treatment plant effluents. (Special issue: endocrine disruptors). Analytical and Bioanalytical Chemistry. Springer-Verlag, Berlin, Germany. 378: 3, 688–696.

Auger, J., 1995. "Decline in semen quality among fertile men in Paris during the past 20 years." New England Journal of Medicine 332: 281–285.

Burger, J., Gochfeld M., Burke, S., Jeitner, C., Jewett, S., Snigaroff, D., Snigaroff, R., Stamm, T., Harper, S., Hoberg, M., Chenelot, H., Patrick, R., Volz, C. D., Weston, J., 2006. "Do scientists and fishermen collect the same size fish? Possible implications for exposure assessment." Environmental Research 101(1): 34–41.

Blom, A., Ekman, E., Johannisson, A., Norrgren, L., Pesonen, M., 1998. Effects of Xenoestrogenic environmental pollutants on the proliferation of a human breast cancer cell line (MCF-7). Achieves of Environmental Contamination and Toxicology 34: 306–310.

Braga, O., Smythe, G. A. et al., 2005. Steroid estrogens in primary and tertiary wastewater treatment plants. Water Science & Technology 52(8): 273–278.

Cargouet, M., Perdiz, D., Mouatassim-Souali D., Tamisier-Karolak A., Levi S., 2004. Assessment of river contamination by estrogenic compounds in Paris area (France). Science of The Total Environment 324(1–3): 55–66.

Carballa, M., Omil, F. et al., 2005. Behaviour of pharmaceuticals and personal care products in a sewage treatment plant of northwest Spain. Water Science and Technology. IWA Publishing, London, UK. 52: 8, 29–35.

Carballa, M., Omil, F. et al., 2004. Behavior of pharmaceuticals, cosmetics and hormones in a sewage treatment plant. Water Research (Oxford). Elsevier, Oxford, UK. 38: 12, 2918–2926.

Carlsen, E., 1992. Evidence for decreasing quality of semen during the past 50 years. British Medical Journal 305: 609-613.

Colborn, V. S., Soto, A. M., 1993. Developmental effects of endocrine-disrupting chemicals in wildlife and humans. Environmental Health Perspectives 101(5): 378–384.

Davis, D., Webster, P., Stainthorpe, H., Chilton, J., Jones, L., Doi, R., 2007. Declines in sex ratio at birth and fetal deaths in Japan, and in U.S. Whites but not African Americans, Environmental Health Perspectives 115(6): 941–946.

EPA, 2001. Removal of endocrine disruptor chemicals using drinking water treatment processes. Washington, DC, Office of Research and Development.

Hemond, H., Fechner-Levy, E., 2000. *Chemical Fate and Transport in the Environment.* 2nd ed. San Diego: Academic Press.

Jobling, S., Williams, R. et al., 2006. Predicted exposures to steroid estrogens in U.K. rivers correlate with widespread sexual disruption in wild fish populations. Environmental Health Perspectives 114 Suppl. 1: 32–39.

Liney, K. E., Hagger, J. A. et al., 2006. Health effects in fish of long-term exposure to effluents from wastewater treatment works. Environmental Health Perspectives 114 Suppl 1: 81–9.

Mellanen, P., Petänen, T., Lethimáki, J., Mákela. S., Bylund, G., Holmbom, B., Mannila, E., Oikari, A., Santti, R., 1996. Wood-derived estrogens: studies in vitro with breast cancer cell lines and in vivo in trout. Toxicology and Applied Pharmacology 136: 381–388.

National Research Council, 2005. Regional Cooperation for Water Quality Improvement in Southwestern Pennsylvania. Washington, D.C.: National Academies Press.

Petrovic, M., Sole, M., Lopez, de Alda, M., Barcelo, D., 2002. Endocrine disruptors in sewage treatment plants, receiving river waters, and sediments: integration of chemical analysis and biological effects on feral carp. Environmental Toxicology and Chemistry 21(10): 2146–2156.

Soto A. M., Lin, T.-M., Justicia, H., Silvia, R. M., Sonnenschein, C., 1992. An ''in culture'' bioassay to assess the estrogenicity of xenobiotics (E-screen). In: Colborn T, Clement C (eds.) Chemically-induced alterations in sexual and functional development: the wildlife/human connection. Princeton Scientific Publishing Co, NJ, pp. 295–309.

Southwestern Pennsylvania Commission, accessed September 4, 2007 http://www.spcregion.org/pdf/datalib/units.xls: original United States Census 2000 Summary File 1 Supplement released June 2003; data are municipal summation of census block data.

Toppari, J., Larsen, J. C., Christiansen, P., Giwercman, A., Grandjean, P., Guillette, L. J. Jr, Jégou, B., Jensen, T. K., Jouannet, P., Keiding, N., Leffers, H., McLachlan, J. A., Meyer, O., Müller, J., Rajpert-De Meyts, E., Scheike, T., Sharpe, R., Sumpter, J., Skakkebaek, N. E., 1996. Environmental Health Perspectives 104 Suppl 4: 741–803.

Three Rivers Wet Weather, 2002. *THE REGIONALIZATION REPORT: An initial study on options for regionalizing the management of sewage collection within the ALCOSAN service area.* 3 Rivers Wet Weather Demonstration Program, 3901 Penn Avenue, Building #3, Pittsburgh, PA 15224. http://www.3riverswetweather.org/f_resources/regionalization_report.pdf accessed 8/26/07.

United States Department of Justice, 2007, Consent Decree between United States of America, Commonwealth of Pennsylvania, Department of Environmental Protection, and Allegheny County Health Department vs. Allegheny County Sanitary Authority, Appendices A&B http://www.alcosan.com/Consent%20Decree/Alcosan%20-%20Consent%20Decree%20for%20Signatures.pdf, accessed August 15, 2007.

Van Waeleghem, K., 1996. Deterioration of sperm quality in healthy Belgium men. Human Reproduction 11: 325–329.

West View Water Authority, 2007. http://www.westviewwater.org accessed 8/30/07.

White, R., Jobling, S., Hoare, S. A., Sumpter, J. P., Parker, M. G., 1994. Environmentally persistent alkylphenolic compounds are estrogenic. Endocrinology 135(1): 175–182.

Sorption and Desorption of Nitrogen and Phosphorus by Zeolite and Shale

Johnsely S. Cyrus and G.B. Reddy

Abstract Wastewaters from animal houses contain excess of nutrients namely nitrogen (N) and phosphorus (P) apart from suspended solids and organics. Therefore, cost-effective technologies to remove N and P need to be developed. Sorption methods using the appropriate solid sorbent can be promising since this would serve as a nutrient trap which could be recycled as a nutrient source or soil mulch. Several sorbents have been investigated so far. Due to the cost effectiveness, stability and possibility of regeneration of shale and zeolites as filter media, these materials have been chosen as sorbents in the present investigation. The study focuses on the removal of N and P from wastewater using zeolite and shale and the feasibility of using the nutrient-sorbed material as slow-release fertilizers. Sorption experiments were conducted by using shale (2 mm and 2–4.7 mm) and zeolite (1 mm and 2–4 mm). Varied concentrations of N and P solutions ranging from 0 to 1000 mg L^{-1} were prepared from NH_4Cl and KH_2PO_4 respectively. Results indicate that Shale 1(particle size = 2 mm) showed the highest sorption of P (170 ± 10 mg kg^{-1}). But, the percent sorption of P was found to have a decreasing trend with increasing concentration of P in solution. Shale 2 (particle size = 2-4.7 mm) of larger particle size showed comparatively lower sorption. Both Shale 1 and 2 showed significant ($R^2 = 0.9566$ and $R^2 = 0.9333$) sorption of P over the P concentration in solution. Zeolites showed approximately 90% sorption of NH_4, over a wider range of feed concentrations ranging from 50 to 1000 mg L^{-1}. The percent desorption was about 30 % of initial phosphate concentration ranging from 200 to 1000 mg L^{-1} when shale 1 was used as the sorbent. Shale 2 showed a percent desorption of more than 30% after 24 h desorption in deionized water. Zeolites 3 and 4 show very low desorption of less than 10%. Even though the percent sorption of NH_4 in 24 h was high, the release of NH_4 was slow and hence, zeolite would serve as a substrate with a property of slow ammonia release.

J.S. Cyrus (✉)
Department of Natural Resources and Environmental Design, North Carolina A&T State University, Greensboro, NC 27411, USA

G.A. Uzochukwu et al. (eds), *Proceedings of the 2007 National Conference on Environmental Science and Technology*, DOI 10.1007/978-0-387-88483-7_48,

Introduction

Untreated wastewater such as domestic sewage, animal waste from livestock (Dairy, Poultry and Swine farms) (Knight et al., 2000) and run-off from agricultural fields contains high concentrations of nitrogen (N) and phosphorus (P). Mostly animal waste is flushed into an anaerobic lagoon and/or applied to the fields. In the state of North Carolina, swine farming is a major industry contributing to the State's economy. Wastewaters from swine farms contain excess of nutrients (N and P) apart from suspended solids and organics. Therefore, cost-effective technologies need to be developed to remove N and P in swine wastewater.

Conventional treatment processes for the removal of nutrients include chemical precipitation, biological treatment methods and sorption processes. Among these techniques, sorption methods using the appropriate solid sorbent can be promising since this would serve as a nutrient trap which could be recycled as a nutrient source or soil mulch (Bolan et al., 2004). Several sorbents such as Limestone (Baker et al., 1998; Hussain et al., 2006), Sand (Bubba et al., 2003; Rodgers et al., 2005), Bittern(Lee et al., 2003), gravel (Szogi et al., 1997; Mann and Bavor, 1993) Shale (Drizo et al., 1997; Drizo et al., 1999; Patel et al., 2005), Zeolite (Bolan et al., 2004; Cooney et al., 1999a; Cooney et al., 1999b; Du et al., 2005; Sarioglu, 2005; Wang et al., 2006) have been investigated so far.

In the present investigation, shale and zeolite sorbents have been selected due to the cost effectiveness, stability and regeneration possibility. These sorbent materials will be used to treat swine wastewater to remove N and P prior to discharging into either constructed wetlands for further cleansing or applied to the land. Preliminary laboratory batch experiments were conducted to study the sorption/desorption of nitrogen and phosphorus using zeolite and shale sorbents, respectively. This is necessary to find out the maximum concentration of ions removal and amount of regeneration prior to conducting the swine wastewater treatment through these selected sorbents by using column studies.

Objectives

The objectives of the present investigation are to (i) investigate the sorption capacity of N and P by zeolite and shale and (ii) study the feasibility of the sorbed N and P for use as a slow nutrient release fertilizer.

Materials and Methods

Materials

Chemicals used in the experiments on sorption of N and P were analytical grade NH_4Cl and KH_2PO_4 procured from Fisher Scientific, USA. All solutions and dilutions were made in deionized water.

Table 1 Physical properties of shale

Media	Particle size description (mm)	Coefficient of Uniformity (Cu)	Hydraulic Conductivity (cm/s)	Specific Surface Area	
				SBET(m^2/g)	MIP
Shale 1	(0.95, 2.00, 2.36)	2.48	0.9	1.41	18.68
Shale 2	(2.75, 4.75, 5.40)	1.96	7.6	1.32	27.04

Table 2 Physico-chemical properties of shale

Bulk Density (g/cm^3)	Particle Density (g/cm^3)	Specific Gravity	Porosity			Degree of Saturation (%)	Exchangeable cations (mg kg^{-1})	
			Intraparticle	Interparticle	Total		Ca	Mg
0.72–0.9	52.55	1.24–1.40	~0.24	~0.48	~0.70	~5–25	~15000	~2500

Expanded shale of varied mesh sizes were procured from Utelite Corporation (Coalville, UT) and their physical properties are shown in Tables 1 and 2.

Zeolite samples were procured from GSA Resources Inc., Cortaro, AZ 85652. Zeolite 2 used in the experiments were ZK406SMZ which is a clinoptilolite modified with HDTMA surfactant to enhance the zeolite's ability to pick up heavy metals and organic contaminants in ground water remediation. These Cabsorb 500 series products were based on the natural zeolite mineral chabazite of varied mesh sizes available as Zeolite 3 (1 mm) and Zeolite 4 (2–4 mm). These provide a high level of performance in cation exchange, sorption and odor control applications at a low cost comparative to the more expensive synthetic zeolite products. The product has a large surface area, a high porosity and permeability, and a high rate of sorption and cation exchange (Tables 3 and 4).

Batch Experiments on Sorption of Nutrients

In the batch studies, the effects of particle size of the sorbent and adsorbate concentration on NH_4 and PO_4 sorption were examined. Experiments were conducted in triplicates in 50 ml centrifuge tubes with samples of 1 g of the sorbent (Shale or Zeolite) for N and P sorption. N and P solutions of varied concentrations(C) ranging from 0 to 1000 mg L^{-1} were prepared from NH_4Cl and KH_2PO_4, respectively. The solutions were prepared in deionized water. The solutions (10 mL) with the sorbent were continuously stirred on a horizontal shaker (Eberbach Corporation Cat.No. 5850) for 24 h at constant room temperature. After 24 h, the samples were centrifuged for 15 min at 5000 rpm and filtered through Whatmann 42 filter paper and the initial and final concentrations of PO_4-P and NH_4-N of the samples were

Table 3 Properties of Zeolite used in the present investigation (GSA Resources, 2000a)

Form	Powder or Granules
Color	Dark Brown (Dry Brightness 40)
Ring Members	8
Crystal Size - Chabazite	Less than 1 micron
Crystallinity	+ 90%
Density	1.73 g/cm^3
Pore Size	4.1 by 3.7 Angstroms
Effective Pore Diameter	4.3 Angstroms
Cavity Size	11.0 by 6.6 Angstroms
Total Pore Volume	0.468 cm^3/g
Surface Area	520.95 m^2/g
Crystal Void Volume	0.47 cm^3/cm^3
Packing Density	Approx. 513 kg/m^3 (32 lbs./ft^3)
SiO_2/Al_2O_3 Ratio	Approx. 4:1
MOH's Hardness	4–5
Moisture as packaged	Less than 3% by weight
pH of 1% Dispersion	8.5
Stability	pH of 3 through 12
Ion Exchange Capacity	2.50 meq/g
Sorption Capacity	Greater than 15 wt.% H_2O at 10% R.H.

Table 4 Chemical Analysis (Anhydrous Basis)

Component (%)							
SiO_2	Al_2O_3	Fe_2O_3	CaO	MgO	Na_2O	K_2O	Dominant Cation
68.10	18.59	2.84	0.27	0.75	8.32	1.12	Na

analyzed by Lachat Flow Injection Analyzer/Ion Chromatograph (QuikChem IC / FIA – 8000 Series). The amount of N and P sorbed (S) was calculated as the difference between the amount of N or P in original concentration and the amount in solution.

Desorption of N and P - Feasibility Studies on Slow Nutrient Release

Sorption studies indicated that shale was a good sorbent for phosphorus while zeolite showed better sorption of ammonia. Previous research studies support these results (Drizo et al., 1999; Du et al., 2005). Hence in the present investigation, P desorption experiments were conducted on Shale and NH_4 desorption experiments were conducted on zeolite. Desorption studies were conducted on N and P solutions ranging in concentration from 50 to 1000 mg L^{-1}. After the sorption phase of the experiment, supernatants were replaced with 10 mL of deionized water and the suspensions were shaken for 24 h at 25± 2°C (room temperature). The concentrations of NH_4-N and PO_4-P in the solutions after desorption were determined. The total desorbed amount, the NH_4 and PO_4 remaining on the sorbents and the percent desorption were calculated.

Results and Discussion

Sorption of Phosphorus

Sorption experiments were conducted with solutions of concentrations ranging from 0 to 1000 mg L^{-1} of P (Fig. 1). It was observed that Shale 1 showed the maximum sorption of P (170 ± 10 mg P kg^{-1}). But, the percent sorption of P was found to have a decreasing trend with increasing concentration of P in solution (Fig. 2). Shale 2 of larger particle size showed comparatively lower sorption, because of its less surface area (Zhu et al., 1997) and high hydraulic conductivity than smaller particle sized material. Both Shale 1 and Shale 2 showed significant ($R^2 = 0.9566$ and $R^2 = 0.9333$) sorption of P.

Sorption of Ammonia – N

Zeolites 3 and 4 which are natural and unmodified, showed better performance (800–900 mg kg^{-1} with 1000 mg L^{-1} feed concentration) compared to Zeolite 2, which is a surfactant modified zeolite as shown in Fig. 3. Higher initial NH_4

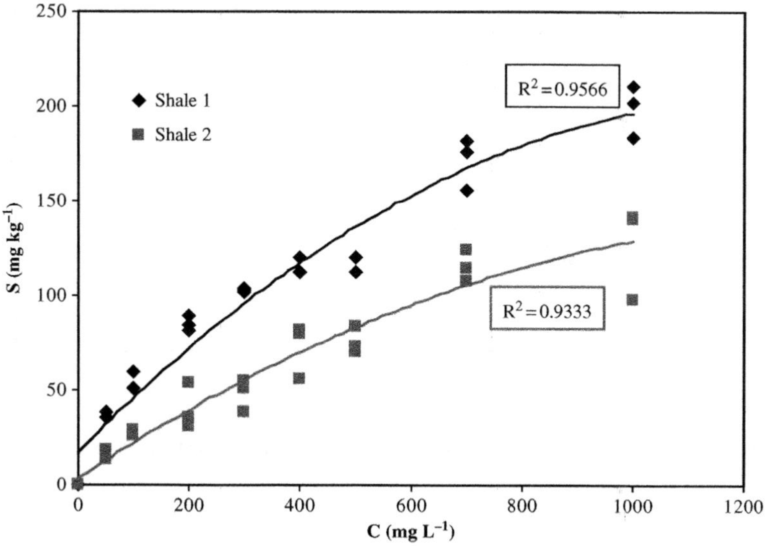

Fig. 1 Phosphorus sorption for Shale showing polynomial trend lines and their associated r^2 values

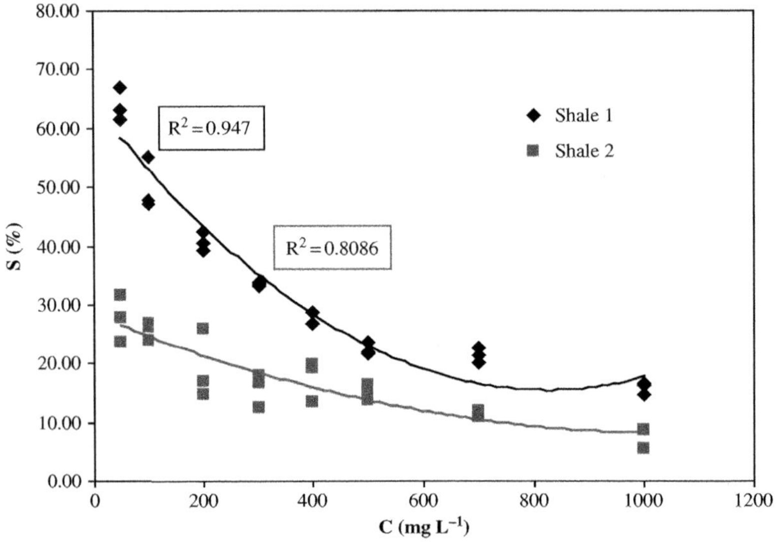

Fig. 2 Phosphorus sorption efficiency for Shale showing polynomial trendlines and their associated r^2 values

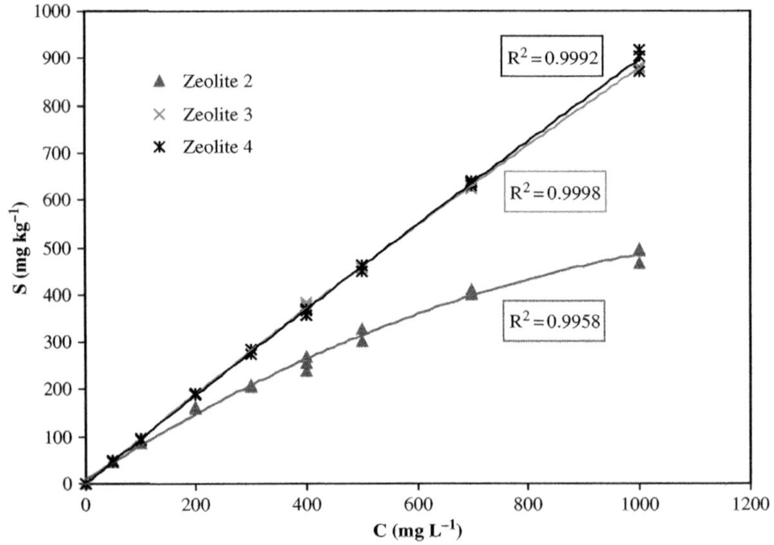

Fig. 3 Ammonia sorption of Zeolites showing polynomial trend lines and their associated r^2 values

concentration in solution resulted in a larger amount of sorbed NH_4. Similar results have been reported by Jørgensen et al. (1979) and Du et al. (2005). It can be interpreted by the fact that higher the NH_4 concentration in solution is, higher the solute concentration gradient is. This provides the necessary driving force so that NH_4 ions could take the place of cations on the surface of the internal micropores of zeolite within given contact time (Du et al., 2005).

The sorption efficiency of the zeolites is expressed in Fig. 4. As the initial concentration of ammonia increased, more sites on the surface of the sorbent were taken up, leading to reduced sorption efficiency (Hussain et al., 2006). The natural zeolites (Zeolites 3 and 4) showed approximately 90% sorption over a wider range of feed concentrations ranging from 50 to 1000 mg L^{-1} (Fig. 4). Zeolite 4 of higher particle size showed slightly lesser sorption efficiency as compared to zeolite 3, since uptake of NH_4 decreases with increase in sorbent particle size. Similar observation has been reported by Wang et al., (2006). Zeolite 2 which is a surfactant modified form, showed lesser efficiency as compared to the unmodified natural zeolites. The surface modification on this zeolite, by the sorption of quaternary amine profoundly alters the chemistry of the zeolite's external surface, causing the charge on the surface to change from negative to positive. Some of the zeolite's original cation exchange sites are retained which helps in the removal of cations (GSA Resources, 2000b). This explains the decline in the sorption efficiency by Zeolite 2.

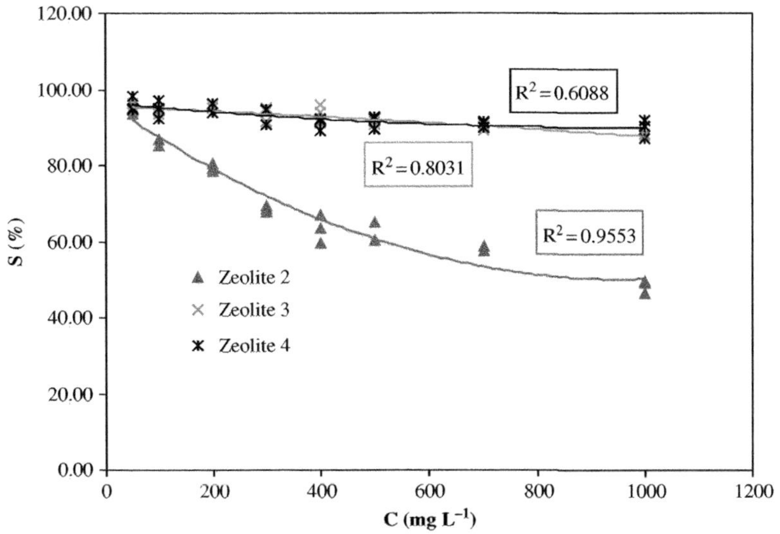

Fig. 4 Ammonia sorption efficiency of Zeolites showing polynomial trend lines and their associated r^2 values

Table 5 Desorption of phosphate-P by Shale

Initial Concentration added (mg L^{-1})	Shale 1			Shale 2		
	P desorbed mg kg^{-1}	P retained mg kg^{-1}	P desorbed %	P desorbed mg kg^{-1}	P retained mg kg^{-1}	P desorbed %
50	3.64 ± 0.16	32.85 ± 1.71	9.99 ± 0.84	4.63 ± 0.42	11.16 ± 2.62	29.92 ± 6.41
100	7.35 ± 0.27	46.47 ± 4.57	13.7 ± 0.81	8.47 ± 1.42	19.12 ± 0.94	30.59 ± 3.71
200	24.76 ± 0.5	60.58 ± 3.73	29.05 ± 1.51	20.69 ± 3.27	19.53 ± 8.92	52.87 ± 7.42
300	33.53 ± 1.28	69.34 ± 1.43	32.6 ± 1.21	29.91 ± 2	18.17 ± 6.54	63.12 ± 7.73
400	41.29 ± 1.5	76.22 ± 4.95	35.18 ± 2.02	37.01 ± 1	35.38 ± 13.9	52.59 ± 10.73
500	45.14 ±1.04	69.99 ± 5.77	39.28 ± 2.46	48.35 ± 3.67	27.42 ± 6.25	64.04 ± 6.05
700	52.98 ± 2.01	118.12 ± 14.57	31.13 ± 3.19	29.54 ± 5.7	85.82 ± 6.65	25.55 ± 4.24
1000	72.82 ± 3.04	126.07 ± 16.31	36.79 ± 4.04	44.17 ± 8.37	82.07 ± 25.89	36 ± 9.68

Table 6 Desorption of ammonia-N by Zeolite

	Zeolite 2			Zeolite 3			Zeolite 4		
Initial concentration added (mg L^{-1})	N desorbed mg kg^{-1}	N retained mg kg^{-1}	N desorbed %	N desorbed mg kg^{-1}	N retained mg kg^{-1}	N desorbed %	N desorbed mg kg^{-1}	N retained mg kg^{-1}	N desorbed %
50	14.78 ± 0.70	32.86 ± 0.41	31.02 ± 1.01	3.30 ± 0.19	44.21 ± 0.61	6.96 ± 0.45	3.27 ± 0.17	44.76 ± 1.09	6.81 ± 0.45
100	27.39 ± 2.61	58.17 ± 3.46	32.03 ± 3.35	7.37 ± 0.95	86.88 ± 0.81	7.81 ± 0.32	5.92 ± 0.32	88.82 ± 2.38	6.25 ± 0.98
200	47.65 ± 0.55	111.07 ± 2.34	30.03 ± 0.57	13.48 ± 2.10	177.16 ± 1.63	7.07 ± 0.27	12.34 ± 0.36	176.84 ± 3.31	6.52 ± 1.08
300	65.54 ± 2.79	140.21 ± 5.14	31.87 ± 1.72	19.51 ± 1.84	260.94 ± 5.10	6.95 ± 0.94	18.65 ± 2.45	261.12 ± 7.63	6.67 ± 0.61
400	84.95 ± 2.37	167.94 ± 12.41	33.63 ± 1.05	25.79 ± 0.77	348.97 ± 9.57	6.89 ± 1.18	27.47 ± 3.93	337.00 ± 9.81	7.55 ± 0.29
500	99.71 ± 1.73	209.29 ± 14.72	32.32 ± 1.65	34.60 ± 2.85	421.53 ± 4.43	7.59 ± 0.89	32.99 ± 4.54	424.91 ± 5.27	7.20 ± 0.65
700	136.37 ± 1.36	267.81 ± 4.96	33.74 ± 0.52	59.36 ± 4.15	567.40 ± 1.56	9.47 ± 0.74	44.01 ± 4.44	592.35 ± 9.31	6.92 ± 0.58
1000	177.28 ± 6.63	306.28 ± 22.70	36.72 ± 2.63	84.49 ± 5.59	795.73 ± 5.82	9.60 ± 0.10	72.40 ± 1.19	825.71 ± 22.34	8.06 ± 0.63

Desorption of N and P

The release of P by Shale 1 and 2 with varied initial P solution concentrations ranging from 50 to 1000 mg L^{-1} is shown in Table 5. Observations indicate that about 126 mg kg^{-1} of P remained and 72.82 mg kg^{-1} was desorbed by shale 1 after 24 h desorption in deionized water when the initial concentration of P in solution was 1000 mg L^{-1}. The percent desorption was about 30% for initial PO_4 concentration ranging from 200 to 1000 mg L^{-1} when shale 1 was used as the adsorbent. Shale 2 showed a percent desorption of more than 30% after 24 h desorption in deionized water. An increased amount of P desorbed with increased initial solution concentration of P. Also similar trend was observed for P retention on shale materials. However the P desorption were approximately two to three folds higher in shale 2 than shale 1 in concentrations <500 mg L^{-1}. This may be explained by the fact that shale 1 being of lesser particle size has more surface area and hence more available sorption sites to hold the sorbed P than in the case of Shale 2, which is of comparatively larger particle size. It has been observed by Dayton and Basta (2005), that most of the desorbable P was removed within 2 h of rinsing from drinking water treatment residuals.

The release of N by Zeolite 2, Zeolite 3 and Zeolite 4 with varied concentrations ranging from 50 to 1000 mg L^{-1} is given in Table 6. The release of N on equilibration with deionized water for 24 h was comparatively lesser mainly in the case of Zeolite 3 and 4. About 306.28 mg kg^{-1} of NH_4 remained on Zeolite 2 and 177 mg kg^{-1} was desorbed after 24 h desorption in deionized water when the initial concentration of the solution was 1000 mg L^{-1}. Zeolites 3 and 4 retained more NH_4 after desorption in deionized water for 24 h. About 30 % of the sorbed N was desorbed in deionized water in 24 h in the case of Zeolite 2. This may be because it is surfactant modified and the bonding of the sorbed N is weaker. Zeolite 3 and 4 show very low desorption rate of less than 10 % desorption over the chosen concentration range. This shows that the sorption of N on to these zeolites were stronger as compared to the surfactant modified form.

The release of NH_4 in 24 h was observed to be less. Hence desorption experiments were continued in deionized water. Results showed a slow release of N indicated by a gradual release of the sorbed N (unpublished data).

Conclusion

On the basis of the experimental results Zeolite was found to be an effective sorbent for N and Shale was effective in sorbing P. Smaller size of the sorbent yielded better sorption efficiency due to greater available surface area. Similarly the percent desorption was lower when the particle size of the sorbent was smaller resulting from higher availability of sorption sites. Thus, Shale 1 and Zeolite 3 were found to be better sorbents, showing higher sorption efficiency and lower percent desorption. Further, the results from serial desorption studies show that zeolite 3 and 4, would serve as a substrate with a property of slow NH_4 release (unpublished data). Therefore, these sorbents may be applicable as soil amendments.

Acknowledgment Southern Water Quality, Texas A&M and USDA CSREES

References

Baker M.J., W.B. David, and C.J. Ptacek (1998). Laboratory Development of Permeable Reactive Mixtures for the Removal of Phosphorus from Onsite Wastewater Disposal Systems. Environmental Science and Technology 32, 2308–2316.

Bolan, N.S., L. Wong, and D.C. Adriano (2004). Nutrient removal from farm effluents. Bioresource Technology 94, 251–260.

Bubba M.D., C.A. Arias, and H. Brix (2003). Phosphorus adsorption maximum of sands for use as media in subsurface flow constructed reed beds as measured by the Langmuir isotherm. Water Research 37, 3390–3400.

Cooney E.L., N.A. Booker, D.C. Shallcross, and. G. W.Stevens (1999a) Ammonia removal from wastewaters using natural australian Zeolite. I. Characterization of the zeolite. Separation Science and Technology 34(12), 2307–2327.

Cooney E.L., N.A. Booker, D.C. Shallcross, and G.W. Stevens, (1999b). Ammonia removal from wastewaters using natural Australian zeolite. II. Pilot-Scale Study Using Continuous Packed Column Process. Separation Science and Technology 34(14), 2741–2760.

Dayton E.A., and N.T. Basta (2005). A method for determining the Phosphorus sorption capacity and amorphous aluminium of aluminium-based drinking water treatment residuals. Journal of Environmental Quality 34(May-June 2005), 1112–1118.

Drizo A., C.A. Frost, K.A. Smith, and J. Grace (1997). Phosphate and ammonium removal by constructed wetlands with horizontal subsurface flow, using shale as a substrate. Water Science and Technology 35(5), 95–102.

Drizo A, C.A. Frost, J. Grace, and K.A. Smith (1999). Physico-chemical screening of phosphate-removing substrates for use in constructed wetland systems. Water Research 33(17), 3595–3602.

Du Q., S. Liu, Z. Cao, Y. Wang (2005). Ammonia removal from aqueous solution using natural Chinese clinoptilolite. Separation and Purification Technology 44, 229–234.

GSA resources (2000a)http://www.gsaresources.com/pdsZS500a.htm Last modified: June 22, 2000.

GSA resources (2000b) http://gsaresources.com/pdsk406smz.htm Last modified: June 22, 2000.

Hussain S., A.A. Hamidi, M.H. Isa, M.N. Adlan, and F.A.H. Asaari (2006). Physico-chemical method for ammonia removal from synthetic wastewater using limestone and GAC in batch and column studies. Bioresource Technology 98, 874–880.

Jørgensen S.E., O. Libor, K. Barkacs, and L. Kuna (1979). Equilibrium and capacity data of clinoptilolite, Water Research 13, 159–165.

Knight R.L., V.W.E. Payne Jr., R.E. Borer, R.A. Clarke Jr., and J.H. Pries. (2000) Constructed wetlands for livestock wastewater management. Ecological Engineering 15, 41–55.

Lee S.I., S.Y. Weon, C.W. Lee, and B. Koopman (2003). Removal of nitrogen and phosphate from wastewater by addition of bittern. Chemosphere 51, 265–271.

Mann, R.A., and H.J. Bavor, (1993). Phosphorus removal in constructed wetlands using gravel and industrial waste substrata. Water Science and Technology 27(1), 107–113.

Patel D.M., J.R. Hauser, J. Johnston, J.G. Curtis, (2005). Pilot filtration studies for turbidity and nutrient removal at Lake Tahoe Journal of the Nevada Water Resources Association Spring 2005 Lake Tahoe Edition 2(1).

Rodgers M., M.G. Healy, J. Mulqueen, (2005). Organic carbon removal and nitrification of high strength wastewaters using stratified sand filters. Water Research 39, 3279–3286.

Sarioglu M. (2005) Removal of ammonium from municipal wastewater using natural Turkish (Dogantepe) zeolite Separation and Purification Technology 41, 1–11.

Szogi A.A., F.J. Humenik, J.M .Rice, P.G. Hunt, (1997). Swine wastewater treatment by media filtration. Journal of Environmental Science Health B32(5), 831–843.

Wang Y., S. Liu, Z. Xu, T. Han, S. Chuan, T. Zhu (2006). Ammonia removal from leachate solution using natural Chinese clinoptilolite. Journal of Hazardous Materials B136, 735–740.

Zhu T., P.D. Jenssen, T. Maehlum, T. Krogstad (1997). Phosphorus sorption and chemical characteristics of lightweight aggregates (LWA) -potential filter media in treatment wetlands. Water Science and Technology 35(5), 103–108.

Index

A

F

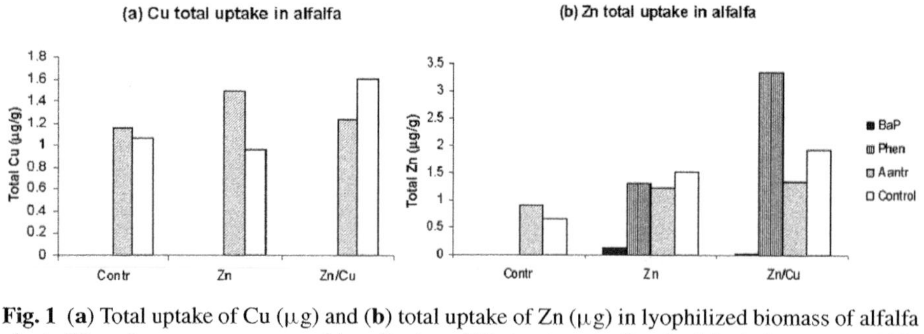

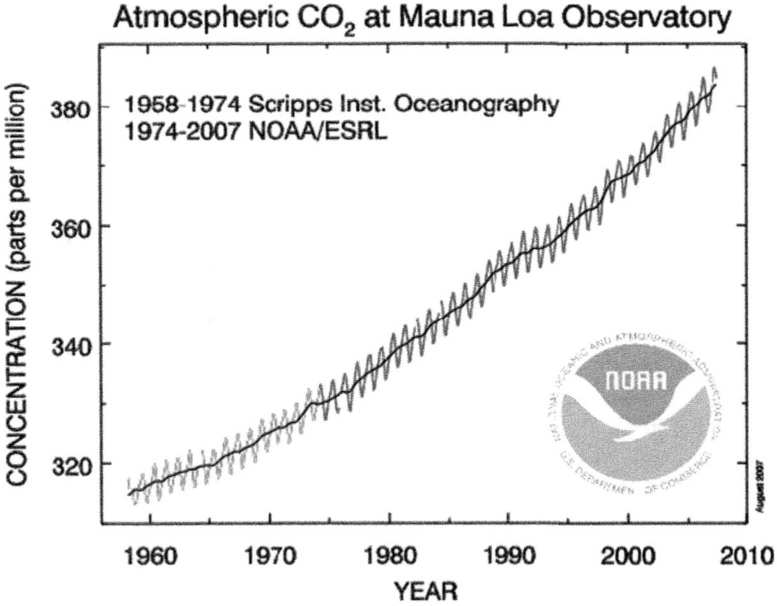

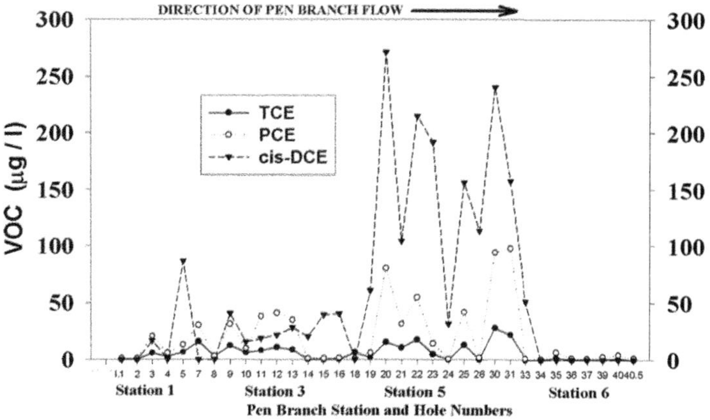

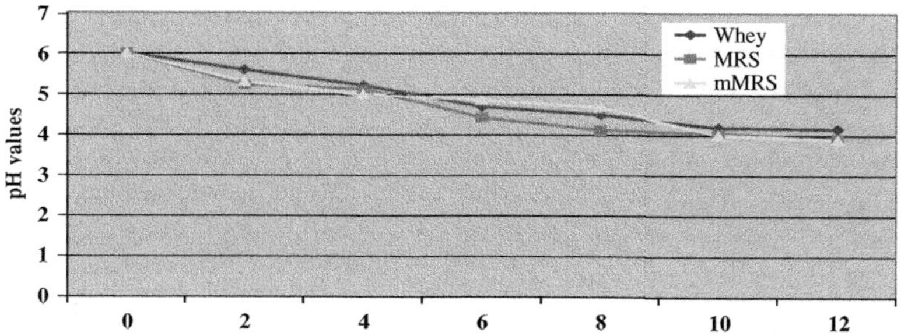

Table 6 Desorption of ammonia-N by Zeolite

Initial concentration added (mg L^{-1})	Zeolite 2 N desorbed mg kg^{-1}	Zeolite 2 N retained mg kg^{-1}	Zeolite 2 N desorbed %	Zeolite 3 N desorbed mg kg^{-1}	Zeolite 3 N retained mg kg^{-1}	Zeolite 3 N desorbed %	Zeolite 4 N desorbed mg kg^{-1}	Zeolite 4 N retained mg kg^{-1}	Zeolite 4 N desorbed %
50	14.78 ± 0.70	32.86 ± 0.41	31.02 ± 1.01	3.30 ± 0.19	44.21 ± 0.61	6.96 ± 0.45	3.27 ± 0.17	44.76 ± 1.09	6.81 ± 0.45
100	27.39 ± 2.61	58.17 ± 3.46	32.03 ± 3.35	7.37 ± 0.95	86.88 ± 0.81	7.81 ± 0.32	5.92 ± 0.32	88.82 ± 2.38	6.25 ± 0.98
200	47.65 ± 0.55	111.07 ± 2.34	30.03 ± 0.57	13.48 ± 2.10	177.16 ± 1.63	7.07 ± 0.27	12.34 ± 0.36	176.84 ± 3.31	6.52 ± 1.08
300	65.54 ± 2.79	140.21 ± 5.14	31.87 ± 1.72	19.51 ± 1.84	260.94 ± 5.10	6.95 ± 0.94	18.65 ± 2.45	261.12 ± 7.63	6.67 ± 0.61
400	84.95 ± 2.37	167.94 ± 12.41	33.63 ± 1.05	25.79 ± 0.77	348.97 ± 9.57	6.89 ± 1.18	27.47 ± 3.93	337.00 ± 9.81	7.55 ± 0.29
500	99.71 ± 1.73	209.29 ± 14.72	32.32 ± 1.65	34.60 ± 2.85	421.53 ± 4.43	7.59 ± 0.89	32.99 ± 4.54	424.91 ± 5.27	7.20 ± 0.65
700	136.37 ± 1.36	267.81 ± 4.96	33.74 ± 0.52	59.36 ± 4.15	567.40 ± 1.56	9.47 ± 0.74	44.01 ± 4.44	592.35 ± 9.31	6.92 ± 0.58
1000	177.28 ± 6.63	306.28 ± 22.70	36.72 ± 2.63	84.49 ± 5.59	795.73 ± 5.82	9.60 ± 0.10	72.40 ± 1.19	825.71 ± 22.34	8.06 ± 0.63

GPSR Compliance

The European Union's (EU) General Product Safety Regulation (GPSR) is a set of rules that requires consumer products to be safe and our obligations to ensure this.

If you have any concerns about our products, you can contact us on ProductSafety@springernature.com

In case Publisher is established outside the EU, the EU authorized representative is:

Springer Nature Customer Service Center GmbH
Europaplatz 3
69115 Heidelberg, Germany

Batch number: 10755284

Printed by Printforce, the Netherlands